GERARD P. KUIPER AND THE RISE OF MODERN PLANETARY SCIENCE

Gerard P. Kuiper (December 7, 1905, to December 23, 1973).
Courtesy of Matthew des Tombe.

GERARD P. KUIPER
AND THE RISE OF MODERN
PLANETARY SCIENCE

DEREK W. G. SEARS

THE UNIVERSITY OF
ARIZONA PRESS

TUCSON

The University of Arizona Press
www.uapress.arizona.edu

ISBN-13: 978-0-8165-3900-0 (cloth)

Cover design by Leigh McDonald
Cover photo of Gerard P. Kuiper, courtesy of Matthew des Tombe

Library of Congress Cataloging-in-Publication Data
Names: Sears, Derek W. G., 1948– author.
Title: Gerard P. Kuiper and the rise of modern planetary science / Derek W. G. Sears.
Description: Tucson : The University of Arizona Press, 2019. | Includes bibliographical references and index.
Identifiers: LCCN 2018038445 | ISBN 9780816539000 (cloth : alk. paper)
Subjects: LCSH: Kuiper, Gerard P. (Gerard Peter), 1905–1973. | Astronomers—Biography. | Planetary
    scientists—Biography. | LCGFT: Biographies.
Classification: LCC QB36.K9 S43 2019 | DDC 523.4092 [B] —dc23 LC record available at
    https://lccn.loc.gov/2018038445

Printed in the United States of America
♾ This paper meets the requirements of ANSI/NISO Z39.48-1992 (Permanence of Paper).

Dedicated to my wife

Hazel

without whom nothing is possible

CONTENTS

PREFACE

THE DUTCH AMERICAN ASTRONOMER Gerard Kuiper thought he was living in the third era.[1] The first era was the time of Copernicus, Galileo, Brahe, Kepler, and Newton, when the basic architecture of the solar system was being figured out and the movements of the planets came to be understood in terms of gravitational theory. The second era was the era of Percival Lowell, who in the nineteenth century conjured up a Mars full of inhabitants busily building cities and irrigation canals all over the planet. It was a boon to science fiction and the popular imagination but a sterile time for planetary astronomy, which fell into a decline as far as the professional astronomers were concerned. Then in the third decade of the twentieth century came Gerard Kuiper, who through his hard-nosed observational programs and confident writings created the third era. Kuiper demonstrated that doing real science of the planets was possible by ignoring traditional disciplinary boundaries and throwing oneself wholeheartedly into the study of the solar system. "It is not astronomy," complained the astronomers, and they were right. It was a new discipline we now call planetary science.

I spent thirty years as a professor at the University of Arkansas. When I reached the appropriate level of seniority, the deans of the Arts and Sciences College and of the Honors College asked me to offer an honors colloquium. I am sure most universities offer these types of courses. Aimed at the brightest and best students from all departments in the university, honors colloquia are an opportunity to stimulate the growth process by making the students think and talk about ideas and techniques they would not normally run into. These students would also get involved in discussions with other students of very different backgrounds from their own. I was not sure what response to give the deans—this was something new to me—but my son David, who was a student at Arkansas and had taken several of these courses, urged me to do it, saying that it was exactly the kind of teaching I would like. He turned out to be right.

I chose "The Exploration of Space" as the theme and taught it through a dozen biographies: William Congreve, Tipu Sultan, Konstantin Tsiolkovsky, Robert Goddard, Hermann Oberth, Wernher von Braun, Sergei Korolev, James Webb, Robert Gilruth, James Van Allen, Gerard Kuiper, and Barbara Mikulski. Warriors and politicians, engineers and scientists, administrators and dreamers. They represent at least five nations. They represent individuals, teams, and governments. They reflect hot wars and cold wars, old and new politics. It worked well. I had art and law students confidently talking about the chemistry of rocket fuels, engineering students discussing astronomy, and chemists talking about international politics during the Cold War. I loved the course and taught it many times. It was never dull.

I gathered a library of relevant books: about a dozen on Goddard and von Braun; several on Korolev, Oberth, Webb, Congreve, and Tipu Sultan; and one each on Van Allen and Gilruth. Senator Mikulski had an elaborate website and she was often in the news, so there was no shortage of information on her contributions to space exploration and the role of Congress that she represented.

Kuiper had no book. He did not even have a decent Wikipedia page the first time I taught the class. I resolved that when I could find the time, I would write a book-length biography of Gerard Kuiper with my honors colloquium students in mind. I felt I owed it to them. This is that book.

As far as I am aware, this is the first biography of Kuiper written by someone who did not know him or who was not associated with his institutions, and I hope thereby it has a perspective that is new.

Derek Sears
Hayward, California
November 7, 2015

ACKNOWLEDGMENTS

A FTER I MOVED FROM ARKANSAS to the NASA Ames Research Center in Mountain View, California, in the summer of 2011, I met Dale Cruikshank. It turns out that he had written the official biography of Kuiper for the National Academy of Sciences, which publishes a series of biographies of fellows. That article has been a help to me in putting together this book, but more especially, Dale has been an enormous help. Not only did he open his files to me and provide me with all sorts of information, but he also gave me the greatest encouragement of all, which was to be interested. He introduced me to Kuiper's daughter, Sylvia des Tombe, who has also offered every kind of support and encouragement. Dale and Sylvia allowed me to conduct interviews, Dale's being published in *Meteoritics and Planetary Science*. Sylvia put me in touch with Gédo Kuiper, Gerard's nephew, who generously guided me through Gerard's footsteps in the Netherlands, from his birthplace, to his high school, to Leiden University and its observatory. I am also grateful to Matthew Kuiper, Gerard's grandson, for supplying large numbers of photographs that greatly helped me to visualize people, places, and events I was writing about. I wonder how many biographers have had the joy and privilege of talking to so many of their subject's family members.

A lot of other people have helped. Tim Swindle, the current director of the Lunar and Planetary Laboratory (LPL) and chair of the Department of Planetary Science at the University of Arizona, has been very encouraging, provided access to records and people, and, with his wife Kerry, offered their hospitality when visiting Tucson. Through him I met Ewen and Beryl Whitaker. While Beryl reminisced with my wife about growing up in England, Ewen reminisced about Kuiper and the lunar work they did together and kindly provided a great variety of documents. Maria Schuchardt of the LPL's Space Imagery Center also supplied much information, including audio and video files. The university librarians at the archives I have consulted (University of Arizona; University of Chicago; University of California, Santa Cruz; University of California, San Diego), especially Robin Roggio at the library of the University of Arkansas, my own institution, have been very supportive. Robin has been a steadfast supporter of my writing efforts, going that extra mile at every opportunity and doing so fast enough to take my breath away.

Melissa Sevigny, author of *Under Desert Skies: How Tucson Mapped the Way to the Moon and Planets*, was especially generous when she made available to me fifty-six oral histories of almost everyone associated with the LPL. Similarly, David van Noortwijk generously made his collection of documents about Kuiper available to me. David DeVorkin kindly provided translations of some early correspondence in Dutch while we discovered that we had a common alma mater, the University of Leicester.

I also owe a debt of gratitude to the staff of the observatories I visited during the preparation of this book: Steve Bramlett and David Doss at McDonald Observatory, Tony Misch and Elinor Gates at Lick Observatory, and Steve Larson at the University of Arizona. Steve gave me and my wife a fascinating tour of the Tucson observatories and a lengthy discussion of their history and his time with Kuiper.

I must return to the colloquium course, "Exploration of Space," I gave at the University of Arkansas that I described in the preface. To the deans who asked me to do this, Don Bobbitt and Bob McMath, to David, who convinced me to do it, to the students who took the course and taught me so much, to them all I owe a special thank you.

As the first rough draft of this book came together, I asked Dale Cruikshank, William Sheehan, and Timothy Swindle to review my progress. Their responses were far more thorough than I had any right to expect; not only did they catch errors and suggest improvements, they also greatly enhanced my efforts by providing additional information from their unique perspectives as Kuiper's student and biographer, historian of astronomy, and current director of the LPL, respectively. This book profited enormously by the suggestions made by two anonymous external reviewers. I also appreciate the contributions of my copyeditor, Timothy Clifford, and the staff of the University of Arizona Press for their help in preparing this book for publication. I am very grateful to them all.

Finally, a lot of additional people have helped me put together this sketch of Kuiper and his times through publications, interviews, correspondence, or other means, and I have made many friends in the process. I have consulted many books and appreciate the efforts of their authors.

Throughout this task my daily companion was my wife, Hazel. She helped me find documents, copied and recorded their details, found the necessary financial resources, advised me on how to present the material and how to write the book, and managed my calendar so I had time to do this. She reviewed the text, and she proofed the multiple versions. Yet she declined to be a coauthor. To her I owe the greatest debt of all.

However, while I have received considerable amounts of help from others, the mistakes are entirely mine, and I would appreciate hearing about them.

GERARD P. KUIPER AND THE RISE OF MODERN PLANETARY SCIENCE

CHAPTER 1

Haringscarspel to California

THERE IS A CHARMING PHOTOGRAPH given to me by Gerard Kuiper's daughter. It is an old photograph, not so much sepia, just browned with age. It is a family group, the Kuiper family, and to me it says everything. Mother Antje and father Gerrit are of course at the center of the family group, seated on a bench. Gerrit wears a peaked cap and holds a pipe in his mouth while his relaxed right arm rests gently in his lap. Antje wears a long, plain, dark dress buttoned to the neck and sits upright. Between them are two of their sons, Pieter and Nicolaas, wearing clothes appropriate to their young ages. Leaning against her mother, standing, is Augusta, hand on hip, legs crossed. The mood of the family group is relaxed, happy with their lot in life, with pleasant, comfortable smiles for the camera. Then there is thirteen-year-old Gerrit (he only became Gerard after coming to the United States). He stands upright, suit, tie, and knee-length boots, heels firmly together. He appears to be a serious, somewhat stern young man. To me the photograph is an indication of the man he would become. He was destined to leave the Netherlands fifteen years later for California. His turbulent rite of passage was to be a story of disappointment, persistence, good fortune, and an extraordinary fascination with the stars.

Birth to Entrance to Leiden University

The Netherlands into which Gerard Kuiper was born was still building up its empire in the East, where the colonial power could reap the benefits of a climate so well suited

The Kuiper family at their home circa 1918. From left to right, Gerard, Gerrit (father), Nicolaas (brother), Pieter (brother), Antje (mother), and Augusta (sister). Courtesy of Matthew des Tombe.

to the production of coffee. A strict bureaucracy and strong military meant that before Kuiper left the Netherlands, its colonies of Java and Sumatra would soon be joined by Borneo, Sulawesi, and New Guinea. At home, universal male suffrage came into being when Kuiper was twelve, and when he was seventeen women were given the vote. The country escaped the destruction of population and property during World War I by remaining neutral, and it emerged without war debts. However, soon afterward the Netherlands was plunged into the worldwide depression and, because of inept internal policies and its dependence on international trade, the economic downturn was particularly long and painful. Wages fell and unemployment rose. Those unfortunate enough to depend on the limited welfare programs were stigmatized by a class-ridden society that insisted on attaching badges of dishonor to their vehicles and clothing. It was hard being poor in the Netherlands between the wars.

Gerrit Pieter Kuiper was born on December 7, 1905, in the village of Tuitjenhorn, which has undergone various administrative changes but was then part of the municipality of Haringscarspel, later changed to Harenkarspel. The region largely consisted of moats and ditches for digging peat but is now pumped dry and more accessible.[1] Nevertheless, maintaining the water channels is still a requirement for anyone owning property in the area. Gerrit was the eldest of four children. His father was a tailor. His younger brothers, Pieter and Nicolaas, were to become engineers and he and his sister Augusta were expected to become teachers. Augusta did go into teaching until she married. After an elementary education in the village, with tutoring by the local priest, Gerrit entered high school in Haarlem, the capital city of North Holland. According to the Dutch American astronomer and fellow student Bart Bok, Kuiper did not go to a regular high school—he attended the teachers' college and was expected to become a primary school teacher.[2] While he attended high school Gerrit would stay with an aunt during the week and take the bus fifty kilometers (thirty-one miles) to Haringscarspel for the weekends.

An interest in astronomy was encouraged by his father and by his grandfather, who gave him a small telescope.[3] Kuiper made sketches throughout an entire winter, recording the faintest members of the Pleiades star cluster. He sent his results to Leiden astronomers, who found that the faintest stars Kuiper could see had a magnitude of 7.5, which is much fainter than those visible to most people. Astronomers define the brightness of a star in terms of magnitudes, which increase in value as the star becomes less bright, so that the Sun is −27, Sirius (the brightest star after the Sun) is −1.46, and the dimmest star most people can see is about 6.0.[4] Even in his later years, Kuiper's visual acuity was exceptional.

At the end of his high school years Gerrit passed both the graduation test and the teacher qualifiers and was well equipped for a career as a Dutch school teacher. However, he wanted to take the test for university entrance, presumably to pursue his interests in astronomy. This was a test he was not prepared for. Entrance to university required going to a selective high school, the lyceum or gymnasium, rather than a teacher's college. But Kuiper had considerable confidence in himself, even as a school student—a trait that partly explains his later success. He therefore took the entrance examination for the highly prestigious Leiden University and, contrary to all expectations, he passed. He entered Leiden University in September 1924.

Leiden University

Bart Bok was another student who entered Leiden University that fall. Many years later he recalled that on the first morning there was little to do and the incoming astronomy students gravitated toward the reference library and to the card catalog of astronomy books. Bok remarked, "After introduction formalities were over, Gerard asked me what my special field of interest in astronomy was. I promptly replied that I was most interested in the Milky Way and the Cepheid variable stars. He responded, 'That is not an uninteresting field. But I expect to study a more fundamental area, the problem of three bodies and related questions about the nature and origin of the solar system.'"[5] True to his words, this topic was to dominate his life.

In class-conscious Holland, Kuiper's humble background and teachers' college training meant that he was never really accepted as an equal by his fellow students at Leiden. He could not join the student corps, populated by wealthy students. He was snubbed and called *knor*, meaning "oink," or "pig-like."[6] We can easily imagine the negative effects this treatment had on him, but it also made him independent and self-reliant. He could survive without popularity. It dictated much about the way he behaved and the way he treated others, and it lined up enemies against him. However, it is also the secret of much of his success.

Kuiper completed his first degree in astronomy in 1927 and immediately went on to graduate studies. At that time the science departments at Leiden were at the peak of their fame, known for such prominent scientists as Ejnar Hertzsprung, Antonie Pannekoek, Paul Ehrenfest, Willem de Sitter, Jan Woltjer, and Jan Oort. In the 1920s the professors at Leiden were enormously respected by their students. Reason was their religion, and there was no place for subjectivity. Human affairs and human emotions were inferior to facts and knowledge. In this Victorian environment Kuiper received an excellent education and developed a close association with some of his teachers, relationships that would last a lifetime. His teachers must also have had confidence in him because at one time he was a tutor to Ehrenfest's son, Paul.

Kuiper's advisor for his PhD research was Ejnar Hertzsprung.[7] Born in Copenhagen, Hertzsprung also inherited an interest in astronomy from his father. Initially, he was persuaded by his father to become a chemist for its better financial prospects. After obtaining a degree at the Polytechnical Institute of Copenhagen he spent several years as a chemist in St. Petersburg and then at Leipzig, where he worked with the famous Wilhelm Ostwald. He returned to Copenhagen to take up astronomical photography and then moved to Göttingen, Germany, where he was soon appointed as a professor. At the outbreak of World War I Hertzsprung moved to Leiden to assume the position of adjunct director of the observatory. He then set out on a distinguished career as an astronomer, using the Cepheid variable stars to measure distance. Cepheid variables go through their cycles of brightness in a way that is proportional to their true brightness, so their apparent brightness as seen from Earth is an indication of their distance from Earth. Hertzsprung also classified stars by their spectrum and brightness. He was the first person to distinguish red giants and red dwarfs. The classic Hertzsprung-Russell diagram, which plots brightness against spectral class, derives from this.

Eclipse Expedition and the Beginnings of a Love for Indonesia

Kuiper made two interruptions to his PhD research at Leiden. The first was six weeks in 1928 he spent at the Astrophysical Observatory in Potsdam, Germany.[8] The second was an eight-month trip to the Dutch East Indies to observe the 1929 eclipse of the Sun. Eclipses have always been important in astronomy because they afford a chance to look at the outer atmosphere of the Sun. After publication of Einstein's general theory of relativity in 1916, eclipses also became an important means to test the theory. The leader of the expedition was the prominent Dutch astronomer Jan Van der Bilt. Seven additional astronomers, seven management and technical support staff, and three students, including Kuiper, joined the trip.[9] Most were assigned specific duties during the eclipse.

Van der Bilt and Minnaert published a technical article in 1929 describing the eight-month expedition.[10] They raised 20,900 florins ($150,000 in 2018 dollars) from the state and various corporations and individuals. What the Dutch astronomers wanted to observe was the spectrum produced just before total eclipse, called the flash spectrum, which cannot be seen at any other time. The flash spectrum contains information on the lower atmosphere of the Sun, such as pressure, temperature, and composition. They also wanted to observe the outer atmosphere—the corona—which also can only be seen during eclipse. Therefore the group took two major pieces of equipment with them: a spectrograph to record spectra and a camera designed to record the corona. A site measuring 70 by 80 meters (77 by 87 yards) near the town of Idi in Sumatra was established, with the instruments on a north–south line. Light from the Sun was reflected into the instruments by a system of mirrors. On the eve of the eclipse Kuiper discovered that another astronomer had incorrectly assembled the spectrograph, but he was able to correct the matter just in time.[11]

The spectrometer, like the camera, has been an essential tool for astronomy since the mid-nineteenth century. It is a means of separating radiation into its various energies—light, infrared, radio, X-ray, and gamma-ray radiation. Visible light radiation is well known because we are all familiar with how prisms (or water drops in the case of a rainbow) split the light into its colors, blue radiation being higher in energy than red radiation.

The efforts of the Dutch team in Sumatra in 1929 were to no avail. The clouds closed in and the effort was largely a failure.[12] A few scientific papers were published, but their contents were far short of the expected flood of new information. The group did obtain a spectrum for the extreme limb of the Sun shortly before totality and were able to show that there was no difference in the spectrum at limb and center, contrary to some predictions.

It was not all astronomy on the island of Sumatra for Kuiper. Kuiper had a passion for art, and as a student growing up in Holland he produced several fine sketches. These sketches were highly meticulous renderings, some of which were lightly tinted with watercolors. There were sketches of a church with trees lining a canal, a windmill nestled in the trees on the side of a lake, a row of houses, street views with houses, more windmills, and more lines of trees. He painted on paper and on rough-cut cloth. While preparing for the eclipse, he made a series of beautiful drawings that reflect his fondness for Indonesia. The young astronomical protégé had an artistic eye, another element of his later success that was emerging. Years later he hired an artist to make large plaster models so he could better visualize the surface of the Moon.

Contact with Aitken

Kuiper had made it clear from the outset that his interests now centered on double stars, an interest he shared with Robert Grant Aitken, director of Lick Observatory. On Christmas Eve, 1929, Kuiper wrote to Aitken. "As a student here at Leiden," he wrote, "I started about 3½ months ago measures of double stars with the 10½ inch refractor, formerly used by Dr. W. H. van den Bos."[13] Van den Bos was an observer who started at Leiden and who moved to Cape Town in 1920 to make a career in locating double stars. The young student went on to say that he had measured thirty pairs of stars on at least three nights each. He attached his data and hoped for criticism and advice from the senior astronomer. He also asked for suggestions for any "special pairs" that he could measure.

The third and final paragraph of his letter describes his plans for the future. It was a roadmap for his coming decade of double-star research, or to be more precise, binary-star research. Double stars are often just accidental alignments of two stars, but Kuiper was interested in binaries, stars that orbit each other. He wanted to make enough observations to perform a statistical analysis, and he wanted to focus on stars that are closest to Earth so he could obtain better data. Nearby stars appear to move back and forth as Earth revolves around the Sun (an observable parallax) or they may be seen moving among the background stars (proper motion). Kuiper was interested in ascertaining how many nearby stars were binary and what was the nature of the relationship between the two stars, how far apart, how bright, and what the classes of the stars were. Many astronomers were working on such topics in the 1920s.[14] Aitken responded to the Christmas Eve letter with encouragement, enclosing suggestions and looking forward to the results of the statistical analysis.

What Kuiper did not say in his Christmas Eve letter was why he was interested in double stars. This was not to emerge until many years later when in 1971 he received

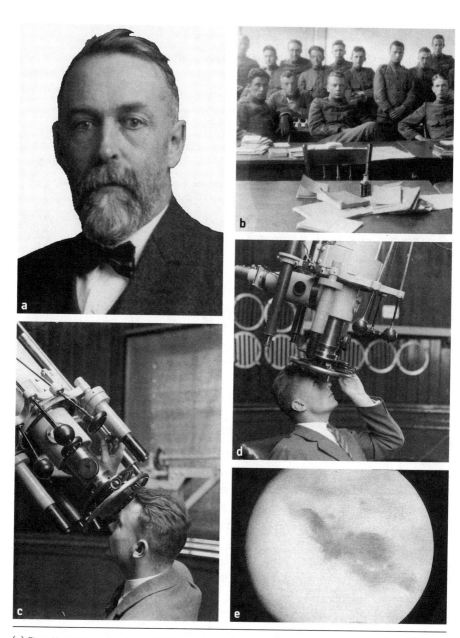

(a) Ejnar Hertzsprung. Courtesy of Alchetron, https://alchetron.com/Ejnar-Hertzsprung. (b) Kuiper during his nine months as a military trainee. Kuiper is in the center of the image. Courtesy of Dale Cruikshank. (c) Kuiper using the telescope at Leiden Observatory while a graduate student. Courtesy of Matthew des Tombe. (d) Kuiper using the telescope at Leiden Observatory while a graduate student. Courtesy of Matthew des Tombe. (e) One of Kuiper's sketches of Mars. G. P. Kuiper, "Visual Observations of Mars, 1956," *ApJ* 125 (1957): 307–17, figure 2.

the Kepler Gold Medal from the American Association for the Advancement of Science. In his acceptance address, Kuiper said, "I felt that I had come to understand the problem of double-star origin, at least in outline; that it was identical to the general process of star formation, from slightly turbulent prestellar clouds upon contraction, with conservation of angular momentum."[15] The solar system was no more than an unsuccessful double star, he felt, with the second star smeared out into a disk that was to become the planets. He also felt that the statistics of double stars would provide an indication of how common planetary systems are. Kuiper's approach explains why his binary research came to an end around the time he started working on the origin of the solar system. However, as we shall see, other important events also were happening when he gave up double-star research.

As with all Dutch youth of college age, Kuiper and his peers were required to serve a year in the military. Most probably they served in the First Force Strategic Reserve Corps that was made of conscripts from The Hague, Leiden, Haarlem, and Westland. Several photographs of Kuiper and his fellow conscripts in military uniform survive, showing him and his cohorts standing proud in their stiff-collared army uniforms.

Thus when Kuiper wrote to Aitken on April 16 he began, "Through absence in military service my response to your letter of February 7th has been greatly delayed."[16] He then went on to thank Aitken for his advice and suggestions. He continued, "I am also convinced that for the general conditions of seeing it is better to go not too far to the small distances [referring to the apparent distance between the two stars in the binary]."[17] "Seeing" is a term used by astronomers to describe the steadiness and sharpness of their stellar images in the face of the blurring and shimmering caused by the atmosphere.[18] Kuiper continued, "Only when the circumstances were favorable I permitted me to go further, partly because it would be possible to compare the results with real measures, made at large telescopes. Should I in future ever be able to work with a large refractor this training might be of some value." Was this the first hint at his career plans?

Kuiper wrote again in October 1931 and December 1931, describing new results and confirming some of Aitken's data.[19] By the time his thesis was complete in 1933, the young astronomer had obtained data for about 830 double stars. Kuiper's first paper, published in 1933, reported relative positions and magnitudes of 225 binary stars determined with Leiden's 10.5-inch refractor telescope.[20] The result of his correspondence and exchange of data with Aitken was that in January 1932, while Kuiper was in the middle of his graduate research, Aitken began to make moves to bring Kuiper to Lick Observatory. These efforts would find success in the summer of 1933.

Kuiper published several research papers while at Leiden: a paper on Mars, a paper on the Titius-Bode law relating planetary distances, and several papers on the orbits of binary stars. He spent considerable time at the observatory and documented his fondness for telescopes through many formal photographs depicting him standing by, sitting alongside, and, of course, viewing through the telescope.

Move to Lick Observatory

Lick Observatory, which was to become the next home for the young astronomer, was half a world away. It had become the first permanently occupied mountaintop observatory in the world in 1887.[21] James Lick, a carpenter and piano maker who made his fortune on property speculation during the gold rush, bequeathed $700,000 (about $22 million today) for the creation of an observatory on Mount Hamilton. The observatory was built, at a site Lick identified, between 1876 and 1887 in the Classical Revival style. It consisted of a long hall lined with offices and an observatory dome at each end—a large dome housing the 36-inch reflector and a small dome housing a 12-inch refractor. A refractor telescope, like the common hand-held telescopes, concentrates the light using a lens, whereas a reflector uses a concave mirror. A year after completion, the observatory was donated to the University of California and run by the astronomy department at the Berkeley campus. The warm and growing relationship between Kuiper and Aitken meant there was never any doubt that Kuiper was destined for Lick.[22]

Aitken controlled the funds for the Martin Kellogg Fellowship and the Morrison Fellowship. He had three prospects for these positions, but one had declined and Aitken was unenthusiastic about the others. Then there was a candidate Aitken was inclined to consider "with some care," G. F. (mistake for P.) Kuiper of Leiden. In January 1932 he wrote to Armin Otto Leuschner about incoming researchers at the observatory. Leuschner, then the dean of the graduate school at UC Berkeley, had to approve all graduate admissions to Lick Observatory. In fact, many years earlier he was the first graduate student at Lick, although a conflict with his adviser meant that he finished his PhD at the Humboldt University of Berlin. He then became a professor at Berkeley and developed a reputation as an astronomer and educator. Aitken wrote to Leuschner about Kuiper, "Hertzsprung has written of him in very complimentary terms and I know something of the work he has done in the observation and the theoretical investigations of binary stars. I am waiting until I have seen Dr. de Sitter before making any decision as to whether or not I offer him the Morrison Fellowship."[23] He was expecting Willem de Sitter, another distinguished astronomer at Leiden, to visit Lick soon. Then, somewhat ominously, especially in view of later events, Aitken wrote, "I quite agree with you in your suggestion that we should be cautious about offering fellowships to our European friends."

On January 27, 1932, Aitken wrote to "My dear Mr. Kuiper," mentioning that he had spoken to de Sitter about Kuiper's desire to come to Lick. Unfortunately, he had learned that the young man would not complete his doctoral work before the academic year started and he might have to wait until the following year. Aitken discussed various options but preferred that Kuiper not delay an entire year. The stipend was $1,200 per year, there was no charge for room and board, and living expenses were moderate, "Aside from purely personal expenses—trips to town, clothing, tobacco (if you smoke), etc.—your monthly expenses need not be $45.00, and may be less. The rooms are plainly furnished, but bed room linen and blankets are not provided. If convenient, it will be well to bring these with you. Otherwise, they can be purchased in San Jose at reasonable cost."[24] Aitken stressed that there was no commitment beyond a year and asked for a cable with Kuiper's decision: "accept July," "accept October," or "delay." It is interesting to note that the information that would now be the basis of an application—birthplace, age, name, academic record, and publications—were asked for in a postscript "for our University records." The real decision was made on the basis of personal knowledge and the opinion of mutual acquaintances. But the young astronomer needed to complete his research program and the senior astronomer would have to wait a year. This was obviously a disappointment to Aitken, who was just a few years short of retirement.

In January 1933 Aitken wrote to tell Kuiper that the deadline for selecting fellows for the next academic year was close.[25] In the days of the Great Depression income from investments was down and the stipend for the fellowship was smaller, only $950 for 1933–34. Again, there was no guarantee of support for a second year. "I am not at all sure you will find this adequate," Aitken wrote. After more exchanges, Aitken made a formal offer to Kuiper for a Kellogg Fellowship on March 7 "on the basis of your credentials and my conversations with Professor de Sitter and Professor Hertzsprung." Paraphrasing, Aitken went on to say that it would be desirable for Kuiper to start July 1, but August 1 was also possible. The 12-inch would not be as available as in the past year, but Kuiper would have some time on this and the 36-inch, and if he was interested in photographic work, that would be possible with the 36-inch. "We can discuss your program when you arrive."[26]

The months rolled by, and on June 14, 1933, a bright red "Radiogram," pronouncing "World Wide Wireless" by RCA Communications, arrived on Aitken's desk. "Because doctors degree June thirtieth prefer arrival August first Kuiper."[27]

The newly graduated Dr. Gerrit Kuiper left Rotterdam on the S.S. Black Eagle to New York where on July 20, 1933, he wrote to Aitken about his travel plans. He apologized for the delay in his departure for Lick. "The observations required had been done for some time already, but the discussions of the results took some more time than I had estimated to be necessary."[28] The full data are in his Dutch thesis, but a summary was due to appear in the Annals of the Leiden Observatory. "I hope you will find that the conclusions arrived at are of sufficient importance to excuse the repeated delay in my coming to Lick Observatory." A second publication in the *Bulletin of the Astronomical Institutes of the Netherlands* (*BAN*) was to contain measurements for about 830 double stars. He went on to explain how because of the letter of introduction from the president of the University of California that the "quota immigration visa" was no problem, which gave him unlimited time in the United States. He had brought his books, tables, manuscripts, and calculations with him but was having his objective gratings sent by mail. "The first two days in the United States have been extremely pleasant and most interesting."

This optimistic account of his time in New York contrasts with the account remembered by his daughter. According to Lucy, Kuiper started his relationship with the new nation by leaving his papers and his wallet on the New York subway.[29] After an anxious few hours, he reconnected with his property in the subway lost-property office. However, he saw a great deal of New York. The next day an American teacher took him by car to Chicago, where he took the train to Oakland and then another train to San Jose. Finally he traveled to the observatory by the local bus. On August 1, Gerrit Kuiper reached Lick Observatory on top of Mount Hamilton. When he passed through it in 1933, San Jose was a town of 58,000. Now, in 2018, it is just over 1 million, having grown with Silicon Valley.

CHAPTER 2

Double Stars, White Dwarfs, a Nova, and Controversy

BINARY STARS HAVE ALWAYS been important in the history of astronomy. One of the pioneers in the use of the astronomical telescope, and a giant figure in the history of astronomy, is William Herschel, who was surprised to find that a large fraction of the stars he observed were double stars.[1] In subsequent years Herschel realized that while some of these double stars were accidents of alignment, some, called binaries, orbited each other. From these much could be learned, especially if Newton's law of gravitation was applied to the observations. The precise orbit and the relative masses of the two stars could be deduced.[2] Sometimes one of the stars would be a white dwarf that was, in the first few decades of the twentieth century, a newly identified star type that was causing much excitement among astronomers. The young Gerrit Kuiper, known to the staff of Lick Observatory as Gerry, took full advantage of his new position, starting early and finishing late, observing when other astronomers thought the conditions too poor, taking little time off. He easily earned another fellowship for a second year. However, his time at Lick was not without its difficulties. First, his insensitive manner of reporting his research led to animosity with Willem Luyten, a Dutch astronomer with a pedigree not unlike his, that was to last a lifetime and foreshadowed enmity with many others. Second, when his second fellowship ended he was not offered the expected position on the permanent staff. It was said by historian David DeVorkin that prejudice against foreign nationals had prevented this, but it may have been a blessing in disguise.[3]

When Kuiper arrived at Lick he found a group of astronomers who were either well known or would soon become so.[4] In addition to Aitken, there were three astronomers, five associate astronomers, an assistant, four Fellows, and six permanent staff (a

(a) Robert Aitken. Courtesy of Special Collections, University Library, University of California Santa Cruz, Lick Observatory Records. (b) William H. Wright. Courtesy of Special Collections, University Library, University of California Santa Cruz, Lick Observatory Records. (c) Robert Trumpler. Courtesy of Special Collections, University Library, University of California Santa Cruz, Lick Observatory Records. (d) Nicholas Mayall. Courtesy of Special Collections, University Library, University of California Santa Cruz, Lick Observatory Records. (e) Willem Luyten in 1935. Courtesy of University of Minnesota Media Digital Archives. (f) Joseph Moore. Courtesy of Special Collections, University Library, University of California Santa Cruz, Lick Observatory Records.

photographer, a foreman/carpenter, an instrument maker, an electrician, and a janitor). William Hammond Wright, Joseph Haines Moore, and Robert Julius Trumpler were full astronomers, and Nicholas Ulrich Mayall was an associate astronomer. All were to play important roles in Kuiper's life.

Wright graduated from the University of California in 1893 to become assistant astronomer at Lick Observatory. Between 1903 and 1906 he worked on establishing a southern observatory for Lick at Cerro San Cristobal near Santiago de Chile. In 1908 he was promoted to astronomer. In 1918 and 1919 he worked at the Proving Ground for the U.S. Army. In the year Kuiper left Lick, Wright became director, a post he held until 1942. He is best known for his work on radial velocity of stars—the speed at which they appear to be moving away from us—and for the design of a spectrograph he used in this work.[5]

Moore obtained his first degree at Wilmington College in 1897 and his PhD in astronomy at Johns Hopkins University in 1903. He immediately joined Lick Observatory as an assistant to William Wallace Campbell, working with him on radial velocities and eventually publishing a catalog. He also published catalogs of the orbital data for binary stars. Early in his time at Lick he was assigned to their southern observatory in Chile. In 1936 he became deputy director and then director in 1942, retiring in 1948 and dying a year later.[6]

Trumpler obtained his PhD at the University of Göttingen in 1910. During World War I, he emigrated to the United States, taking a position at Allegheny Observatory and then Lick Observatory. He is most noted for observing that the brightness of the more distant open clusters was lower and redder than expected because of interstellar extinction. He also devised a system for the classification of star clusters that is still in use today.[7]

Mayall obtained his PhD at Berkeley in 1934 and then worked at Lick Observatory, where he remained until 1960, except for a brief period during World War II when he worked at the Massachusetts Institute of Technology's Radiation Laboratory. He was interested in nebulae, supernovae, the internal motions of spiral galaxies, the redshifts of galaxies, and the origin, age, and size of the universe. In 1960, Mayall was appointed director of the Kitt Peak National Observatory, retiring in 1971.[8]

The intellectual environment and the facilities Kuiper found waiting for him at Lick Observatory were well suited to building a successful career in astronomy. The colleagues he met were to be important to him, and the friendships he made were to last a lifetime.

The Telescopes of Lick Observatory

Kuiper observed with a variety of telescopes during his two years at Lick Observatory. For his double-star measurements he used 12-inch and 36-inch refractors, and for his white-dwarf observations he used the Crossley 36-inch reflector fitted with objective gratings. A diffraction grating placed over the telescope objective lens means that instead of seeing just the stars as points of light, each star appears as its spectrum.

The 12-inch refractor at Lick Observatory. Courtesy of Special Collections, University Library, University of California Santa Cruz, Lick Observatory Records.

It is said that Kuiper's favorite telescope for observing binary stars was the 12-inch refractor. This telescope, the first installed at Lick Observatory, was a secondhand purchase. The telescope and its optics were made by the premier telescope makers of the time, Alvan Clark and Sons of Cambridge, Massachusetts. The dome that housed the telescope was the first structure built on Mount Hamilton. In fact, the bricks used for the structure were fired in a kiln on the site. The 12-inch had been used by E. E. Barnard to make many valuable observations; for instance, he observed an eclipse of Iapetus by the rings of Saturn on November 1, 1889.[9] The dome still exists, and the 12-inch telescope remained in use until the 1970s, its mechanical drive clock having been replaced with an electronic drive. The telescope is now being restored in Lick workshops at the University of California, Santa Cruz. The Santa Cruz campus assumed responsibility for Lick Observatory in 1966.

In 1934 Kuiper used the 12-inch with a grating across the objective to allow him to obtain the spectrum of the bright star Sirius and its small companion, a white dwarf. He also made an estimate of the companion's magnitude. His 1934 paper acknowledges the Lick Observatory technician for making the gratings,[10] although some gratings were brought to Lick from Leiden after Aitken arranged for the import duties to be waived.[11] The gratings were still in the dome that housed the 12-inch telescope until at least 1981.[12]

The large dome of the observatory housed the 36-inch refracting telescope, the largest refracting telescope in the world until the 40-inch was constructed for Yerkes Observatory in 1897. Warner and Swasey designed and built the telescope mounting and Alvan Clark produced the lens. The telescope was used by Barnard in 1892 to discover a fifth moon of Jupiter, Amalthea,[13] and by Campbell for his pioneering determinations of the radial velocity of stars using their Doppler shift.[14] Doppler shift is the well-known phenomenon of changes in the spectrum caused by movement of the sources—visible light moving away from the observer reddens the spectrum whereas objects moving toward the observer are blue-shifted.

Like most large telescopes, the telescopes at Lick Observatory had equatorial mounts, meaning they were attached to an axis that was parallel to the axis of Earth. Stars appear to move around this axis as Earth spins from east to west. A second axis perpendicular to the first allows the telescope to move north–south. The two axes are termed "right ascension" and "declination," analogous to Earth's longitude and latitude. Many other types of mount exist, such as the altazimuth mounts used by Herschel and commonly used with small telescopes because of their simplicity. In this case one axis is perpendicular to the surface of Earth and the second is horizontal, so two motions are necessary to track a star. Some telescopes have fixed mounts (zenith mounts); others are fixed in a north–south plane (transit mounts).

In 1895 a 36-inch reflecting telescope was donated to the observatory by British politician Edward Crossley. Crossley had bought the telescope from Andrew Ainslie Common, who between 1879 and 1886 had used the telescope to demonstrate the value of long-exposure astronomical photography. The instrument that came to be known as the Crossley telescope had also demonstrated the viability of metal-coated glass mirrors. Until then astronomical mirrors were made entirely of metal.

When the Crossley telescope was acquired by Lick Observatory it was in terrible shape. Barnard said he wouldn't have paid five dollars for it. James Keeler, who was director from 1888 to 1891, organized the complete refurbishment of the telescope so that by Kuiper's time hardly a scrap of the original instrument remained. A dome was constructed for the telescope a few yards down the hill from the main building. The Crossley was important in establishing the value of large reflectors. It was also one of the first telescopes to receive a slitless spectrograph when Mayall, a longtime user of the telescope, fitted this important attachment.[15] The spectrograph, which is an instrument for recording a spectrum photographically, was essential for Kuiper's identification of white dwarfs. Normally such instruments let the light in through a slit, so a single "normal" spectrum is produced. However, as mentioned above, by removing the slit and looking at a field of separated stars through the telescope, each star produces a spectrum. The stars may then be quickly characterized. (By contrast, a spectrometer records the spectra in the form of an electrical signal either using a strip recorder or, since the early 1980s, digitizing the data and storing them in a computer.)

The Lick Years

Upon arriving at Lick, Kuiper launched himself into two years of intense observing. He observed from sunset to sunrise. The weather had to be truly terrible before he would give up. In the boarding house, he was first in and out for dinner. He rarely left Mount Hamilton, although he had a standing invitation from Kay and Nick Mayall to ride into town with them. Leaving the mountain meant enduring the observatory road, which is said to have 365 turns. He was sure to be carsick. "When he went with us, we had to make repeated stops while he retched, and we suffered silently for him," Mayall wrote many years later.[16] As a consequence, he seldom made the trip. "He would let his hair grow for months until he looked like a long-haired hippie or a bona fide professor," Mayall added.

Double Stars

Kuiper continued his observations on double stars, searching out faint visual doubles, determining their colors and spectra, and extending his observational program to white dwarfs. His work in Holland was published in *BAN*, but at Lick he published in *Publications of the Astronomical Society of the Pacific (PASP)*. As with his student work, everything was published. To the young Kuiper, publication gave him a voice, a way to map his progress, and it enabled him to participate in the grand discussion that was astronomy research.

By 1934 the double-star count for Lick observations was 533, which included 81 newly discovered binaries. On January 31, 1934, Kuiper sent Aitken a request for a second year of funding that included a detailed report on the progress he had made.[17] In the first year he had completed a survey of the brightest stars; in the second year he wanted to survey the fainter stars with high parallax. He wrote, "Together they cover the greatest range in absolute magnitude possible." By taking long-exposure photographs he could detect a binary star not visible to the naked eye. He found a star system consisting of two M (cool) stars, which required the 36-inch, and using the larger telescope he reported five new close binaries, then a sixth a few months later. He found the fastest orbiting binary pair,[18] a binary system in which one of the stars was a binary (making a three-star system),[19] and finally he published a paper on probably the most famous binary star system, Sirius A and B.[20] While his Lick data were appearing in the journals, Kuiper reported the results of his PhD research in a two-part article in *PASP* in 1935, with a special emphasis on the statistics.[21]

As Kuiper's reputation grew, he began to receive invitations to speak at meetings. On June 26–28, 1935, Frederick C. Leonard, an astronomer at the University of California, Los Angeles (UCLA), organized a three-day joint meeting of the Astronomical Society of the Pacific and the Pacific Division of the American Association for the Advancement of Science (AAAS) to be held at his university. The morning session consisted of six invited talks on the terrestrial and cosmic timescale that included a presentation by Kuiper.[22] In the afternoon and for the next few days there were talks on a variety of astronomical topics, including a session on Nova Herculis, and Friday afternoon and evening there was a formal dinner in Kerckhoff Hall and bus tours to Griffith Observatory and Mount Wilson. Leonard published a report of the meeting in *PASP*, and abstracts of the talks were published in both the proceedings and *Science*.[23]

White Dwarfs

One or two of the binary star systems Kuiper observed contained a white dwarf star, which sparked Kuiper's interest in this newly discovered type of stars. Kuiper realized that stars this faint would only be seen if close to Earth, so he focused on stars with large proper motion. When he started this work, two were known (Sirius B and Van Maanen's Star), but his efforts resulted in five more about which he published in *PASP* in 1934 and 1935.[24] One of the stars Kuiper discovered was A.C. 70° 8247, the smallest star then known, half the size of Earth with 2.8 solar masses and spectral class O0. He also discovered that Wolf 219 was a white dwarf and very similar to Van Maanen's Star in luminosity and spectrum. By 1941, when Kuiper summarized the known white dwarf stars, thirty-eight were known. Kuiper claimed responsibility for fourteen and credited Luyten for eleven, although he added that he had spectroscopically confirmed Luyten's work. We now know that these stars, generally comparable with the Sun in mass but only equal to Earth in size, are formed when a star collapses after the exhaustion of its nuclear energy source. In 1983, Subrahmanyan Chandrasekhar received the Nobel Prize with William Fowler for his theoretical explanation for the evolution of stars, which included the formation of white dwarf stars.

Nova Herculis

In 1935 Kuiper was distracted from his double-star and white-dwarf research by a spectacular nova, an exploding star often visible to the naked eye. In fact, 1935 was the year of "the Great Nova," the brightest nova of the twentieth century. The nova in the constellation Hercules was first observed by a British amateur astronomer, J. P. M. Prentice, on December 13, 1934.[25] The nova reached its peak brightness on December 22, with an apparent magnitude of 1.5, and was visible to the naked eye for several months.[26] The event caused a public stir. Recently Professor Brad Ricca of Case Western Reserve University has suggested that the creators of Superman, writer Jerry Siegel and artist Joe Shuster, were inspired by Nova Herculis. However, by the time they sold the idea to DC Comics, the object bestowing superpowers on Clark Kent had changed from the nova to the planet Krypton.[27]

In April, Kuiper took a break from his work and went to Bryce Canyon. He stayed at Ruby's Inn, which is now a cowboy-themed hotel operated by Best Western. Kuiper kept a menu as a souvenir. A cheap breakfast was 40 cents; a fancy breakfast was 65 cents.[28] From the inn, Kuiper wrote to Aitken that he had visited Mount Wilson and Flagstaff, Arizona, where he saw "many interesting things."[29] However, his main topic was that his travels were delaying his latest paper on the nova. Kuiper published his observations of the nova steadily through the first half of 1935, and on July 4 he reported that it was now a double star.[30] He postulated that the system was made by the fission of a single star. He then drafted a press release: "Mr. Gerrit Kuiper of the Lick Observatory found that the new star, or 'Nova,' in the constellation of Hercules had split up into two parts."[31]

Kuiper went on to publish two more papers on the Great Nova, one in 1937 and one in 1941. The first showed that the separation of the two objects was increasing, an observation confirmed by several astronomers. The second paper was a summary of current data. It included plots of light intensity against time (light curves), estimates of the separation of the two components, and sketches. Nova Herculis is now seen as a prototype of a cataclysmic variable in which an apparent star is a binary star consisting of a donor star that is losing mass to the accretion disk of a white dwarf.[32] The apparent separation observed by Kuiper and others may be a result of the evolution of the system rather than the splitting of a single star.

Kuiper, Mars, and Astrobiology

Kuiper had published a large three-part paper on Mars as a graduate student, in which he reviewed the latest information on the nature of its surface.[33] While at Lick Observatory, he observed clouds on Mars with the 12-inch and wrote a brief note about them for *PASP*.[34] The paper was coauthored by Wright, a rare example of Kuiper having a coauthor in those days. Wright was a pioneer in the use of filters in astronomical photography, and presumably the images that appear in the article were taken by him. The clouds were bright and similar in color to the polar caps. There is a significant reference to "canals" (in quotes in the paper) emanating from the north pole.

The reference to canals on Mars was sensitive in the astronomical community and it played an important part in the Kuiper story. When the Italian astronomer Giovanni Schiaparelli saw lineations on Mars during the 1877 opposition (when it is at its closest to Earth and its details clearest), he referred to them as "canali."[35] Literally translated, this means "channels," but Schiaparelli meant them to be nothing more that straight lines. We now know that these features are an optical illusion: the human eye tends to draw straight lines on weak blurry patterns.[36] But to Percival Lowell at the observatory he built in Flagstaff, the canals were of the most profound importance.

Lowell was born to a wealthy Boston family. He graduated from Harvard University in 1876 with a distinction in mathematics. He later received honorary degrees from Amherst College and Clark University. Upon graduation he ran a cotton mill for six years and then launched himself into extensive travels in the Far East while serving as a U.S. diplomat. He wrote books on Japanese religion, psychology, and behavior, and on various aspects of Japanese life, language, economics, and geography. He wrote on the development of personality, postulating that human progress is a function of the qualities of the individual and imagination. In 1892 he was elected a Fellow of the American Academy of Arts and Sciences. At about that time, Lowell became captivated by Mars. He had found a copy of Camille Flammarion's *La planète Mars* and consequently dedicated the last twenty-three years of his life studying the planet. In 1894 Lowell chose Flagstaff, a town in Arizona Territory with an altitude of 6,900 feet (2,103 meters), for the location of his observatory.

The canals, Lowell argued, were clearly made by an advanced civilization busily pumping water from the poles to the lower latitudes. He published three books on this topic: *Mars* (1895), *Mars and Its Canals* (1906), and *Mars as the Abode of Life* (1908). While astronomers published paper after paper debunking Lowell's assertions and were astounded that such nonsense was coming from a highly educated man, the public lapped it up. The excitement defeated common sense because it was an idea that people wanted to believe. The same effect is alive and well today.

It was no coincidence that H. G. Wells's work of science fiction, *The War of the Worlds*, appeared at the same time. The story appeared first as a series in popular U.S. and U.K. magazines in 1897, and then as a book in 1898. It has never been out of print, and at least six movie versions have been produced. The astronomers drew in their ranks. Solar system science, Mars in particular, was tainted, and it was not a suitable subject for real astronomers. Kuiper must have known this when he published his papers on Mars, first as a student,[37] and then as a Fellow at Lick, yet he mentioned canals. He may have meant it in the serious scientific sense of the day—straight lines—but what he said was "canals."

The Luyten Complaints

Among the string of papers that flowed from Kuiper's pen in 1935, and between the first and second part of a review paper on double stars, was what appeared to be a noncontroversial report, "A New White Dwarf of Large Parallax."[38] Interesting, but nothing one would think to inflame the anger of rivals. But that is exactly what happened. The review paper on double stars and the announcement of another white dwarf in the early days of Kuiper's career were the origins of an acrimony that persisted long after Kuiper's death.

Wolf 219 was the third white dwarf discovered by the young postdoc at Lick, doubling the known number of such stars to six. Kuiper reported that the magnitude of Wolf 219 was 14.8 and its parallax was 0.068 ± 0.008 arc seconds. ("Parallax" refers to the distance a star appears to move across the night sky as Earth moves from one side of its orbit of Sun to the other. The arc in the sky corresponding to this distance is measured in seconds, thus "arc-seconds.") Wolf 219 was identical in its color and luminosity to Van Maanen's Star. But then came the statement that enraged Luyten. Kuiper brought up a recent article by Jaako Tuominen that had appeared in *Lund Meddelande* ("Lund Messenger"), a publication of Lund Observatory in Sweden.[39] Tuominen was a young Finnish astronomer who was educated in Turku and worked in Europe and the United States between 1933 and 1950 (war years excepted) prior to obtaining a professorship at the University of Helsinki in 1951. At Helsinki he became distinguished for introducing astrophysics and radio astronomy to Finland.

In his *Lund Meddelande* article Tuominen described eight new white dwarf stars of large parallax. Kuiper wrote that "this amazing result calls for a closer examination of the basic data." After studying the star with the quartz slitless spectrograph attached to the Crossley reflector, Kuiper claimed that the first of Tuominen's white dwarfs, Wolf 1056, has a spectrum of an M3 star. This puts it on the main sequence. Stars are essentially classified by their size and color using the Hertzsprung-Russell diagram, which will be discussed below. On the diagram the star's brightness, corrected for its distance from the Earth, is plotted against its temperature as determined from its spectrum, blue (hot) stars on the right and cool (red) stars on the left. The temperature axis is divided into regions with letter-labels (O, B, A, F, G, K, M, and N), and within these letter divisions they are further subdivided by numbers 1–9. For example, the Sun is G2 and Wolf 1056 is M3. Stars in the top right of the diagram are red giants, stars on the bottom right are red dwarfs, and stars in the bottom left are white dwarfs. Most stars plot along the diagonal top left to bottom right, and this is referred to as the "main sequence." Kuiper claimed that the second, Kapteyn's Star, is a normal red dwarf. Using parallax measurements, to determine distance, that Kuiper considered superior, he found that the other six have luminosities too large for white dwarf stars and were also on the main sequence. Kuiper's last sentence in the note was, "Our conclusion is that all eight stars considered very probably belong to the main sequence, and are in no way exceptional." To one white dwarf astronomer, Willem Luyten, this was all too much.

Willem Jacob Luyten was born in the Dutch East Indies, now Indonesia, and educated in Holland, with a first degree at the University of Amsterdam and a PhD at Leiden University with Ejnar Hertzsprung as his advisor. His enthusiasm for astronomy was stirred, as with so many astronomers, by witnessing a comet in 1910. The identity of this comet is unknown; it could have been Halley's Comet. As well as sharing an alma mater and mentor with Kuiper, Luyten's early career path was similar. From 1921 to 1923 he was at Lick Observatory, and from 1923 to 1930 he was at Harvard College Observatory, spending much of the time at their observatory in Bloemfontein, South Africa. In 1931 he obtained a position at the University of Minnesota, where he spent the remainder of his career. Although he retired in 1967, he remained an active astronomer until he died in 1994. Like Kuiper, Luyten was interested in the proper motions of stars and white dwarfs. After the unusual faintness of these stars was first recognized in 1910, it was Luyten who coined the name "white dwarf" in 1922.

On May 7, 1935, an inflamed Luyten wrote to Aitken,

> I have watched with interest the several notes and articles which G. P. Kuiper has published in the A.S.P. [*PASP*] the last few months. He has done a lot of good work (although I was sorry to see that he did not think it worthwhile to mention the fact that Hertzsprung put him up to doing these things). I have also watched with alarm, however, the fact that he is so careless in putting his articles together, and does not take the trouble of looking up the literature. His last two things, e.g. on the "white dwarfs" which Tuominen talks about was completely unnecessary, there existing much better material than he obtained at the same observatory where he is working and there having been published a much more complete discussion, some of it in the same journal to which he now contributes.[40]

Criticism of Kuiper's rebuttal of Tuominen's work was not enough. Luyten went on to get upset about Kuiper's summary of double-star research. "I strongly object to the way in which he lays down the law." Luyten goes on to write,

> Insofar as I know no one has as yet appointed him sole arbiter of what is and what is not important in connection with double stars. Especially, however, since the article is published in the A.S.P. of which you are the editor, and from Lick Observatory, of which you are the director, there would be justification in the belief that the article has your

sanction, yours as the foremost author on double stars. Now I hope that this is not so, and for that reason I am submitting two papers for publication in A.S.P. The one on double stars could go either as an article or as a letter to the editors whichever way you prefer.

In other words, Luyten thought it unfair how Kuiper trashed the work of this young Finnish colleague in his white-dwarf paper and disliked the arrogant attitude of his double-star paper.

Six days later Aitken responded.[41] He appreciated that Luyten thought Kuiper did good work, and stressed that, as it stated in the journal, the words of the authors were their responsibility. He also wrote that "some of the points you make are quite in harmony with my own views" but "if I were to decline an article simply because I did not agree with the statements made in it I should assume a dictatorship." On the subject of the two papers Luyten had offered him, Aitken was quite willing to print them, but Kuiper's double-star paper was to be published in two parts and he suggested Luyten hold his criticisms until the second part had been published. In the meantime, Aitken suggested that Luyten talk to Hertzsprung and Dutch South African astronomer Willem Hendrik van den Bos—who would attend the next International Astronomical Union (IAU) meeting in July—about presenting a "dispassionate review" of the three Kuiper papers.

Luyten wrote back on May 17, agreeing to Aitken's suggestions for publishing his two papers but remarking that "I believe he [van den Bos] is a better judge of these matters than Hertzsprung. When I was at Leiden last summer I felt that both he [Hertzsprung] and Oort were inclined to take the view that Kuiper could do no wrong. Hertzsprung especially is apt to attack my criticisms either because part of them are obvious and hence not worth printing, and the remainder because there might be some technical slip in the language, which is really in itself trivial."[42]

Luyten had the bit between his teeth and was not going to let go. "One may be pardoned for inferring that when a beginning student of double star astronomy publishes articles from your observatory that he had shown these articles to you, and taken your advice about them." And he repeated his criticism of Kuiper, who he believed had been "extraordinarily lucky in finding 3 white dwarfs," for ignoring everybody else's work. When he was at Lick, Luyten claimed, he worked as hard as Kuiper, looked at as many stars, even though he did not find any white dwarfs. Campbell, director at the time, had censored several of his papers, for which Luyten was grateful since they would have made him look foolish.

Aitken delayed his response to the latest letter, this time on the grounds that he was preparing to move to Berkeley, but on June 8 replied that "Kuiper is doing some very good work and I expect him to do still better work in the future. As I wrote to you, however, he will probably be somewhat less certain of some of his conclusions ten years from now than he is at the present moment."[43] In this last statement, Aitken was to be disappointed.

It is certainly true that Luyten was not without his unattractive character traits. In his biography for the National Academy of Sciences (NAS), Luyten's colleague at the University of Minnesota, Arthur Upgren, wrote,

> Luyten had a talent for alliterative broadsides in his publications. Some of his feistiest papers bore such titles and references to colleagues as "The Messiahs of the Missing Mass," "More Bedtime Stories from Lick," and "The Weistrop Watergate," the last, of course, coming much later. They made for very amusing reading, but they were too disrespectful and too full of negative allusions to his colleagues and their work. They were not in the best interest of science, even though much in them was factually correct. In his later years, Luyten referred to himself as a curmudgeon, an epithet bestowed on him at times by others.[44]

However, it is equally true that this exchange, so early in Kuiper's career, captures much about Kuiper. His colleagues were to be in one of two camps: they either admired his extraordinary abilities as an astronomer and put up with his idiosyncrasies, his self-centered approach, his excess of confidence, and his somewhat insensitive attitude to others in the field, or they could not get beyond his personality and thoroughly disliked the man despite his accomplishments.

In October 1937, after Kuiper's move to Chicago and after Wright had assumed the directorship of Lick Observatory, Wright showed Kuiper the correspondence between Luyten and Aitken. One can only guess at Wright's motives. Kuiper took the letters seriously enough to laboriously copy every word for his own records; his handwritten copies now reside with the Struve correspondence in the Kuiper papers in the University of Arizona. But they did not change him even though he went through life knowing this was how at least one colleague felt, and he must have guessed there were others. It led to further conflicts with Luyten and, more significantly, with an equally important character in twentieth-century planetary science, Harold Urey. Geoffrey Burbidge was another adversary. Ultimately Kuiper's personality was a factor in his leaving Chicago in 1960.

Failure to Get Permanent Position at Lick and an Offer from Harvard

Just months after the Luyten complaint letters, Kuiper left Lick Observatory. The conditions under which Kuiper left Lick have been documented in some detail by

Donald E. Osterbrock in his history of Yerkes Observatory and by David DeVorkin in his biography of Henry Norris Russell.[45] Russell was one of the best-known astronomers of the early twentieth century; DeVorkin describes him as the "dean of American astronomy." After obtaining his bachelor's degree and PhD from Princeton University, Russell spent short periods at Cambridge Observatory and the Carnegie Institution. He then obtained a faculty position at Princeton, where he spent the rest of his career. He is best known as the coinventor of the Hertzsprung-Russell diagram, but his contributions were widespread, mostly focused on the determination of the composition of the Sun and stars from their visible spectra.[46] Russell thought highly of Kuiper from his visits to Lick, and he had heard only good things from Harlow Shapley, chair of the astronomy department, and Bart Bok, an assistant professor, at Harvard University.

As Kuiper completed the second year, there was talk of him replacing Aitken. Aitken liked Kuiper. Even though, according to DeVorkin, Kuiper was brash, impetuous, and a brazen self-promoter, Aitken respected his research and wanted him to stay. However, nativist colleagues like Wright wanted to hire Americans. Hearing in the fall of 1934 that Kuiper might leave Lick, in January 1935 the persuasive Bok encouraged Kuiper to fight it out.[47]

In fact, Aitken made considerable efforts to secure outside support for Kuiper. The university was applying pressure not to hire foreigners. Thus Aitken asked his close friend H. A. van C. Torchiana, the consul general of the Netherlands, if he could help. Kuiper wanted his friend to try to persuade the Holland-America Chamber of Commerce (which was affiliated with the consulate) to find funds for a fellowship. In his four-page proposal to the chamber, Aitken mentions that Kuiper was thinking of taking a position in at the Bosscha Observatory in Java. He argued that it would be a shame to lose such talent to the United States.[48]

Russell frequently wrote a column for *Scientific American* and used it to campaign for Kuiper.[49] Russell also wrote (via Aitken) to Robert Gordon Sproul, president of Berkeley, and to Aitken and Leuschner (who were on the search committee to find a replacement for Aitken), saying that the loss of Kuiper would be a disaster to binary-star astronomy and Lick, and asking Aitken whether nationality was involved in Kuiper's potential departure. He argued, tactlessly, that Kuiper had made discoveries Aitken should have made. Offended, Aitken denied nationality was a factor and told Russell not to worry about Lick.

While Kuiper liked his position at Lick and asked Russell to help make his case, he was always drawn to Bosscha Observatory and the island he enjoyed so much as a student. Bosscha Observatory, named after the owner of a tea plantation who donated the land, is the oldest observatory in Indonesia, having been in operation since 1928.[50] It is located in Lembang, West Java, on a hilly six hectares of land 4,300 feet (1,310 meters) above mean sea level. The observatory has an unobstructed view of the sky while also having easy access to Lembang, once proposed to replace Batavia (present-day Jakarta) as the Dutch colony's capital.

Wright campaigned for the directorship at Lick and was appointed on condition that he hired astronomers trained in modern astrophysics and atomic theory. Aitken and Wright wanted to hire Arthur Bamberg Wyse (Wright's student),[51] Mayall, and Kuiper, but there were budget constraints. Ultimately, Wyse was hired on university funds and Mayall was hired on observatory funds. Trumpler was angered that Kuiper was not appointed. In fact, both Aitken and Wright placed Kuiper higher than Wyse, but Aitken defended himself by saying Kuiper, thinking he was doomed at Lick, had already accepted a position in Java.

DeVorkin points out that while Russell was strongly against nativism, he declined to hire Kuiper at Princeton. Harvard and Princeton were in competition to hire German American astronomer Rupert Wildt, and both had the funds. Wildt, who had interests in planetary atmospheres, particularly those of Jupiter and Venus,[52] went to Princeton, and Shapley used his funds to hire Kuiper on a one-year teaching position at Harvard.[53]

Kuiper's years at Lick nurtured his growth as an observational astronomer, enabled him to establish a career of frequent publication—which facilitated his research as well as broadcast it—and introduced him to the American astronomical community. He was denied an opportunity for a permanent position and looked toward Indonesia for a position at Bosscha Observatory. However, the short-term appointment at Harvard was to extend his stay in the United States for almost a year, but as soon as he set off for the East Coast, another major opportunity unfolded.

CHAPTER 3

Aging Stars, a Young Wife, and Another Move

THE QUIET ASTRONOMICAL COMMUNITY on Mount Hamilton, California, is a continent away from Harvard College Observatory in Cambridge, Massachusetts, but somehow the distance seemed much greater to Gerard Kuiper in the summer of 1935. The weather in Massachusetts was awful compared to California—moisture in the air and sudden changes in temperature affected his health. The Observatory was close to a big city, which was not ideal.[1] Kuiper was hired to teach undergraduate astronomy in the fall and graduate astronomy in the spring. He was expecting to spend nine months in Cambridge and then take up a position in his beloved Java at the Bosscha Observatory. But not long after his arrival at Harvard, Kuiper received an offer from the University of Chicago, where they were just completing one of the largest telescopes in the world in Texas's Davis Mountains.[2] He agonized, but in the end wrote a three-page letter to the authorities in Java explaining why he was going to stay in the United States: "Opportunities to make a contribution are just better in the U.S."[3] Despite his teaching commitments, which he learned to enjoy, he continued his research on double stars and white dwarfs. He also thought about the evolution of stars, combining Trumpler's observations of clusters and Bengt Strömgren's theory for the evolution of stars to explain the behavior of open clusters as they age. He met some of the best-known astronomers of the day, but in the end he went to the University of Chicago and Yerkes Observatory with two of them—Chandrasekhar, who he met at Harvard, and Strömgren, who was hired at Chicago at the same time. At Harvard, he also met a woman who was to be his companion for the rest of his life, Sarah Parker Fuller. Bok, his friend from undergraduate days, said two events of 1936 were major factors in Kuiper's lifetime success, the move to

Yerkes and his marriage to Sarah.[4] Kuiper's short nine months at Harvard College Observatory are described in this chapter. They were a story of aging stars, a young wife, and another move.

Harvard College Observatory and Its Golden Years

At the end of summer, 1935, Kuiper arrived in Harvard in time for the academic year. Although a lecturer on a nine-month appointment, he was accorded faculty status and invited to college faculty meetings, tea at 3:30.[5] But while he found that the experience of being connected with "the great Harvard University" was "interesting," he complained about the weather, especially "the moisture and the sudden temperature changes." The sensitivity that caused him to be sick on the car rides from Lick now made it difficult for him to cope with the Massachusetts coastal climate.

Kuiper had been brought to Harvard by Harlow Shapley, director of Harvard College Observatory and chairman of Harvard's astronomy department. Shapley had been Russell's graduate student at Princeton, where he used the period-luminosity relation for Cepheid variable stars to determine distances to globular clusters. He helped disprove that Cepheid variables were spectroscopic binaries. Instead they were pulsating stars whose rate of pulsation was proportional to their absolute luminosity. Apparent luminosity is the luminosity observed while absolute luminosity is luminosity as it would appear if the star was one astronomical unit (AU) away, the AU being the distance of Earth from the Sun. Thus Shapely was the first to realize that the Milky Way Galaxy was larger than previously thought and that the Sun was out on a spiral arm.[6] In April 26, 1920, while at Mount Wilson Observatory, he participated in the "Great Debate" with Heber D. Curtis on the nature of nebulae and galaxies and the size of the Universe. Shapley believed that the nebulae and galaxies were part of our Milky Way, while Curtis believed they were outside our galaxy and were "island universes." When Cepheid variables were found in the Andromeda Galaxy, Curtis won the argument. In fact, the galaxies were independent equivalents of the Milky Way. After the debate, Shapley was hired to replace the recently deceased Edward C. Pickering as director of Harvard College Observatory. He served as director from 1921 to 1952. During the first half of his tenure, Shapley was responsible for building a powerful group of astronomers at Harvard.[7]

Thus Kuiper arrived at Harvard during its golden age for astronomy. In addition to Shapley, well-known astronomers like Bok, Peter van de Kamp, Abhijit Saha, Knut Lundmark, and Fred Whipple were there. Chandrasekhar and Jerry Mulders, a solar astronomer from the Dutch city of Utrecht, were, like Kuiper, postdocs in the department.

(a) Harlow Shapley. (b) Henry Norris Russell. (c) Chandrasekhar in 1934. (d) Fred Whipple, Subrahmanyan Chandrasekhar, Gerard Kuiper, and Jerry Mulders at a party at the Whipples' home. Courtesy of NASA/CXC/SAO.

Bok had been at Harvard since 1929, completing his PhD in 1932. In 1933 he was hired as assistant professor and was in that position when Kuiper arrived. His research interests focused mostly on the Milky Way. He was married to Priscilla Fairfield, who worked so closely with him that it is often impossible to separate their contributions. By 1936, the Boks had two children, a son and a daughter.[8]

In the 1930s the department at Harvard included one person who was to become a lifelong friend and who was to play a major role in Kuiper's career, from their meeting in Harvard in 1936 until he left Chicago in 1960. This was Subrahmanyan Chandrasekhar. Chandrasekhar (often abbreviated to Chandra) was born in Lahore,

Punjab, India, and studied at Presidency College in Madras from 1925 to 1930, publishing his first paper as an undergraduate.[9] He was then awarded a government scholarship to pursue graduate studies at the University of Cambridge, where his advisor was Ralph. H. Fowler. Chandra was awarded a PhD in 1933, and the following October he was awarded a fellowship at Trinity College for 1933–37. During this period Shapley offered him a lectureship at Harvard College Observatory for the academic year 1935–36. In September 1936 Chandra married Lalitha Doraiswamy, who he had met as a fellow student at Presidency College. After deliberations on the merits of England, India, and the United States, which involved considerable input from his father, he accepted the Yerkes offer of a position there and he started in January 1937.[10]

Chandrasekhar became one of the most successful astrophysical theoreticians of his time, winning the Nobel Prize in Physics (with William Fowler) in 1983. He is probably best known for his studies of the evolution of stars, in particular white dwarfs, but he worked on a wide variety of problems, from radiative transfer to general relativity to gravitational waves. Many book-length biographies have been published describing his highly productive life.[11]

Also at Harvard was Fred Whipple. Whipple was born in Iowa, the son of a farmer.[12] After an education at Occidental College in Los Angeles and UCLA, graduating in 1927, Whipple obtained a PhD from the UC Berkeley in 1931. That year he joined Harvard College Observatory and studied the trajectories of meteors, confirming that they originated within the solar system rather than from interstellar space. In 1933 he discovered or co-discovered six comets and an asteroid. He is best known for his discussions of the nature of the cometary nucleus.

Kuiper made important lifelong friendships at Harvard. Frank Edmondson, in his oral history deposited with the American Institute of Physics, observed that "it was very interesting at the Christmas party to hear Mulders and Kuiper and Bart Bok singing ribald songs—in Dutch—that nobody else could understand!"[13]

Teaching at Harvard

Gerard threw himself into teaching at Harvard, just as he threw himself into everything he undertook. His fall course had forty students, "a lively group," that he "greatly enjoyed working with."[14] In the spring he would teach Donald Menzel's graduate-level astronomy course. Menzel was at that time a member of the Harvard astronomy faculty. Strong, minimalistic lecture notes in large, bold writing, about half the words underlined, six pages per lecture, characterized his lecturing.[15] As might be expected of an observational astronomer, the laboratory portion of his teaching emphasized observing. Throughout the term two observations of Delta Cephei and occasional observations of Algol were required, and there were evening visits to the Oak Ridge

Observatory on the land Sarah's parents had donated. They also observed the position of the Moon over six nights, as well as the constellations, and completed laboratory assignments and reading, with the appropriate reports. He graded with the same care and rigor that characterized his research, deriving letter grades with the help of histograms of test scores. There was little room for subjectivity in Kuiper's grading.

Research at Harvard—White Dwarfs and the Hertzsprung-Russell Diagram

Kuiper published a summary of his work on double stars in the popular magazine *The Telescope* (later to be merged with a similar magazine to become *Sky and Telescope*) in his year at Harvard.[16] He reported the smallest white dwarf (A.C. 70° 8247) and the shortest-period binary.[17] He also began thinking about the Hertzsprung-Russell diagram.[18]

White Dwarfs

Kuiper had published on two new white dwarfs, A.C. 70° 8247 and Wolf 1346, when at Lick Observatory. In the meantime, Walter Sydney Adams and Milton Humason had made new observations of A.C. 70° 8247 that Kuiper wanted to discuss, particularly regarding its classification and the absence of atomic lines in its spectrum. In addition to the continuum of radiation objects emit by virtue of their temperature, atoms in the stars' atmosphere produce sharp lines. However, Kuiper ended his article by confirming his earlier conclusion that A.C. 70° 8247 was the smallest star yet observed.[19] We now know that A.C. 70° 8247 is indeed an unusual white dwarf, noted for its strong and poorly understood magnetic field. Although Kuiper made and reported the first detailed observations of A.C. 70° 8247, most authors cite the two or three sentences describing the work of Rudolf Minkowski, director of Mount Wilson Observatory, in his 1938 annual report.[20] Kuiper detailed serious criticisms of Minkowski's work in a letter to Chandrasekhar dated October 12, 1937.[21] At this point, Kuiper started to power down his work on white dwarfs. His last paper on the topic, published in 1941, five years after leaving Harvard, was a list of known white dwarfs. It included thirty-eight stars, of which Kuiper had discovered ten. Among his list of accomplishments should certainly be included his contributions to the early years of white-dwarf research.[22]

Hertzsprung-Russell Diagram

Kuiper also produced a major paper using the Hertzsprung-Russell diagram during his year at Harvard. The Hertzsprung-Russell diagram is arguably the most important plot in astronomy. In one simple diagram it captures the major properties of the stars and their relationship to each other, and it enables astronomers to track their evolution.[23] The origin of the diagram probably lies in the realization that much can be learned about stars from their visible light spectra. A catalog of stellar spectra had been published by Annie Jump Cannon and Edward C. Pickering at Harvard College Observatory in 1901.[24] They described what became known as the Harvard Spectral Classification Scheme. A few years later, Hertzsprung observed that red stars (the K and M stars in the Harvard scheme) consisted of two distinct brightness groups, the "giants" and the "dwarfs." Still later, Hertzsprung published in *Astronomische Nachrichten* a graph of the absolute magnitude of the stars against spectral class with hottest on the left and coolest on the right. The large band of stars plotting diagonally across the plot, from top left to bottom right, as every schoolboy now knows, he termed the "main sequence."

Starting in 1913, Russell went through the astronomical lecture circuits in Europe and North America explaining Hertzsprung's ideas and elaborating on them, adding considerable amounts of new data. At the same time, he wrote large, high-visibility articles for *Nature*, *The Observatory*, and *Popular Astronomy*. As a popularizer of the diagram, he was extremely successful and at no time, at least in the published articles, did he neglect to mention Hertzsprung's contribution.[25] So it comes as something of a surprise when Kuiper wrote to Chandrasekhar in March 1936, almost in anger, about an event at the Gaposchkin's.

Celcilia and Sergei Gaposchkin were Harvard College Observatory astronomers.[26] The University of Cambridge had refused to give Cecilia (then Cecilia Payne) a PhD for her work on the composition of stars because she was a woman, so in 1923 she moved to the United States. Shapley hired her as an astronomer—other job titles not being open to women—and persuaded her to write a thesis for which she was granted a Harvard PhD in 1925. The thesis was highly praised by Otto Struve, chair of astronomy at Chicago. It showed that stars were predominantly composed of hydrogen and helium. Cecilia met Sergei on a trip to Göttingen and learned that he was keen to leave Germany in the face of the Nazi rise to power. She organized a visa and position at Harvard College Observatory and he moved to the United States in 1933. The couple was married that year.

Apparently, at the event at the Gaposchkin home Russell remarked that "Hertzsprung did not make it!," referring to the Hertzsprung-Russell diagram. Kuiper protested in the letter to Chandrasekhar.[27] There was more. Bok had quoted Russell as saying, "Hertzsprung is a perfect failure as an astronomer, although he has brilliant ideas." "I think he meant that he has good ideas, but does not follow through. That is definitely wrong," Kuiper retorted.[28] Later he wrote, "It is incredible how the astronomers have been hypnotized by the name Russell diagram."[29] Pouring scorn on

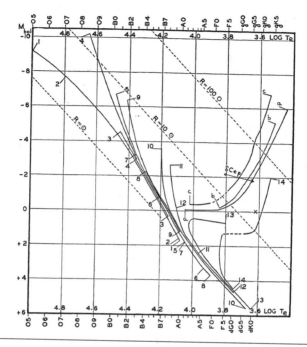

Kuiper's Hertzsprung-Russell diagram for clusters. The vertical axis is absolute magnitude from +6 to −10 and the horizontal axis is stellar class O5 (blue) to K5 (red). G. P. Kuiper, "On the Hydrogen Content of Clusters, Binaries, and Cepheids," *Harvard College Observatory Bulletin* 903 (1936): 1–11.

the attacker of his beloved mentor, Kuiper noted that Russell is wrong even to the point of thinking that Hertzsprung was German! Later in the same letter, Kuiper writes, "When I will have reached sufficient independence I shall speak only of the *Hertzsprung*-diagram, nothing more or less."

But it all blew over. This exchange between Kuiper, Bok, Russell, and Shapley displays the friction between Russell and Hertzsprung and the nationalism of the time, with World War II just a few years away. However, several years earlier Strömgren had published a paper in *Zeitschrift für Astrophysik* with the title "On the Interpretation of the Hertzsprung-Russell Diagram."[30] Kuiper quickly adopted the phrase and so did the astronomical community.

On May 1, 1936, Kuiper submitted a paper to the *Harvard College Observatory Bulletin*. It was not the usual one-or-two-page article describing some detail concerning double stars or a new white dwarf—it was an eleven-page interpretation of Trumpler's cluster diagrams in terms of Strömgren's reinterpretation of the Hertzsprung-Russell diagram.[31]

In 1930, Trumpler, a colleague of Kuiper's at Lick, had published a major paper reviewing the properties of open star clusters.[32] He discussed their distance, dimensions, and magnitudes corrected for interstellar extinction and their distribution in the sky. They were more-or-less spread out in the galactic plane. By assuming that open clusters were of roughly similar size and using their apparent sizes as a proxy measure of their distance, Trumpler showed that the more distant clusters appeared less bright and their stars reddened. It was this observation that led him to discover the extinction of starlight by interstellar dust, which previously had been an unquantified concept. Most importantly for Kuiper, he published a catalog giving the classifications, diameters, and distances of 334 stars in open clusters. Kuiper was able to produce a series of Hertzsprung-Russell diagrams for 14 of the better-characterized clusters.[33]

Strömgren had published calculations for the hydrogen content of stars as they moved across the Hertzsprung-Russell diagram. Originally the stars would plot on the main sequence, but as they aged—the biggest and brightest aging fastest—the stars would move to the upper right corner of the diagram. He presented his results as a series of lines of constant hydrogen content superimposed on the Hertzsprung-Russell diagram.[34]

Kuiper showed that the trends shown by the stars in Trumpler's clusters matched the contours on the diagram for lines of constant hydrogen composition. In other words, he showed that stars of a given cluster have the same composition, an important step in understanding clusters and how stars evolve, and that the point at which stars left the main sequence depended on the age of the cluster.

Kuiper mentioned to Chandrasekhar in a letter dated May 11, 1936, that "in spite of everything" he had finished his paper on the open clusters, noting that it was a preliminary paper pending two larger papers he planned to write for the *Astrophysical Journal*.[35] Why did he say "in spite of everything"? The "everything" was a woman.

Marriage

On April 18, 1936, Kuiper confided in Struve, "I do not know yet how my personal life will develop in the coming months, but it looks quite possible that I may marry before next fall."[36] Three-quarters of the way through his year at Harvard, Kuiper met Sarah Parker Fuller. She had just finished university and had spent a year teaching. She was educated at Cambridge High and Latin (now known as Cambridge Rindge and Latin School) and then she went to Smith College where she was an English major. When she met Gerard she had just spent a summer in England biking around the country and having a wonderful time. Upon her return to Cambridge, and feeling a sense of malaise, she met Gerard.[37] Sarah's father, Charles Sumner Fuller (1871–1943),

(a) Sarah's family in the 1920s. From left to right, sister Eleanor, mother Anna Lillian Hayes, Sarah Parker, brother Sumner, and father Charles Sumner. "Fuller Ancestor Photos," Geocities, archived webpage, October 2009, http://www.oocities.org/hough97/Fuller.html. (b) Sarah in 1936. Courtesy of Matthew des Tombe. (c) The Kuiper marriage party. Courtesy of Matthew des Tombe.

and mother, Anna Lillian Hayes (1873–1968), were socialites in Cambridge, Massachusetts, and had the wherewithal to donate the land on which Harvard's Oak Ridge Observatory is built. They had three children: Eleanor Hayes, Sumner Hayes, and Sarah Parker.

Kuiper was fond of a cousin back in the Netherlands, and he had other sweethearts he had left behind, but his father encouraged the partnership with Sarah, thinking that an American wife, especially one from a respected family, would help him settle and do well in the United States.

(a) Kuiper with baby Lucy in 1947. Courtesy of Matthew des Tombe. (b) When Sarah became pregnant for the first time, Lalitha and Subrahmanyan Chandrasekhar were told to "prepare yourselves to become Aunt and Uncle." Courtesy of Special Collections Research Center, University of Chicago Library. (c) The Kuiper family on a walk in about 1945. Courtesy of Matthew des Tombe. (d) Lucy as a teenager. Courtesy of Matthew des Tombe.

In a letter to Chandrasekhar on May 11, Kuiper explained that in April, "I met a dear girl (whom I had met a few times before), that I am now engaged to her, will marry on June 20, bought a new car, started to take driving lessons, arranged everything about our house at Yerkes." He explained that instead of going to Europe that year he would take a honeymoon to Maine, Quebec, Georgian Bay, and Green Bay. "These things had to be, it was now the right time. Life is a compromise, and the work had to wait."[38]

In early June, Gerard and Sarah visited Williams Bay, Wisconsin, where Yerkes Observatory was located, and were shown their new home, the house originally built by the distinguished astronomer E. E. Barnard and then bequeathed to the University of Chicago upon his death in 1923. The Yerkes astronomer Frank Ross and his wife lived there for a while. The rent was $37.50 per month, unfurnished, with seven rooms and a garage. The university was willing to update and redecorate. Gerard asked that a shower be fitted over the bathtub because showers are faster than baths.[39]

A few weeks later a marriage was arranged. It took place at Sarah's parents' home at 18 Francis Way, Cambridge, just a stone's throw from Harvard University. Sarah, a slender brunette, wore a gown of white silk and lace with a veil held in place by a delicate diamond tiara. Many years later, his life-long friend and colleague Bart Bok noted that Kuiper's marriage to Sarah and the appointment at Yerkes Observatory were "strong positive stimuli" for his significant scientific accomplishments. The wedding announcement for June 20, 1936, said that after the wedding the couple would be "at home" at Yerkes Observatory from September 1.[40] Kuiper and his new wife were destined to become associated with Yerkes Observatory and the University of Chicago, which owned the observatory. The story of how Kuiper came to move from Harvard to Yerkes Observatory begins at the beginning of his Harvard year. In fact, it starts at the very time he was moving to the East Coast.

Overtures from Chicago and a Permanent Job at Yerkes Observatory

In August 1935, as Kuiper was preparing to move to Harvard, Struve was told by Shapley and Bok that Kuiper was about to travel from the West Coast to the East Coast via Chicago. Struve leapt at the opportunity to invite Kuiper to Yerkes Observatory. There were frequent and easy trains that would get him to a destination near Williams Bay where he could be met by a car from the observatory. So on August 30 Kuiper met the man who was to play a major role in his life for the next decade and a half.[41]

Otto Struve came from a distinguished family of Russian scientists, and in late 1921 he began working as a stellar spectroscopy assistant at Yerkes with a monthly salary of $75. After obtaining his PhD in 1923 he was offered a faculty position and rapidly rose through the ranks. Throughout his career he had considerable support from Edwin Frost, the second director of Yerkes Observatory, after George Ellery Hale.[42] When Frost retired in 1932, Struve took over and held the directorship until he left Chicago in 1947.

Struve is widely regarded as one of the most successful twentieth-century astronomers.[43] In 1937, he discovered the Struve-Sahade effect (spectral lines vary in strength as two stars in a binary orbit each other), ionized hydrogen in the interstellar medium, and stellar rotation and the relationship between rotational speed and temperature. He investigated the Stark effect in stellar spectra (splitting spectral lines by an electric field), and turbulence of stellar atmosphere and expanding shells around stars. He published, with Pol Swings, spectroscopic studies of peculiar stars. He also published many articles popularizing astronomy. Yerkes Observatory had a distinguished history but was in decline when Struve assumed the directorship in 1932. By the time Struve contacted Kuiper, Yerkes was fast becoming one of the foremost centers for astronomy in the world.

We must assume that the Kuiper's visit to Struve went well, for a month later, September 30, 1935, Struve wrote to Bok that he wanted to hire to Kuiper.[44] Struve wrote, "While I am aware of the criticisms that emanate from Lick Observatory, and while I am reasonably certain that they will not be pleased with any appointment that Kuiper may get here, I feel that it is quite impossible for me to let my actions be influenced by the likes or dislikes of the astronomers at Mount Hamilton."

Struve did not specify the criticisms from Lick astronomers. He may have been aware of Luyten's views or he might have been thinking of Lick's decision not to offer Kuiper a permanent position. It is also clear that many astronomers at Lick were supportive of the young Kuiper. Mayall wrote fondly of their relationship, Aitken tried hard to retain him at Lick, and relations between Kuiper and Wright seemed cordial enough. In any event, Struve saw only an excellent astronomer and did not want to be distracted by personal comments. He went on:

> It is much more important to make conditions here agreeable for new members of our staff and in that respect I can assure Dr. Kuiper that he will meet with the friendliest possible attitude on the part of all members of our staff. [George A.] Van Biesbroeck, [Christian T.] Elvey, and [William W.] Morgan have independently expressed to me their hope that Kuiper might be induced to come here, and Morgan, especially, has been urging me to proceed with this appointment. Ross met Kuiper during the summer at the Lick Observatory and has written to me in very high terms about him.[45]

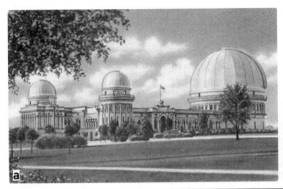

(a) Yerkes Observatory in a postcard. (b) The big dome of Yerkes Observatory. Courtesy of Special Collections Research Center, University of Chicago Library.

Struve then discussed research areas and equipment, mentioning that the mounting for the new 82-inch telescope was built but not yet installed, and that in addition to the dormitory facilities in the dome he was planning to build several cottages for astronomers with families. Interestingly, he was quick to add that he only had funds for astronomers to travel to and from the observatory, not their families! Since most astronomers took their families on their observing periods, this meant some financial burden for them.

Struve approached Robert Maynard Hutchins,[46] the university president, and Henry Gordon Gale, the dean of the Arts and Sciences College, on October 8, 1935, and was authorized to appoint Kuiper at $3,000 per year, about $54,000 in 2018 dollars. The next day, two months after arriving at Harvard, Kuiper was offered a permanent position in the University of Chicago.[47] Struve outlined the nature of the position, vacations, leaves of absence, how he would live in Williams Bay and visit McDonald Observatory on Mount Locke in Texas (which the University of Chicago operated until 1960) for observing runs. There was a suggestion that the 40-inch refractor at Yerkes be moved to Texas, although plans were vague. The idea was floated to make Kuiper assistant director of Yerkes and McDonald Observatories, but this was dismissed because many people (like Joel Stebbins, director of the Washburn Observatory at the University of Wisconsin–Madison, and Wright) objected to putting foreigners in positions of authority. Instead, Yerkes astronomer William W. Morgan was asked to assume the role of deputy director.

Letters flowed freely between Kuiper and Struve while Kuiper anguished over his commitment to Bosscha Observatory. "The authorities on Java have been counting on my coming for almost two years now, and some of my country men would like to see me go there," he wrote to Struve on October 28.[48] In an October 30 letter to Wright, Kuiper tells of his pending move to "a new observatory in this country" but observes that "it will mean a real sacrifice to me not to go to the beautiful and happy island of Java."[49] He also expressed his appreciation for his time at Lick, where he learned to use large telescopes. In response Wright expressed pleasure that Kuiper was to remain in the United States and assured his young colleague that Lick got as much out of their association as he did: "I hope it means that we shall see you here frequently. . . . My wife would join me in kind regards, were she here. We often speak of you, and she is your very good friend."[50]

Then on November 1 Kuiper sent a telegram Java to say that he had received an offer and was considering it. He immediately received a telegram from Joan Voûte, director of Bosscha Observatory: "If you fancy Texas opportunities accept offer stop congratulations and best wishes for future."[51] That same day, Kuiper wrote to Struve, "This friendly attitude towards a change that certainly must upset Dr. Voûte's plans has relieved me very much, and simplifies matters considerably."[52] On November 7, Kuiper formally accepted the offer and, as requested, immediately set about laying out his requirements.[53]

Letters continued to flow between Kuiper and Struve, and Ross was also brought into the discussion because of his technical knowledge of the telescopes. Kuiper wanted to continue his double-star and white-dwarf research, and after his experience at Leiden and Lick, he knew precisely what he required and spared no effort providing details. Mostly he wanted a slitless spectrometer for the 82-inch telescope and a micrometer.

Frank Elmore Ross graduated from the University of California in 1905, was director of the International Latitude Observatory and then physicist for Eastman Kodak, where he developed photographic emulsions and designed lenses for astronomical use.[54] In 1924 he was hired by the University of Chicago as the successor of Barnard. He remained there until retirement in 1939. Using the blink comparator acquired by Barnard,[55] he discovered 379 new variable stars and more than 1,000 stars with high proper motion. A blink comparator is a mechanical way of quickly alternating between two photographic plates so changes in brightness or position stand out as the objects appear to "blink." After one long and highly technical letter in which Kuiper laid out the design for a slitless spectrometer, Ross calmly remarked, "I have done just enough work on the optical problem to learn that this reduction of focal length is a serious matter, and may require months of hard work to solve. Just at present, I haven't the time to put on it, but will, of course, when working for the 200-inch. If I solve it, then the solution will be available for you for the 82-inch."[56]

On November 14 Kuiper confirmed to Wright that his new position would "enable me to work with the 82-inch McDonald reflector, visually, photographically, and spectroscopically" and a few days later sent a postcard to say that his new appointment was official.

Aitken heard of the new position when he received a paper from Kuiper on a new white dwarf in which was stressed the density of the star, 36 million times that of water. "All I, in the befuddled state in which your deductions leave me, can say is that on that star of yours it would be questionable pleasure to have one's sweetheart sit on one's lap. Furthermore she would, I take it, be rather dense. As for myself I doubt whether I would be any denser then than I am here. There are maxima to all things." He goes on, "Many congratulations on your prospects. Quite aside from the advantages provided by the splendid equipment, you will, I think, find it stimulating to work in a brand new place. . . . I know you will make the most of it."[57] These few words, and the words of Wright and Mayall mentioned earlier, speak of a warm and friendly relationship with the astronomers at Lick.

Following on the discussions surrounding the hiring of Kuiper, and the equipment he would need, the discussions between Struve and Kuiper turned to who else Chicago should hire. The level of confidence and the weight given to the opinions of the young astronomer are noteworthy. On December 23, Kuiper's opinion was that for a theoretician the order of priority should be Strömgren, William McCrea, then at Imperial College London, then Chandrasekhar. For the observer, the priority was Marcel Minnaert, then at Utrecht University in the Netherlands, Strömgren, then McCrea.[58] On January 6 an offer went out to Strömgren, who tentatively accepted the offer a few weeks later. However, President Hutchins felt that Minnaert, almost forty-three at the time, was too old for a faculty position at Chicago.[59]

On January 28 Kuiper reported that he had attended a series of talks given by Chandrasekhar and considered him a genius. Kuiper was now a dedicated campaigner for Chandra.[60] But Struve echoed the concerns he had heard from Dean Gale.[61] Would American students accept an Indian professor? How would Chandra handle the prejudice he will surely encounter? What about the rumors that Chandra held unpopular political views that might induce him to ignore the laws of the United States and Illinois? Kuiper responded vehemently to every point, concluding by arguing that theoreticians were rare in the United States, yet every astronomy department needs one.[62] Kuiper argued that, if need be, Chandra could be hired as a researcher at Yerkes rather than as a professor at the university. Struve was well aware of the strong working relationship between Chandrasekhar and Kuiper. He knew Chandra had helped Kuiper interpret Trumpler's cluster diagrams in terms of Strömgren's explanation of the Herztsprung-Russell diagram. In the end, Hutchins said that only scientific issues mattered in hiring staff. He also knew that Chicago was in a race with Harvard to recruit Chandrasekhar, so he pressed Struve to hire him over the protests of Dean Gale.

In early 1936 Chandra visited the campus, and on March 12 he accepted a position at the University of Chicago and Yerkes Observatory. In all probability, Chandra accepted the Chicago offer over the Harvard offer because of his relationships with Kuiper and Strömgren.

On March 15, 1936, Kuiper wrote to Chandrasekhar, mentioning that Struve had told him that Chandra was moving to Chicago. "Struve writes that he is 'very happy about this result.' I feel the same way." In a moment of self-indulgence, Kuiper mentions discussions he had with Leland Cunningham, an assistant of Whipple's who made a career in orbital mechanics and computing. Cunningham expected Chandra to take the position at Harvard. "Well, look at this; he gets a fine job, no obligations, no lectures, is not expected to work, only to study. That should appeal to an Englishman!" Kuiper replied, "But he is no Englishman! C.: Oh. He is! I: we'll see.—So I was very proud that I had won the bet; I said: he is no Englishman."[63]

In March, Struve and Kuiper arranged to be at the Warner and Swasey facility in Cleveland to watch the mounting for the 82-inch being disassembled prior to shipping to Mount Locke.[64] In a follow-up letter on April 18, Kuiper mentioned that he also took advantage of the opportunity to visit Ohio State Observatory, which had books it could donate to the new McDonald Observatory.

In August 1936 Kuiper's year at Harvard was over. He met many new colleagues, he taught students at the undergraduate and graduate level—and thoroughly enjoyed it—and while he did not maintain the output of his Lick years, he certainly made major research contributions. He knew within months of arriving at Harvard that he was destined for Chicago, despite agonizing over his commitments to Bosscha Observatory and the attractions of Indonesia, and during his stay at Harvard College Observatory laid a thorough groundwork for that move, not only ensuring he had the equipment he needed and quickly establishing a close relationship with Struve, but also ensuring that he had the right colleagues to work with. He played a decisive role in persuading Struve to hire Chandrasekhar, who was to be a mainstay during Kuiper's years at Yerkes Observatory. And, not least, he found a critical life partner in Sarah. It was a good year.

CHAPTER 4

Two Observatories, a Child, and War on the Horizon

THE TINY VILLAGE OF Williams Bay, on the side of Lake Geneva in Wisconsin, is the home of about two thousand people. It is almost one hundred miles from the University of Chicago astronomy department, about a two-and-a-half-hour drive; too far for anything but an occasional trip on university business. On the west side of the village is Yerkes Observatory, sitting in an open grassy space surrounded by trees, with an oval drive that connects the observatory to the road. At the periphery of the observatory grounds, among the trees, are the homes of those that serve the observatory. When Kuiper arrived at Yerkes in 1936, there were thirty astronomers and support staff at the observatory, a number that had not changed much when he left in 1960. It was a peaceful, gentle lifestyle. Kuiper was able to go home for lunch and walk over to the observatory in the evenings to observe. Fourteen hundred miles to the southwest, several days' drive, is Fort Davis, Texas, the nearest town to McDonald Observatory. The small, friendly town was made up predominantly of ranchers. When Kuiper arrived at Chicago, McDonald Observatory was under construction, and it, too, would be an important part of Kuiper's life. However, these quiet places in the woods of Wisconsin and in the mountains of the Southwest were not immune to the campus politics that would reach out from the university in Chicago while its astronomers dealt with the mighty astronomical questions of the day. The astronomy department at the University of Chicago was on its way to becoming one of the most productive astronomical arrangements of the day, and at times one of the most turbulent.

Aerial view of Yerkes Observatory.

Yerkes Observatory

Yerkes Observatory was nearly forty years old when the Kuipers moved there.[1] In the 1890s a young astronomy professor at the University of Chicago, George Ellery Hale,[2] heard about two surplus 40-inch lens blanks he could use to build a major telescope, and he took the idea to the University of Chicago president, William Harper. Harper agreed to help find a benefactor willing to pay for the telescope and an observatory. In 1897 a benefactor was found in the form of Charles T. Yerkes, who had made his fortunes from the public transport system of Chicago and, it was rumored, various dubious financial dealings. Toward the end of the nineteenth century there was a move to large, privately funded observatories that were not just domes with a telescope, but full-blown institutes with laboratories, libraries, lecture halls, and offices. Yerkes Observatory is a narrow T-shaped building, with a large dome on bottom and two smaller domes at the ends of the top of the T. Between the domes are the other facilities (Lick Observatory was built on similar principles).[3] The building is splendid. The Greco-Roman facade with intricate stonework and carvings was designed by Henry Ives Cobb, who designed several buildings at the University of Chicago.[4] The telescope, which was to be the largest refracting telescope in the world, was built

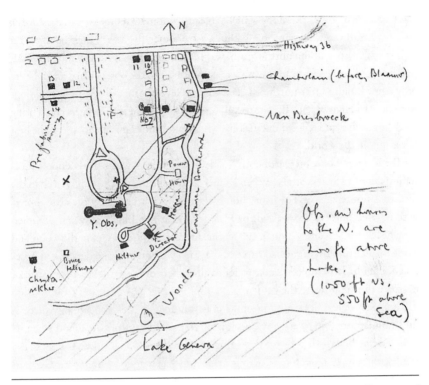

Hand-drawn map by Kuiper of Yerkes Observatory and the locations of the astronomers' homes as of the late 1950s. Courtesy of Special Collections Research Center, University of Chicago Library. The map is reproduced in E. A. Whitaker, "Clash of the Titans (or Why and How the LPL was Born)" (PowerPoint presentation, Symposium to Celebrate the 50th Anniversary of the University of Arizona's Lunar and Planetary Laboratory, Tucson, Arizona, October 2, 2010), https://www.lpl.arizona.edu/sites/default/files/history/lpl50/lectures.pdf.

by Alvan Clark and Sons, the same company that built the 36-inch telescope at Lick. The founder of the firm, Alvan Clark, had died before Yerkes Observatory was built; his son Alvan Graham Clark supervised the completion of the 40-inch lens.[5]

A hand-drawn map by Kuiper shows the Yerkes Observatory situation as of the later 1950s. The observatory sat at the end of an elliptical drive that was connected by a north–south road to Highway 36, known locally as West Geneva Street. Also running north–south and to the east was Parkhurst Place, which hooked behind the observatory and along which observatory housing was located—first Morgan's home, then the director's house, and then the home of William Albert Hiltner (usually known as "Al"). Some distance to the west, and beyond the Bruce Telescope building,[6] was the Chandrasekhar home. Running parallel to Parkhurst Place and farther to the east was Constance Boulevard, along which Yerkes astronomer George A. Van Biesbroeck and Adriaan Blaauw, another young Dutch astronomer, lived with their families. Yet farther east, also running north–south, was Dartmouth Road, along which the Kuipers lived. All were within an easy walk of each other and the observatory, and all were only a few hundred feet north of Lake Geneva. Several other observatory buildings were close to Highway 36. The site was all very cozy. Here the astronomers could bring up their families and enjoy domestic bliss while having no commute to one of the best telescopes in the world.

The Kuipers' house on Dartmouth Road was two stories, with steeply sloping roofs reminiscent of a Swiss cottage. When Kuiper served as director the family lived in the director's house, a large, two-story home, three stories if including the attic rooms, with pillared porches. It was built by Hale for himself and his family and occupied until he left for Mount Wilson in 1903.[7] The director's house was set in the woodland while the Dartmouth home had a splendid garden with an intricate network of hoses that Kuiper had assembled to take water to all corners of the yard. He spent much of his spare time tending to the garden.

Otto Struve was not only a prominent astronomer, he was also a highly successful administrator. He reversed the decline of Yerkes Observatory, he found when he assumed the directorship, by ruthless firing and hiring. In 1936 he hired Kuiper to the faculty, then in 1937 he hired Chandrasekhar. He then added Strömgren; Gerhard Herzberg, a spectroscopist and future Nobel Laureate then at University of Saskatchewan; and Jesse L. Greenstein, then a student of Menzel's at Harvard. This was a powerful group of observational and theoretical astronomers, and their hiring led to what Morgan called "the renaissance of the observatory." In fact, hiring theoreticians at an observatory was a new practice, and perhaps even revolutionary.[8]

One of the astronomers that was on the faculty before Struve, and was to have a long association with Kuiper, was Van Biesbroeck, who had been at the observatory since 1915.[9] He obtained a degree in theoretical astronomy at Belgium's Ghent University and worked at Heidelberg, Germany, before joining Yerkes. His interests were wide, and included double stars, comets, asteroids, and variable stars. He was forced to retire in 1945 at the age of 65 but increased his observing time at Yerkes and at McDonald Observatory, making frequent automobile trips between the two. Later he was involved in several astronomical expeditions to remote parts of the world, including a trip to Khartoum, Sudan. There he observed an eclipse in a test of Einstein's theory of relativity. Shortly after his retirement, Van Biesbroeck and his wife, Julie, converted their home into a boarding house where graduate students, young instructors, research associates, and visiting astronomers could stay. His sister Marguerite was also well known around the observatory as a part-time librarian and for more than fifteen years appeared in the annual departmental photographs.[10]

(a) Otto Struve. Courtesy of MacDonald Observatory. (b) Frank Elmore Ross. Courtesy of Special Collections Research Center, University of Chicago Library. (c) Bengt Strömgren. (d) Subrahmanyan Chandrasekhar in his fifties. Courtesy of Special Collections Research Center, University of Chicago Library. (e) Yerkes Observatory staff in 1938. *Back row*: Perry Nicholas, Hans Sieghold, Otto Struve (Director), Frank R. Sullivan, Edward L. McCarthy, Roy Wickham, O. B. Fensholt, Charles Ridell; *Middle row*: George Van Biesbroeck, Theodosia Belland, Charles W. Hetzler, Frances Sherman, Louis G. Henyey, Edith M. Kellman, William W. Morgan, Carl Rust; *Front row*: Alice Johnson, Subrahmanyan Chandrasekhar, Mary Ross Calvert, Bengt Strömgren, Marguerite Van Biesbroeck, Gerard P. Kuiper, Lalitha Chandrasekhar, Jesse L. Greenstein. Courtesy of MacDonald Observatory.

In the fall of 1937, fifteen months after moving to Yerkes, the Kuipers were back on the West Coast for a few weeks. "Our trip was interesting," he wrote in a letter to Chandra on October 12, 1937. The Kuipers stopped at Berkeley for three days on their way to Mount Hamilton where they met with Hertzsprung, who was visiting Lick. The Kuipers pleaded with Chandrasekhar and his wife, Lelitha, to join them on the West Coast: "The bridges near San Francisco are beautiful and inconceivable in size."[11] Sarah sent a couple postcards, one showing an aerial photograph of the Golden Gate and Bay Bridges and one urging their friends to join them. "The weather has been wonderful the last few days. We did some climbing last Saturday. I spend all my time sitting in the Sun learning Dutch and reading. Do come and see the U.S.A." Chandrasekhar declined, saying that he needed to spend the time on his writing.

In 1940 there were serious discussions of moving the 40-inch telescope to Texas. This was "something that may happen," said President Hutchins to Struve, and Struve quoted this to Kuiper in a letter dated February 20.[12] But the difficulties of the war and the improvements in telescope technology meant the glorious old refractor was on its way to being little more than a museum piece. In November 2007 a group of astronomers and others interested in the observatory recommended selling much of the land and maintaining the observatory as an education and outreach facility.[13] The group's recommendations were adopted by the University of Chicago.

McDonald Observatory

When Kuiper arrived at Yerkes, Struve was busy overseeing the creation of a major observatory in Texas, McDonald Observatory. On February 26, 1926, William Johnson McDonald, prominent Northeast Texas banker, bequeathed his fortune of more than $1 million to the University of Texas for building and supporting an astronomical observatory.[14] The McDonald family came to Texas in the 1830s, opened law offices, invested in land, bought and opened banks, and in 1887 moved to Paris, Texas. At some point, William Johnson McDonald was given a copy of Flammarion's *Popular Astronomy*, which prompted him to buy a telescope. He also amassed a large library that now resides in the observatory. Urbane and intellectual, McDonald died at 81, lonely and in pain from a variety of illnesses. His bequest was a surprise to everyone, and the family was not pleased. According to the will, family members were to receive only $15,000. The family took the matter to court. After five hearings, the matter was settled out of court with the university receiving $840,000 (about $11.8 million today) and the family the remainder.

The University of Texas heard of the will from reporters. Dean of Science Harry Yandell Benedict described it as "like lightning out of a clear sky." He wrote to several observatories, soliciting advice and seeking assurances that bequeathing funds this way was not unusual. Among his correspondents was Edwin B. Frost of Yerkes Observatory. Frost wrote a strong letter of support and long list of suggestions.[15]

When Struve became director of Yerkes in 1932, he picked up the baton. Yerkes was struggling with old equipment and poor skies, and he wanted an observatory at a better location in the United States. He had in mind a site near Amarillo, Texas. Struve and Benedict worked together to agree over management and utilization of the new facility, and they worked to identify a location. Russell, Frank Schlesinger (then director of Yale University Observatory), Shapley, and Hale were all asked for advice. The major problem was that the University of Texas did not have even a single astronomer on their faculty. Eventually, Struve and Benedict came to an agreement whereby while the University of Texas owned the facility, the director of Yerkes Observatory would also be the director of McDonald Observatory, and the Chicago astronomers could use the telescope. The agreement was revolutionary at the time.

Struve arranged for two astronomers, Christian T. Elvey and T. G. Mehlen, to tour the state in search of a site. They took with them a 4-inch refractor to check how well Polaris could be seen in visible light—astronomers call it the checking the "seeing"—and they carried a camera with a prism to check for spectral clarity, using the bright star Vega as a test star. Elvey was a Yerkes astronomer who received his PhD with Struve in 1930 and now worked on spectra of stars, galactic light, aurorae, and the gegenschein, an illuminated region of the sky caused by sunlight scattered from dust orbiting Earth.[16] Mehlen was an Amherst astronomer who agreed to be Elvey's assistant. After crisscrossing Texas from Amarillo to Galveston to El Paso in a little Chevrolet van, they proposed Black Mountain in the Davis Mountains for the site of the new observatory. Struve visited the site and talked to locals, and then moved the location to Mount Locke, also in the Davis Mountains. President Horace W. Morelock of Sul Ross State University, forty miles away, offered to provide offices for the McDonald Observatory astronomers.[17] After some lengthy and complex legal negotiations, the site was acquired.

The observatory was to have three floors, with working and living accommodations on the first two floors and observation facilities on the third. This was an uncommon approach, but it saved the costs of building working and living space elsewhere. There were to be concrete piers, battleship plate floors, and a catwalk. A detailed model was constructed that now resides in the observatory's visitor center. Two years later, a power plant and the first dwelling houses were built: the director's house, the resident astronomer/superintendent's house, and three houses for visiting astronomers. Finally, a garage was constructed. In 1933, with the sites for the dome and houses selected, a contract was awarded to Warner and Swasey to build the telescope. Struve published an article in *Popular Astronomy* laying out his plans for research at the observatory.[18]

(a) Steel skeleton of McDonald Observatory. Courtesy of Special Collections Research Center, University of Chicago Library. (b) The completed McDonald Observatory. Courtesy of Special Collections Research Center, University of Chicago Library. (c) The McDonald Observatory telescope waiting the mirror. Courtesy of Special Collections Research Center, University of Chicago Library. (d) John S. Plascott, C. A. Robert Lundin, and Henry Norris Russell with the final mirror. Courtesy of Special Collections Research Center, University of Chicago Library.

(e) One of the astronomers' residences at McDonald Observatory. Courtesy of Special Collections Research Center, University of Chicago Library. (f) The McDonald Observatory vehicle. Courtesy of Special Collections Research Center, University of Chicago Library. (g) The Davis Mountains, Texas. Courtesy of Special Collections Research Center, University of Chicago Library. (h) Participants attending the dedication ceremony for McDonald Observatory standing on the steps to the observatory. Left to right: H. N. Russell, J. S. Plaskett, O. Struve, H. Shapley, H. P. Rainy (University of Texas president). Courtesy of Special Collections Research Center, University of Chicago Library.

In the fall of 1934 Franklin Roach, a graduate student at the University of Chicago, was appointed as the first resident astronomer, and a 12-inch refractor was brought from Yerkes.[19] After two years Roach graduated and accepted a faculty position at the University of Arizona, where he stayed until the outbreak of World War II.[20] A year later, Elvey replaced Roach. Elvey was soon replaced by Kuiper. Before the formal dedication of the observatory in 1939, Kuiper had already obtained four hundred spectra of stars.[21]

While originally planned to be an 80-inch telescope, design margins inserted at the outset were not necessary, and the telescope at McDonald Observatory ended up being an 82-inch. In fact, Struve and his wife were invited to Corning to watch the pouring of the Pyrex mirror. After four months, cooling cracks appeared and Struve insisted on a new pouring. The grinding and polishing were slow. In March 1937 Struve, Van Biesbroeck, and Kuiper discovered errors in Warner and Swasey's method, so the Yerkes optician, Lloyd McCarthy, was sent to the company's facilities in Cleveland. The mirror was completed October 15, 1938, and marked with a celebration dinner at the University of Chicago's University Club. The following February, the mirror arrived at the nearest railroad depot in Alpine, and the astronomers watched its installation in the dome. That night they observed Sirius, the Moon, and the Orion Nebula. The telescope was formally accepted on March 25, 1939, having made a loss of $85,000 for its manufacturers. The telescope went into operation May 5, 1939.[22]

A dedication ceremony was held that day in conjunction a with three-day meeting of AAAS at Sul Ross University. Kuiper gave a talk titled "Under-luminous Stars," and Sarah made a film of the proceedings. There were 250 visitors, some local, some from far away, including astronomers, administrators, and members of the public. Visitors were taken by bus to the observatory to watch Elvey operate the telescope and Carl Seyfert, a Harvard astronomer hired for McDonald Observatory in 1936, to run the projector; Kuiper spoke to the press. There was a chuckwagon barbecue and a rodeo paid for by Warner and Swasey. There were also speeches by Charles Stilwell, vice president of Warner and Swasey, and Struve. Missing were several D.C. politicians who needed to remain in Washington, concerned about conditions in Europe. The following Monday a technical seminar, chaired by Wright, included talks by many well-known astronomers. Wright also offered a formal appreciation of Warner and Swasey.

One year later it was clear that Yerkes Observatory was no longer capable of supporting modern research, and a campus committee chaired by the distinguished physicist Arthur H. Compton suggested that the astronomy department faculty be moved from Yerkes Observatory to the University of Chicago campus starting July 1, 1942. Compton, Chandra, and Hutchins were enthusiastic about the plan. Kuiper was so pleased with the new facility in Texas that he wanted to go one step further and proposed to Struve that he officially divide his time between the campus and McDonald Observatory.[23]

True to form, his proposal was long, detailed, and forcefully argued. Pro Texas: the telescope, the climate (especially important for Sarah), friends, and a "new institution." Pro Yerkes: strength of faculty, library, house, and medical facilities. This led to a detailed discussion of the facilities at McDonald and how they could be improved, particularly the housing. Kuiper suggested his distribution of time between McDonald and Chicago be 50–50 or even 75–25. Sarah was so excited by the idea of spending most of her time at McDonald that she wrote to the dean's wife about leasing apartments in Chicago, an action that upset Struve.[24] The Yerkes astronomers would eventually move to the University of Chicago campus and, in the case of Chandra, set up a new institute for theoretical astronomy. In one sense, the writing was on the wall for Yerkes, but the move did not happen for many years, and Kuiper's proposal did not go anywhere. In January 1942 Struve heard that the plan had been postponed, probably overtaken by the course of events during the war. This was not the war's only impact; in fact, the war was to almost decimate the department.

Sheep Shearing and Barbeques

On June 10, 1940, Sarah wrote to Chandra describing the social life in Fort Davis.[25] She reported that on this trip to Texas she had "new experiences again. . . . The chief one was sheep-shearing at the Merrill ranch." It reminded her of Thomas Hardy's *Far from the Madding Crowd*. The shearing itself was not very "romantic." It was done by eight "Mexicans" with electric sheers. "Mexicans" was a term used at the observatory for anyone of Hispanic origin; many had been U.S. citizens longer than the whites. However, it was the sight of cowboys that excited Sarah, dressed in full regalia and riding "spirited" horses. She intended to buy the Merrills a copy of *Far from the Madding Crowd* as soon as she could reach a bookshop.

Sarah also went to a big barbeque on the Prude ranch, the scene of a rodeo put on by the district Rotary clubs. There was also a picnic, which was inside because of the rain, and she learned to dance the schottische, described by her as an old-fashioned, rather pretty "scotch" dance. She also met one hundred Marfa people at a "Smorgasbord party." Marfa is a town twenty miles to the west that is noted for its dust and mysterious lights (probably glow discharges). Sarah exclaimed, "In these few weeks I have been to almost as many parties as I have attended before in a whole winter!" Sarah was obviously very happy at McDonald Observatory, but it is sad to note that Gerard attended none of these events, meeting with the local people only when they were invited to tour the observatory after the barbeque. This was to change after the war.

Sarah ended her letter by hoping that Struve would allow her husband a few more days at the observatory to compensate for observing time lost to poor weather. She noted in passing that "the storms are very exciting to watch from the mountains." The couple would then return to Yerkes by train via Dallas next Sunday.

The Kuipers next visit to McDonald Observatory was in November 1940, this time driving from Williams Bay to Fort Davis by car. Stopping overnight in Little Rock on Saturday, November 9, Kuiper had time to write to Chandrasekhar and review a manuscript. "We had rain all day, and made only 220 miles. The road No 67 is none too good and many accidents have occurred here lately. So we decided to stop at 6 p.m."[26]

This early stop gave him time to finish reading a manuscript of Leningrad nuclear physicist George Gamow. "I made pencil remarks as before. These remarks may be easily erased, but I hope you won't erase them but let Gamow see them first." He adds, somewhat mischievously, "Sarah saw Gamow's MS and thought, from the appearance of the front page that it was from a crank. Poor Gamow."

Kuiper thought that Gamow's article illustrated the need for good textbooks and summaries, which led him to think about his own publications. "I now have a clear picture of what I should do in the way of publications: Nov or Dec 1940, white dwarfs and Subdwarfs for *Reviews in Modern Physics*; Spring 1941, white dwarfs, for *Astrophysical Journal*; June or July 1941, Dwarfs and Subdwarfs of Types A to M." So in late 1940 white dwarfs were still very much on his mind, but that was soon to change.

A few days later Kuiper wrote from McDonald, where he was living in the dome because the cottage was not finished. His main point was to discuss some of Chandrasekhar's recent theoretical work on white dwarfs. He ended his letter, "Please give our regards to Lalitha." He never finished a letter to Chandra without mentioning Lalitha.

Despite the living conditions, Sarah was still thoroughly enjoying Texas as 1941 unfolded. On January 29 she wrote another long letter to Chandra saying how much she liked Fort Davis. It was a friendlier place and people were more individual, and more social, than in Williams Bay.[27] There were many parties, and even a president's ball, and Marfa had a movie theater. In fact, the Kuipers were thinking of changing their residence to Texas. Among her many activities, she was knitting, which is better than sewing "because you can read while knitting." She continued, "I have been working quite consistently on R.A.F. [Royal Air Force] knitting, for British Relief."

In January 1941 Kuiper heard from Struve that Chandra was thinking of moving from Yerkes Observatory to the campus to start a center for research in stellar structure.[28] "We will miss you and Lalitha in Williams Bay," he wrote to Chandra, "but it would be in your own interest and the interest of the University if you were to do that . . . you would be our intermediary in Chicago." However, a month or so later he heard that Chandra was staying at Yerkes in the face of "unpleasantness" in Chicago. "There are so many boorish and prejudiced people in this country," replied Kuiper, who added, "I have profound hopes that this period of international crisis will clear the minds of many Americans. . . . But one must be perfectly happy to do good work."[29]

However, the main point of Kuiper's January 16 letter was to tell Chandra that he was planning a monograph he wanted to title *Dwarfs, Subdwarfs, and White Dwarfs: Their Frequencies, Motions, Spectra, and Dimensions*, and in this regard has been "having a lot of fun going systematically through some 50 volumes of the Ap.J." There was to be nine chapters and three appendixes, and he had already selected a title for each chapter and thought in some depth about its content. This is another sign of how intense his interest in white dwarfs was at this time. However, despite the enthusiasm, the monograph never happened.

(a) The 82-inch reflecting telescope at McDonald Observatory. Author's photograph. (b) George Van Biesbroeck using the 40-inch refracting telescope at Yerkes Observatory.

The January letter provides insights into developing relationships at Chicago and the astronomical community. "I'd like to have your comments on this project, but should like to ask you not to discuss the matter as yet with others at Yerkes, because there will be some personal relations to think about, which have to be handled separately and diplomatically."[30] Kuiper mentions overlap with Morgan's work and need to get data from Van Biesbroeck. He also mentions "problems with Luyten to straighten out (!)" and that "[McDonald Observatory resident astronomer Daniel] Popper has still much to do." There is even a competitiveness to his relations with Chandra: "I hear you are writing your second monograph: dynamics! Fine. I have my second one *planned*; on double stars. (1942–1943). But you will have your third by then."

Kuiper continued his work on double stars, white dwarfs, and other nearby stars throughout his first few years at Yerkes, mainly publishing old results.[31] He clearly intended to keep up these efforts, but the war was to change this. He made major advances in understanding the relationship between stellar mass and luminosity and how the components of binary stars interact with each other, publishing a remarkable paper with Struve and Strömgren on Epsilon Auriga and another on Beta Lyra.

Epsilon Aurigae

Epsilon Aurigae is a naked-eye star in the northern constellation of Auriga. About every twenty-seven years its brightness drops from an apparent visual magnitude of +2.92 to +3.83, more than a factor of two, the dip lasting somewhere between 640 and 730 days. In 1821, the German astronomer J. H. Fristch suggested that it was a variable star, and this was soon confirmed. In 1912, Hans Ludendorff, of Potsdam Astrophysical Observatory, concluded that it was an eclipsing binary, which means the two stars orbit each other but line up with Earth, taking turns concealing each other.[32]

With his expertise in binary stars, Kuiper was aware that Epsilon Aurigae was not a normal eclipsing binary. Something was wrong. When he arrived in Williams Bay he not only observed the light curve but also how its spectrum changed during the eclipse. The result was confusing. First, such a large difference in brightness between the two components required different spectral types, but they were seen to be the same. Second, if the two stars were same type, then spectral lines should become doublets because of the Doppler effect as one component comes forward and one goes back. Third, there was spectroscopic evidence for movement in one component other than the regular orbital movements. Finally, the spectrum from the star on the far side was seen during eclipse whereas it would have been hidden by the nearer star in a normal eclipsing binary.

Working with Struve and Strömgren, Kuiper thought they had come up with an answer, and in 1937 the three published their idea in *Astrophysical Journal*.[33] For nearly thirty years, it was widely quoted as the best explanation for this remarkable star system, and their ideas became known as the Kuiper-Struve-Strömgren model for Epsilon Aurigae. Kuiper concluded that the orbital inclination relative to Earth was sixty-five to seventy degrees rather than the ninety degrees usually assumed. This meant the two components underwent a grazing eclipse—the farther star was only partially concealed and could be seen during eclipse. Kuiper also suggested that there is a shell of absorbing material around the eclipsing star and that the upper layer of this star is highly ionized. Struve took this model and made a detailed analysis of the spectra to be expected during their orbit. Finally, Strömgren discussed the required astrophysics for the proposal.

The Kuiper-Struve-Strömgren explanation of the strange properties of the Epsilon Aurigae system remained popular until 1965, when Su-Shu Huang, a Chinese American astronomer then at the Goddard Space Flight Center in Maryland, pointed out that the outer-shell idea was not realistic and proposed instead that one of the components of the binary system was a large disk seen edge-on from Earth.[34] But astronomers are still restless with respect to Epsilon Aurigae and plan to bring more new technology to the problem by utilizing space-borne telescopes.

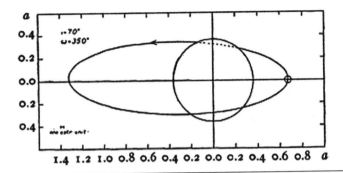

The orbit of Epsilon Aurigae as proposed by Kuiper and his coauthors in 1937. G. P. Kuiper, O. Struve, and B. Strömgren, "The Interpretation of ε Aurigae," *ApJ* 86 (1937): 570–612.

Beta Lyrae

Beta Lyrae is another binary system that caught Kuiper's attention.[35] The variable nature of this star was first observed in 1784 by English astronomer John Goodricke, who was best known for discovering that Algol is a variable star—and for being deaf.[36] The change in Beta Lyrae's apparent magnitude is from +3.4 to +4.6 over an orbital period of thirteen days. Kuiper found that the two stars were in contact but enveloped in a common gas. He coined the term "contact binary." Such stars are intrinsically unstable, and matter will stream from one to the other, forming a giant pinwheel with a streamer. This leads to a complex spectrum with a lot of details reflecting the individual components of the system. Even a shadow thrown by the streamer on the spiral arms produces observable changes on the spectrum. Both the transfer of mass from one star to the other and the ejection of the streamer will cause the period of the binary system to change. This conclusion has been widely accepted and astronomers now believe that the equivalent of the Sun's mass is transferred between the stars every fifty-thousand years, which results in an increase in orbital period of about nineteen seconds each year. In 2008, the primary star and the accretion disk of the secondary star were resolved and imaged.[37]

The Mass-Luminosity Relationship

The large database of high-quality observations on binary stars he had been collecting since his student days in Leiden placed Kuiper in a strong position to address one of the hottest topics of the time: how the luminosity of a star varies with its mass. The history of this fundamental question in astronomy was traced by Kuiper in the

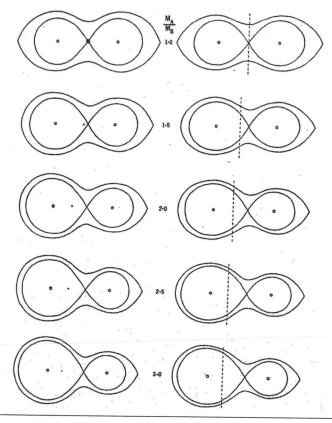

A drawing from Kuiper's 1941 paper on Beta Lyrae, where he proposed that the binary pair of stars were in contact with each other with an envelope around both stars. G. P. Kuiper, "On the Interpretation of β Lyrae and Other Close Binaries," *ApJ* 93 (1941): 133–77.

opening of his article on this topic published in *Astrophysical Journal* in 1938.[38] He thought that the idea developed gradually with knowledge of parallax and that the fact that Sirius B is off the main sequence delayed the discovery of a relationship. It was probably J. K. E. Halm, a German astronomer then at the Royal Observatory, Cape of Good Hope, who in 1911 first reported a relationship between intrinsic brightness and mass in a paper in the *Monthly Notices of the Royal Society*; since he used spectroscopic binaries to derive masses and distance, he included only main sequence stars.[39] Russell reported a relationship in 1914, as did Hertzsprung in 1918. Hertzsprung's equation, published in *Astronomische Nachrichten*, was that $\log m = -0.06 (M_v -5)$ where m is mass and M_v is visual magnitude.[40] Importantly, Adriaan

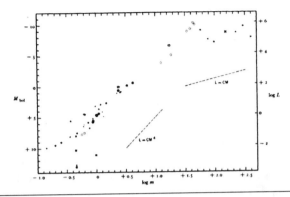

Kuiper's relationship between the magnitude of the star and its mass taken from his 1938 paper. The symbols indicate the source of data. The broken lines indicate the slope of the trends. G. P. Kuiper, "The Empirical Mass-Luminosity Relation," *ApJ* 88 (1938): 472–507.

van Maanen observed that the mass-luminosity stretched across the spectral types.[41] Many investigations followed, which were reviewed by Lundmark in 1933 in *Handbuch der Astrophysik*, but best known at the time was a paper by Arthur Eddington.[42] In view of the criticisms Kuiper was to suffer later (that he did not sufficiently cite the work of others), it is interesting to note the care with which he recorded previous work in his mass-luminosity paper.[43]

Kuiper collected all the best data on binary stars and from their orbit calculated their relative masses and the maximum separation of the stars as they orbited their mutual center of mass. During his observations Kuiper had systematically recorded their maximum angular separation. He also added data for stars for which mass and luminosity could be determined by other means, such as those whose parallax enable distance to be calculated. He thus produced the most comprehensive database on which to determine the mass-luminosity relationship. His publication remained the standard reference on the topic for many years.

As Kuiper prepared his work for publication, a very similar publication appeared in a Soviet journal by P. P. Parenago, an astronomer at Moscow State University.[44] Kuiper carefully documented the similarities and differences. The methods used by both authors were very similar, but Kuiper thought he used improved orbits and mass ratios, and he also used Trumpler's data for large stars in clusters. The mass-luminosity curve that Kuiper played such a major part in developing is now part of the fundamental understanding astronomers have of how stars work, and it plays a critical role in using the Hertzsprung-Russell diagram.

Prepare Yourselves to Become Aunt and Uncle

While Kuiper worked busily through a stream of important findings, his personal life continued to develop. On June 13, 1941, Sarah wrote to Lalitha and Chandra, telling them to "prepare yourselves to become Aunt and Uncle in November!"[45] Later, "Yerkes plus McDonald really seem to be 'turning out' babies at a great rate, don't they?," referring to an imminent birth expected by the Poppers. Daniel Popper was the resident astronomer at McDonald Observatory, being hired in 1939.[46] "We hope it will be twins, but whatever it is we will adopt one of the other kind." Sarah reported problems with choosing a name, especially one that goes with "Kuiper," and was anxious about schooling because in the isolated communities around the observatory pupils either had to live away from parents or be homeschooled and miss the socialization. On November 23, 1941, the Kuipers became the proud parents of their first child, a son named Paul after Ehrenfest's son.[47] Gifts and congratulations came in from all the Williams Bay community of astronomers. Kuiper wrote to Chandra on December 10, thanking him for a gift and noting that Paul was doing well, slowly putting on weight.[48]

After a series of joyful, happy letters about the wonderful and busy lifestyle in West Texas, a little melancholy set in. Sarah wrote to Chandra that "seven months on the mountain is too long."[49] She was getting restless. After the birth of Paul, melancholy gave way to something more serious. In his December 10 letter to Chandra, Kuiper wrote, "Sarah, unfortunately, is in a seriously upset condition, not physically (in which regard she is fine) but as a result of emotional strain. She could not remain at home, and is being taken care of by a specialist." Initially, Kuiper was confident in the specialists, but as their daughter has observed in her oral history, he eventually lost patience and removed Sarah from hospital against the wishes of the doctors.[50] He took her to Fort Davis, where among her rancher friends, whose company she loved so much, she slowly recovered.

War on the Horizon

In a letter to Chandra mentioned earlier, Sarah wrote of knitting scarves for the RAF. The United States was not in the war, but the Kuipers were. The Nazis had invaded the Netherlands and there were stories of atrocities. Sarah's comment about scarves for the RAF was not the only sign that war was creeping up on the astronomers of Yerkes and McDonald. First, several visiting astronomers could not return to Europe because of German naval activity. Then fear of an invasion from Mexico caused Struve to enquire from Warner and Swasey about burying the mirror to protect it; after all, Germany did try to persuade Mexico to invade the United States during World War I.[51] Then Elvey reported to Struve that Theodor Immega (an engineer from Germany who was hired as an observatory assistant) was spreading rumors that the foreigners at McDonald Observatory were pro-Nazi, a problem fixed by a stiff warning from Struve.[52] Then a photoelectric device for measuring the light of the night sky at various elevations that was built by Roach and Elvey for use during construction of the observatory was requisitioned by the National Defense Research Committee for the British war effort.[53]

While the Kuipers were watching the progress of the war in Europe and grappling with the nature of binary stars and Sarah's depression, the war came violently and without warning to the United States. Writing three days after the Japanese bombed the U.S. fleet in Pearl Harbor, Kuiper ended his letter to Chandra, who was visiting Princeton at the time, with the words, "The war must have shaken Princeton as it did us here. When do you expect to be back?" The talk then became about whether and how scientists should directly support the war effort.[54] The scientists would have to decide what to do. To the Kuipers, there was no debate. The Germans had been occupying the Netherlands since May 1940.

Shortly after Pearl Harbor, Kuiper was approached by Albert Tucker of Princeton University with a request that he join the war effort.[55] Professor Tucker was a distinguished mathematician who was working on target-location and gunnery-direction problems for the Office of Scientific Research and Development (OSRD) at the Frankford Arsenal in Northeast Philadelphia. In view of Sarah's condition, Kuiper declined. "Thank you for your very full explanation of the unusual difficulties standing in your way. You have been most cooperative in trying to make yourself available. I am sorry that we have added to your worries at such an inopportune time," replied Tucker.[56] He went on: "We shall not have to press you further in the matter. We are making note of your willingness and desire to be of service under more favorable circumstances."

Through January 1942 a series of letters flowed from Kuiper to Struve documenting Sarah's progress and describing the effect on Kuiper's activities. By mid-January, six weeks after Paul's birth, Kuiper wrote to Struve, "With Mrs. Kuiper on the road to recovery I am regaining my equilibrium; I have found it pretty hard to think straight during the past weeks."[57] Struve's wife Mary replied, "We were pleased to receive your letter with the fine news of the great improvement in the health of Mrs. Kuiper. We can now really look forward to having her back home before long."[58] But Struve continued to have great concern with respect to future trips to McDonald Observatory. From MacDonald he wrote,

> I quite understand your reluctance to leave Mrs. Kuiper in the north. Although there is from our point of view no compelling reason why you should not bring her here (even the house H will be vacant, since Swings will go north in early March), but in view of

the whole situation brought about by the war, the severe restrictions in travel by truck, the absence of any kind of entertainment without automobile travel, and the difficulty of getting to a doctor, I still advise very seriously that you try to find an agreeable way of either leaving her in Williams Bay, or sending her home to her parents in the east. . . . I think I must, in fairness to yourself and ourselves, advise utmost caution."[59]

To this Kuiper replies, "Thank you very much for your kind letter of Jan 22. As to whether or not Mrs. Kuiper will join me in Texas I agree that we should proceed cautiously. I have an open mind on the subject except that prolonged separation would be very hard on both of us."[60]

In March, Struve wrote to Kuiper, "I am extremely pleased to learn that Mrs. Kuiper is coming along so well, and I sincerely hope that the improvements will be permanent. The course of her illness shows how much patience one has to have in cases of this sort and I think you should not be disappointed if there are temporary relapses. The important thing is to keep up the gradual improvement, irrespective of small depressions."[61] Kuiper replied, "Mrs. Kuiper seems almost entirely recovered. She is doing her normal share of house work, and is feeling fine, regaining lost weight. While small relapses may be expected, the whole development since our departure on March 4 has been very favorable." In April, Kuiper wrote to Struve, "Mrs. Kuiper's condition has continued to develop favorably. For all practical purposes here she may be regarded as recovered; but I would like to see her stability still increase before she assumes responsibilities, i.e. before putting her in charge of the baby. It may be difficult for me to persuade her to stay here through May; she wants to go North badly."[62]

By June 1942, Struve assumed, with Sarah recovered, Kuiper's mind would inevitably turn to war work. He wrote to Kuiper, "I am fairly convinced, and have been for a year or longer, that we must all be ready to accept calls for government work, provided two conditions are met: (1) that the nature of the work is of sufficient importance to use our training and knowledge to such extent as to make it, in our minds, appear advantageous to the country to sustain the loss of our scientific and educational work: (2) that the work is of such a character as to justify the assumption that our training is adequate."[63] In other words, Struve was counselling Kuiper to make sure the work justified his time and that he had the relevant expertise. Reading between the lines, Struve was advising that Kuiper not search out war work but wait for it to come to him. He goes on to explain the importance of maintaining the university's programs and how the university will support the war effort by providing free courses on navigation to the Chicago community.

While Kuiper had no reservations about being engaged in war work, this was not the universal attitude shown by the astronomers. Mayall agreed with Kuiper; in fact, he regarded war work as a privilege.[64] Chandra complained about the interference with his research but accepted a position with the army's Aberdeen Proving Ground, where the prime interest was shell ballistics and solving the infamous "ballistics equation," a method of calculating the trajectory of a shell fired from a canon. Chandra was very critical of the work at Aberdeen. It was boring, and the atmosphere was too militaristic for him. He encouraged Kuiper, who had also been invited to Aberdeen, to take the position he had recently been offered supporting the war effort at Harvard, not that he needed much encouraging.[65] Morgan refused to leave Yerkes because, having just been appointed chair, he felt he was indispensable.[66]

The wartime difficulties to be faced by McDonald Observatory were apparent soon after Pearl Harbor and Kuiper's move to Harvard for the war effort. On May 14, 1942, Struve wrote that at the government's request they were converting the oil furnaces at the observatory to coal.[67] Before that, on January 13, 1942, Struve noted that the tire rationing was going to impact the movements of the observatory's truck such that it could make the short drive down the mountain only one day in five. This meant that Kuiper's furniture in their home in Williams Bay was to be moved into storage so others could use their house; Kuiper would have to leave his family in Cambridge while he slept in the dome during visits to McDonald. This was a real sacrifice for Kuiper, who rarely went to the observatory without his family. "Undesirable," wrote Struve, but one of the "lighter inconveniences of war time." Apart from accommodation issues, food was in short supply and the lack of gas meant difficulties with private cars.

On May 21, 1943, Kuiper heard from Struve that he had been promoted to full professor; this was just before he joined the war effort. In fact, the loss of faculty was to strain the department at Chicago so much so that, as Struve observed, "one illness among the astronomers at McDonald Observatory and we should mothball the facility for the duration of the war."[68] To his credit, that never happened, and it was during these wartime measures that Kuiper discovered the atmosphere on Saturn's largest moon, Titan.

Particularly hard hit by the war were the many computers, who were always women. Prior to the invention of the electronic computer, routine calculations were performed by hand using large sheets of paper ruled out in columns and rows. While the male astronomers were assigned to war work, there was no work for the computers. The war effort paid for ten months of Kuiper's time while the university covered the remaining two months, which he spent at McDonald. This was causing real hardships.[69] Struve did his best to find work for the computers; for example, when Kuiper made his first wartime visit to McDonald in December 1943, he was asked by Struve to find work for Mary Calvert. Struve also decided that women might be considered for "simple tasks," such as some photographic work and even some observational work with bright targets. As with industry, gender equality was being forced on astronomy by the war, albeit in small increments.

Mary Calvert in 1927.

Most of the computers were the wives of students (e.g., Mrs. Williamson), faculty (e.g., Mrs. Johnson), or women who lived locally (e.g., Mrs. Rubenstein or the six "Mexicans" hired in June 1941). However, it was common for Yerkes computers to spend time at McDonald. Of note is Mary Calvert, who started working in 1905 for her uncle, E. E. Barnard.[70] When Barnard died in 1923, she became curator of the photographic-plate collection and worked with Frost to publish her uncle's classic *A Photographic Atlas of Selected Regions of the Milky Way* in 1927. Similarly, she worked with Frank Ross to publish *Atlas of the Northern Milky Way* in 1934. She was a high-level assistant for Kuiper and the other astronomers until her retirement in 1946. There is no doubt that in a later age she would have been a distinguished astronomer in her own right.

In October 1943 the Kuipers wrote to Chandra about the status of the war and their war efforts. Kuiper learned that, in view of the wartime conditions, Chandra was not going to the annual meeting of the American Astronomical Society (AAS). At Harvard, Shapley argued that the English war effort was stronger than the U.S. effort, but the English had suffered more. Swings was working at Mount Wilson on the war effort, but, Kuiper noted, he still used the library for his own research. Using the library meant getting on with German astronomer Walter Baade despite Swings's hatred of Germans, but fortunately Baade was a "decent German." He ended his letter to Chandra by assuming Chandra had heard what the "Hun" did to the library at Naples.[71] In another letter, Sarah remarked after one social gathering that "here at Harvard, Don Menzel looked very elegant in his white uniform."[72]

The scene was set for Kuiper's war years. The seven-year run-up to the World War II, while Kuiper was a professor at the University of Chicago and astronomer at Yerkes and McDonald Observatories, enabled him to help in the construction of a new telescope and start research. He was making observations of binary stars well before the telescope was officially opened. During this period he established productive relationships with other astronomers at Chicago, and he enjoyed a lifestyle of living and working in Williams Bay while commuting to Fort Davis once or twice a year with his young family, now consisting of himself, Sarah, and his son, Paul. Aware and sympathetic to the war efforts against Nazi Germany, the Chicago astronomers closely followed news of the war in Europe, especially Kuiper's homeland. However, when the bombing of Pearl Harbor brought the war to the United States, there was universal shock. As tragic as it was, and as painful as if was for Kuiper in many ways, World War II was to make profound changes to his life and to the astronomy of the planets.

CHAPTER 5

War in Europe and a Moon with an Atmosphere

AS AMERICA ENTERED THE WAR, Kuiper was asked to join the war effort. There was never any doubt he would; the Netherlands had been overrun and his own family was painfully involved in the struggle. Kuiper chose to go to the Radio Research Laboratory that was being organized at Harvard. The next thirty months were to be the most turbulent of his life. Helping to develop radar countermeasures with Fred Whipple, he went to Britain to install the apparatus on the bombers that were flying sorties over Germany. After D-Day, being fluent in Dutch, German, French, and English, he was tasked with reading and analyzing reports, debriefing French and Dutch scientists as they were liberated, and interviewing Germans to gain intelligence and determine their role in the war. Most importantly for our story, he was to search for information on technologies of value to the Allies. Despite this busy agenda, he arranged passage for his brothers and others to England for desperately needed recuperation. Amid all this, Kuiper made one of the most important discoveries in twentieth-century planetary science, which propelled the subject in a new direction.

Joining the War Effort

Just before Kuiper started teaching his spring-quarter course in navigation at the University of Chicago in 1943, he was visited by someone who was to profoundly change his life. Clarence N. Hickman is a fascinating man.[1] He got his degree at the college where Robert Goddard built the world's first liquid-fueled rockets, Clark University in Worcester, Massachusetts, and he learned rocketry from Goddard. During World War I Hickman helped develop the bazooka. After the war he used his expertise in physics to design better pianos for the American Piano Company and became well-known for it. When the company collapsed during the Great Depression he moved to Bell Laboratories, eventually becoming its director of research. In that capacity he was asked to be part of the nation's science leadership for the war effort by heading the rocket-ordinance section of the National Research Defense Committee, then an advisory committee to the Office of Scientific Research and Development, the government agency for managing and funding the nation's scientific war effort.

Hickman was helping to assemble a group of the country's best minds to work on radar countermeasures. He visited Yerkes Observatory, where he had met Sarah and asked Morgan to join the war effort. Morgan declined, so Hickman approached Kuiper. Kuiper had already been told by Chandra that he was about to be approached by Edwin Hubble, who was assembling a team to work on military ballistics at the army's Aberdeen Proving Ground in Maryland. Kuiper told Hickman about the pending invitation from Hubble, but Hickman was not deterred and asked that they discuss the matter more in about a week when he expected to be in Chicago.[2] Kuiper mulled it over, noting that Sarah would welcome a chance to live in Cambridge for a few years. They would live with her parents. It was clear that Kuiper was going to accept Hickman's invitation, although he lamented that it was going to interfere with his navigation course, which promised to be first-class. Fifty students had enrolled in the course, including naval officers up to the rank of lieutenant commander. "Real ones, old timers," he added.[3]

The Radio Research Laboratory

The Radio Research Laboratory at Harvard was a spin-off of the Radiation Laboratory established in October 1940 at the Massachusetts Institute of Technology by Alfred Loomis, a philanthropist-physicist who played a leading role in the development of radar.[4] Between 1941 and 1945, while the MIT laboratory focused on radar development, the Harvard laboratory focused on radar countermeasures. The director of the Harvard laboratory was Frederick E. Terman, an electrical engineer from Stanford University now known as one of the fathers of Silicon Valley. Between 1925 and 1941 he designed research programs and courses for Stanford and in 1932 published the textbook *Electronic and Radio Engineering*, which is still in use today, after multiple editions.[5] Terman counted William Hewlett and David Packard among his former students. At its maximum, the Radio Research Laboratory placed more than 850 staff members under Terman's directorship. The Laboratory is best known for supplying

(a) Kuiper's identification photograph taken by the military as he joined the war effort. Courtesy of Matthew des Tombe. (b) Clarence N. Hickman. "Automatic Musical Instrument Collectors' Association. A. E. Werolin, Clarence N. Hickman," *AMICA* 13, no. 5 (1976), http://www.amica.org/Live/Organization /Honor-Roll/Hickman.htm. (c) Frederick Terman. Image from O. G. Villard Jr., "Frederick Emmons Terman, June 7, 1900–December 19, 1982," *BMNAS* 74 (1998), https://www.nap.edu/read/6201/chapter /18. (d) Samuel Goudsmit. Image public domain. (e) James Hopworth Jeans. Image public domain. (f) Fritz Houtermans. From his K. Landrock, "Friedrich Georg Houtermans (1903–1966)—Ein bedeutender Physiker des 20. Jahrhunderts," *Naturwissenschaftliche Rundschau* 56 (2003): 187–99.

radar jammers to the Allies (one placed in Southeast England could jam every radio transmitter in Europe) and for developing methods for detecting enemy radar signals. It also developed chaff, which was to occupy Kuiper for nearly two years.

On June 20, 1943, during their preparations for moving to Harvard, Sarah and Gerard learned that Sarah's father had suddenly died. This gave Sarah and her husband greater incentive to move to Cambridge. Sarah was close to her parents and Kuiper was highly sensitive to the grief in their family, but moving to Cambridge also made it easier to support the war effort and gave him an opportunity to be closer to Harvard College Observatory and the weekly astronomy colloquia.

(g) Kuiper and Hertzsprung at the IAU in 1958. Courtesy of Owen Gingerich. (h) Max Planck. Image public domain.

Kuiper's appointment at Harvard officially started July 1, 1943, the hottest summer in Cambridge since 1911. Conditions in the laboratory were difficult. With new building construction surrounding him, "it sounds like 10 dentists working on 10 of one's teeth." The secretary, with her typewriter and Dictaphone two feet behind him, were minor ripples in the noise.[6] He set to work relearning his basic physics and electronics, and he tried to connect with scientists at the large MIT group. Chandra had written to Kuiper to complain about the boring and tedious nature of the work at Aberdeen and the lethargic nature of the people there, but things were different at Harvard. Most people "are working like dogs," Kuiper reported, from 8:30 a.m. to 11:00 p.m., but his working day was eight hours; he was just too tired to do more. Kuiper did note the poor understanding of science that most engineers had, but perhaps, he said, this is inevitable.

During this important time at Harvard, the fall of 1943, Kuiper's thoughts clearly were never far from astronomy. He wrote to Chandra about his recent work on binary stars, and he was confused about why Sirius's companion white dwarf had no hydrogen lines. He had asked Walter Sydney Adams, director of Mount Wilson Observatory— and also a white dwarf observer—for some plates of Sirius B, and in October 1943 received fifteen. Still no hydrogen lines. Knowing this would upset Eddington, who predicted that these stars would have the lines of heavier elements, he asked Chandra for discretion. "I do not want to offend; I just want to get facts straight."[7]

Settling in to his Harvard duties, Kuiper remarked how while he would enjoy returning to astronomy, he could not regard the war work as a waste of time, as some of his contemporaries did. "Indeed, it was a handful of scientists that won the Battle of Britain," he wrote, presumably thinking of radar.

Titan: A Moon with an Atmosphere

For nearly a decade, Kuiper had made it his practice to visit McDonald Observatory for about a month every winter. Grappling with weather and a heavy research program, he would frequently write to Struve for permission to extend this period. The winter of 1943–44 was to be no exception. With Sarah and Paul, he left Cambridge, drove across country to Fort Davis, moved into the housing at the observatory, and set about continuing his long-term program of looking for white dwarfs among the stars of high parallax and high proper motion, recognizing them from their spectroscopic properties as described by Adams. On one night that winter, as he finished observing the stars he had planned (he referred to them as his "program" stars), he found the telescope pointing at a region of the sky where the planets and the ten brightest satellites were located.[8] He moved from object to object, noting the magnitudes of the moons and the lines of various gases in the spectra of atmospheres of the planets and their absence for the satellites. All was normal—the spectra were just as the textbooks had explained for many years. Then he came to Titan, and our understanding of planetary atmospheres was about to be turned on its head.

Titan is the largest moon of Saturn, half as big again as Earth's Moon. It is the second largest moon in the solar system after Jupiter's Ganymede. What caught Kuiper's eye was that Titan's spectrum contained the unmistakable line of gaseous methane at 6,190 angstrom (0.619 micrometers). This was remarkable. Titan had an atmosphere. But more remarkable yet, Titan's atmosphere contained methane, CH_4, a hydrogen-rich molecule. This discovery went totally against current thinking about planetary atmospheres. It refuted the arguments of some of the greatest astronomical authorities of the day.[9] Aware of the implications of the discovery, Kuiper sent a telegram to Russell: "Found satellite Titan to have atmosphere with strong methane bands and possible ammonia. Have not seen result published. Please advise whether result new in which case observations continued." Russell replied, "Spectroscopic observation new and important."[10]

Word spread rapidly. Congratulations came from many directions. But while it was to completely change Kuiper's life as an astronomer, he refused to see it as a cause for praise. In response to a congratulatory letter from J. H. Moore, now the director at Lick, Kuiper described the circumstances of the discovery: he was merely conducting his routine measurements and just happened to look at Titan.[11] It was not worth commendation because it was completely unplanned. He was to use the same argument

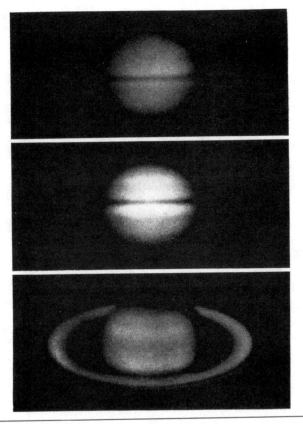

Saturn as it appeared through the best telescopic images before the advent of the space age. A. Dollfus, "Visual and Photographic Studies of Planets at the Pic du Midi," in *Planets and Satellites*, ed. G. P. Kuiper and B. M. Middlehurst, 534–71 (Chicago: University of Chicago Press, 1961).

for not accepting Chandra's congratulations in discovering Uranus's smallest moon, Miranda, a decade later. According to Kuiper, not only did he not deserve Moore's compliments, but in view of the beautiful work Moore had been doing on Saturn, which included spectroscopy, it would have been more fitting if Moore had made the discovery and been enjoying the world's adulation. In the face of remarkable success, Kuiper was sticking to one of his own heartfelt convictions: success should not come by accident.

The reason why Titan's methane atmosphere was so important is illustrated by an influential review paper written by Russell in 1935.[12] Russell argued that large planets have atmospheres containing hydrogen compounds, middle-size planets have atmospheres containing oxygen compounds, and small planets have no atmospheres. It was

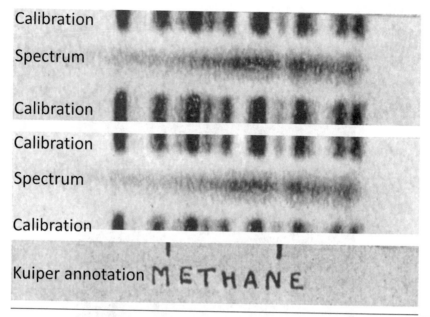

Calibration

Spectrum

Calibration

Calibration

Spectrum

Calibration

Kuiper annotation METHANE

The negative spectrum of Titan showing the presence of methane. Taking the negative means that the spectra are dark and the backgrounds are white. Astronomers often invert images this way as it is usually easier to see the areas of interest as the darker markings. Adapted from G. P. Kuiper, "Titan: A Satellite with an Atmosphere," *ApJ* 100 (1944): 378–83 (with 5 plates).

not a complicated argument, but one based on the simple laws of gas behavior; lighter gases needed larger objects for their retention as an atmosphere. Russell relied on the work of another giant of turn-of-the-century astronomy, Sir James Jeans, who had written the definitive *Dynamical Theory of Gases*.[13] The book was published in 1904, and every physics student had a copy. It had become the Bible for understanding the behavior of gases, planetary atmospheres included. Russell based his arguments about planetary atmospheres on Jeans' theories of gas behavior, which required knowledge of the mass, density, and absolute surface temperatures of the planets. It was all very straightforward.

Kuiper's discovery of the methane-bearing atmosphere on Titan was clearly serendipity—he had been searching for white dwarfs. However, his mind may have been primed for his discovery by Wildt's June 1942 symposium at the University of Chicago in which he summarized current knowledge on planetary atmospheres.[14] Kuiper had also shown some prior interest in the topic of atmospheric compositions when he looked unsuccessfully for an atmosphere on Pluto and for differences in hydrogen abundance from pole to equator on Jupiter and Saturn.[15] After the war his interest was further stimulated by contacts with astronomer Bernard Lyot in France and the planetary work in progress at Pic du Midi Observatory in the French Pyrenees.

Very little that Kuiper did was without controversy. Within weeks of his discovery, Jeans was commenting that he had predicted that Titan and two Jovian moons would have atmospheres and suggested that the observation of an atmosphere on Titan was not new.[16]

Kuiper wrote to Chandra on January 28, 1944, "I am still puzzled about that remark by Jeans . . . (he said that two Jupiter satellites also might have atmospheres; that is not confirmed). I presume it was just a lucky guess by a visual observer. Russell told me that my result was new; he has not yet replied to my enquiry as to the source of Jeans' remark. Jeans often banks on wrong observations (as the rotation of the spiral, etc.), but somehow was right here. Still, . . . the significant point is not that Titan has an atmosphere, but that the atmosphere is hydrogen rich."[17] In the last paragraph of his discovery paper for Titan's atmosphere in *Astrophysical Journal* Kuiper acknowledges a "doubtful" earlier observation of an atmosphere of Titan.[18]

Jeans was a prominent man.[19] With an Oxford, Princeton, Cambridge background, he had proposed a popular theory for the origin of the solar system in which the planets formed from a string of matter drawn out from the Sun by a passing star. With Eddington, he was the first to postulate the steady-state theory of the universe. He formulated the Jeans length (the size required for an interstellar cloud to collapse) and co-discovered the Rayleigh-Jeans law (the relationship between the energy of radiation and the temperature of its source). By 1944 he had long since retired but had become a well-known author of popular books describing complex and modern theories for the day, such as relativity.[20]

The truth behind Russell's skepticism of the old observation is not hard to locate and has echoes in the disputed discovery of an atmosphere on Venus.[21] When Russell replied to Kuiper's request for more information on the earlier work he recollected some old visual observations that he thought were without value. He wrote, "I strongly suspect that someone thought the satellite looked fuzzy and reasoned from that to an atmosphere—which is obviously nonsense."[22]

José Comas i Solá was a Spanish astronomer and director of the Fabra Observatory in Barcelona. In 1908 he published a short paper in *Astronomishe Nachrichten* on the major satellites of Jupiter and Saturn.[23] The final paragraph read, "On August 13, 1907, with a very beautiful image and using the magnification of 750 times, I saw Titan with very dark edges, standing in the darkness of the sky (something similar to what is observed in the disc of Neptune), whilst towards the central part, much clearer, were seen two round and whitish spots which resembles a diffuse double star. We can legitimately assume that this great darkness of the edges indicates that a very absorbing atmosphere exists on Titan." In other words, he had observed the phenomenon of limb darkening, where the edges of the disk are darker than the rest, which is characteristic of a thick atmosphere.

The first edition of Jeans book appeared before Solá's observation, but his second edition appeared afterward, in 1916. Chapter 15, "Aerostatics and Planetary Atmospheres," contains the statement, "An atmosphere has been observed on Titan, for which the critical velocities are about the same as the Moon, but this [the discovery of Titan's atmosphere] is explicable in view of its greater distance from the Sun, and the same consideration is probably adequate to account for the suspected atmospheres on two of Jupiter's satellites." Jeans believed his kinetic theory of gases, as it is now called, could account for atmospheres on Titan and two of Jupiter's moons if allowance is made for their low temperatures.

Solá had observed and Jeans considered plausible the existence of an atmosphere on Titan before Kuiper discovered gaseous methane using visible spectroscopy. They had a right to dispute Kuiper's priority, and the authors of some modern reviews do likewise.[24] Kuiper would dismiss Solá's work as not conclusive and Jean's words as speculative, and many modern writers agree. The kinetic theory is of limited value, Kuiper believed, because what was really needed is a comprehensive theory for the origin of the solar system. However, there is no doubt that Kuiper's discovery of a methane-bearing atmosphere convinced the community more surely than the earlier work, although Solá published in a most prestigious journal and Jeans was a highly distinguished scientist. Kuiper is usually credited with the discovery of an atmosphere on a moon; however, more important is the effect of the discovery on Kuiper. His work not only drew attention to the possibilities of spectroscopy in the study of the planetary system, but also set the scene for Kuiper to create planetary science as we know it today. Kuiper published his discovery in the *Astrophysical Journal* in 1944. Embedded in the paper was also an idea that was to be central in Kuiper's later work, which is that planets and moons formed hot, therefore all atmospheres must be secondary.[25]

Chaff or Window?

By March 1943, the small Kuiper family was back at Francis Street in Cambridge, and Gerard had returned to war work. He had become involved in chaff, a radar countermeasure in which small strips of aluminum foil are scattered ahead of bombers to confuse enemy radar.[26] Whipple had made the detailed analytical studies of this and devised a formula giving radar cross section at a given wavelength per kilogram of chaff. The British originally came up with the idea; the main challenge for their U.S. allies was to devise a method for deploying the strips. Whipple played a key role in designing such a device.

After almost a year of development, the chaff and its dispenser were ready for field trials, and it fell to Kuiper to represent the Harvard effort in those trials. From August 17 to October 1, 1944, Kuiper was in the U.K. to discuss the procedures for deploying the chaff. Kuiper wrote a detailed report in the form of a diary.[27]

Air Transport Command flew Kuiper into Prestwick, near Glasgow, where he was met by a car to take him to his center of operations in the U.K., the Telecommunications Research Establishment in Malvern, where chaff was referred to as "windows." The drive probably took eight to ten hours under wartime conditions. The Telecommunications Research Establishment was the center of British research on radio navigation, radar, and infrared detection for heat-seeking missiles. The facility was established under conditions of great urgency in Malvern College, an independent boy's boarding school in Worcestershire.[28]

In the six weeks Kuiper was in the U.K., he reported to several organizations in London, the British headquarters of OSRD, the U.K. War Ministry, the Supreme Headquarters of the Allied Expeditionary Forces (a tent encampment in Bushy Park, Richmond, with the codename "Pinetree"), and the U.S. Eighth Air Force (also located in Bushy Park, but with the codename "Widewing").[29] He interviewed U.S. bomber pilots in Brampton, supervised the installation of chaff into cannon shells at Porton Down, and installed chaff dispensers in de Havilland Mosquito bombers in Bovington Camp. Going all over the country by car, he once broke down in Oxford and had to call on the local military for help. After returning to the United States months later, Kuiper reported that the development of the dispensing machines for chaff had been satisfactory and the dispensers had been tested over Berlin. It had not been worth installing chaff dispensers on the bombers themselves, but they could be used ahead of the bombers when needed, with three divisions of aircraft of different sorts with different size chaff bombs.

Kuiper remarked that he would leave within a week with a small U.S. Navy mission for Liège, Belgium, and Paris to obtain a stereoplanigraph, an instrument for making topographic maps from stereo images.[30] Few details exist of this "small Navy mission," but one can easily imagine a nighttime sortie by Navy SEALs dressed in black and breaking into top-secret laboratories for the purpose of recovering equipment that could only be recognized by a top scientist. On the other hand, the mission may have been a simple delivery service in broad daylight. I prefer to believe the first.

Back in the United States, Kuiper was surprised to hear senators saying that the war would be over "within a few days." "It would be wiser to worry about the chance we still have of losing if certain high explosives are developed in time," he reports.[31] This possibility may be one reason why the Germans are not giving in. Few people in Cambridge expected an immediate collapse. He presages his next role in the war, as part of the program to determine whether Germany has a nuclear bomb and how advanced their research is in this direction. Kuiper had played an important wartime role in the development of radar countermeasures, so one might expect that the war would be over for the professor of astronomy from Chicago, but one would be wrong. The worst was yet to come.

The Alsos Mission

On December 10, 1944, Kuiper sent a V-Mail to Chandra informing him of a change of address.[32] No longer was it "Headquarters, 8th Air Force, A.P.O. 634," but it was now "U.S. Army, Alsos Mission, A.P.O. 887." In November 1943 Lieutenant General Leslie Groves, chief of the Manhattan District atomic bomb project, formed a team that would search Germany for information on Nazi progress with a nuclear bomb. He named the project "Alsos," Greek for "grove," a play on his name. The leader of Alsos was to be Colonel Boris Pash, a counterintelligence officer, and the chief scientist was to be Samuel Goudsmit, director of Brookhaven National Laboratory.[33] Kuiper planned to return to the United States for a few weeks, but instead of returning to Yerkes as he expected, he would have to return to Europe. In his early days with Alsos, Kuiper met with his brothers, Pieter and Nicolaas, in Eindhoven and arranged for them to spend time in England recuperating from their wartime experiences. For Nicolaas in particular, the war had been tough. He was a member of the Dutch resistance and had spent a year in hiding from the Nazis.

As the Allies moved across Europe on their way to Berlin, teams of scientists followed the front line looking for information that the Allied war effort could use. It was a highly successful program. Project Paperclip brought German rocket scientists to the United States who would became the nucleus of the program that put a man on the Moon.[34] Project Alsos also looked for information on several high-priority science and technology topics, most notably Germany's progress with a nuclear bomb.[35] After several months scouring France and the Netherlands for equipment and information, in April of 1945 Kuiper was dispatched to Alsos headquarters in Aachen, the westernmost city in Germany and the first German city occupied by the Allies.

Samuel Goudsmit, Alsos's chief scientist, had been a senior participant in MIT's Radio Research Laboratory.[36] Goudsmit was born in The Hague and attended Leiden University, where he earned his degree with Paul Ehrenfest. Goudsmit was best known for suggesting that the electron has a spin and as a result was known to his military colleagues as "Dr. Spin." In a letter dated April 1, 1945, Goudsmit outlined Kuiper's duties as a member of Alsos: assist the senior scientist (W. F. Colby, a spectroscopist at the University of Michigan), contact enemy astronomers to determine their support for the war effort, contact research institutions within reach of Aachen to determine the nature and organization of German war research, and contact his colleagues in Holland to obtain information on German research.[37]

The Briefing Sheets

Kuiper received the details of the sought-after technical information laid out in twelve tri-fold sheets (roughly eleven inches by five inches before folding) that neatly fit a breast pocket.[38] The sheets were produced in London by a committee charged to identify strategic German technology advances during the war. Many were based on reports from the front and from undercover agents. Broadly speaking, information was sought on directed missiles, proximity fuses, jet engines, and optical instruments. Under the first two headings was equipment, techniques, locations, and personnel. Under jet engines was high-thrust devices, turbines, metals and alloys, fuels, the Messerschmitt Me 163 rocket-powered aircraft, and rocket propulsion. The fourth topic, optical instruments, had the greatest level of detail and mostly involved infrared technology. The list included image-forming and non-image-forming infrared devices, photocells, bolometers, filters, transmitters and beacons, phosphors, rangefinders, gun sights, periscopes, binoculars, cameras, lenses, and so on. Kuiper must have been struck by the German interest in infrared technology.

A Personal Agenda

The discovery of Titan's atmosphere caused a lot of work for Kuiper. Mostly he wanted to obtain spectra in the infrared region, where lines that were diagnostic of specific molecules were numerous. However, the photographic films available were not sensitive enough for infrared spectroscopy. This was never far from Kuiper's mind as he sought to locate German infrared technology.

The April 1, 1945, letter from Goudsmit ends by noting Kuiper's desire to salvage scientific instruments relevant to his own research. Apparently, Kuiper made no secret of the fact that he was interested in learning about German infrared technology he could use at McDonald. This is not part of the Alsos directive, says Goudsmit, and Kuiper was to make his own arrangements for collecting and shipping to the United States any technology he thought he could use. Additionally, such efforts were not to detract from Kuiper's Alsos work. Kuiper collected boxes of papers and visited a great many relevant sites. On May 12, 1945, Kuiper wrote to Goudsmit, "Unless you have a better suggestion, I might propose that the papers bearing on infrared work be kept in Paris, to be worked over later by myself. I have still some important documents on infrared, and intend to follow them up by a trip to [the German city of] Jena. I am much interested in infrared work myself for my own use upon return to the U.S.A."[39]

Submitted herewith is the report of Working Party "C" which has had under consideration the following items:

I.   INFRA-RED DEVICES & EQUIPMENT

(a) Image-forming Infra-red Receivers:

These are devices which convert invisible infra-red radiation into visible light, the resulting image appearing on a specially sensitized screen.

Two types of receivers are known. One is the phospher type, the other the cathode ray tube or electronic type. The phospher type viewing device in general has no power supply and must first be exposed to sunlight ot ultra-violet radiation before it is capable of detecting infra-red. The electronic type receiver requires a high voltage power supply which may either be self-contained in the instrument itself or it may be carried in a separate pack and connected to the instrument by means of an insulated power supply cable.

The phospher type receiver will be a monocular telescope. The electronic instrument may be either a monocular or binocular. Both receivers have their peak sensitivity between 0.9μ and 1.3μ

The Germans are believed to possess infra-red devices having unusual qualities in range and sensitivity. They are definitely known to possess a phospher type receiver called "KATER", which is sensitized by exposure to daylight but not direct sunlight. Other image-forming devices of the electronic type are reported to be:- "SEEHUND", "SEEHUND DREI", "SPANNER", "FLAMINGO", and "PELIKAN".

"SEEHUND", "SEEHUND DREI" and "FLAMINGO" are reported to be infra-red detecting systems in use on German submarines.

The "SPANNER" device is most frequently connected with aircraft and tanks. In planes it may be employed as a search or identification device. In tanks, it may be used as a night driving aid.

a

1)

b

(a) One of the sixteen tri-fold "briefing sheets" that listed German technologies the allies were seeking information on during their liberation of Europe. This is the first of several pages describing various infrared technologies. Courtesy of University of Arizona Libraries Special Collections. (b) Two pages from the notebook Kuiper kept of his trip to Jena to investigate German development of infrared technologies. Courtesy of University of Arizona Libraries Special Collections.

His trip to Jena was recorded in a notebook, which now only exists as torn pages.[40] The observatory had been wrecked by the Nazis, but the technician, a man named Delforte, was around for Kuiper to interview. The observatory had a one-meter telescope, but the mirror was taken to Brussels in July 1940. A "monteur" (mechanic) for Zeiss, a manufacturer of optical devices, by the name of Theodore and a lieutenant from Berlin accompanied the mirror. They also took the viewfinder and the guiding telescope. The Nazis clearly thought they could use the equipment for monitoring Allied activity in the North Sea. A similar one-third-meter telescope in Hamburg probably also went to the coast. There is no indication that Kuiper found any information on infrared detectors in Jena, although Zeiss was known to have made major advances in this field.

Interviews with Germans

With Charles Vallas of the U.K.'s Telecommunications Research Establishment, and at the request of the military, Kuiper interviewed many Germans.[41] Ernst Lubcke was interviewed about acoustic antiaircraft fuses, parabolic mirrors for sound, and construction of artificial sources of underwater and submarine-like sounds; professors Schroeter and Magnus were interviewed about radar developments at Telefunken and efforts to hide the radar signature of submarines;[42] Albrecht Eisenberg and Joachim Thilo were interviewed about the acoustic proximity fuses being developed at Akustisches Laboratorium. Kuiper heard about documents that were buried in the woods at Auerhahn. He also learned that German senior officers were very superstitious—using a metal pendulum over a map of the Atlantic to detect Allied submarines, for instance—and that Hitler and Himmler had an official astrologer by the name of Dr. Fürher. He noted the level of support these persons provided to the Nazis and how reliable he considered them. By May of 1945, most of his interviewees were anti-Nazi.

One of the most significant interviews Kuiper conducted during the war was with Fritz Houtermans, who was familiar with German war research on physics and uranium.[43] Houtermans was a Dutch-Austrian-German atomic and nuclear physicist with an interest in geochemistry and cosmochemistry. Among his contributions was his suggestion that several transuranic elements were fissile and could be used as an alternative to uranium in a bomb. The second atomic bomb dropped on Japan by the United States was a plutonium bomb. In 1937 Houtermans, a communist, emigrated to Russia, but two years later was imprisoned during Stalin's Great Purge and returned to Germany under the German-Soviet Nonaggression Pact of 1939. After three months of imprisonment by the Gestapo, Houtermans was released, but he kept records on German nuclear research that he shared with Kuiper. Houtermans indicated that the Germans did not put much emphasis on science and technology until they realized how it was being used successfully by the Allies during most of the war.

Now he had serious concerns. He sent a message to colleagues in the United States warning them about the German work on fission and that Germany would soon have an atomic bomb.

Kuiper had high respect for Houtermans. "Professor Houtermans is one of the few German professors who are mature politically; his judgment seems sound and reliable," Kuiper wrote in his report of the interview. Houtermans also told Kuiper about the level of support for the Nazis among the German scientists. Carl Friedrich von Weizsäcker and Werner Heisenberg had visited Copenhagen to try to convince Niels Bohr that they were anti-Nazi. Max von Laue felt strongly enough to put his anti-Nazi views in writing in an obituary for Fritz Haber in 1940. These German physicists, Houtermans reported, were keen to publish their wartime work, but the Allies would probably not permit this. Kuiper suggested to Houtermans that they send their papers to Bohr to hold until the Allies approve publication.

Kuiper wrote a paper on astronomy in Germany during the war, which Struve presented at a New York meeting, presumably at the New York Academy of Sciences.[44] While some were appreciative, others were hostile, perceiving the article as pro-German. On February 19, 1946, Kuiper wrote to Struve, "I was quite amused at the violent reactions you described on the part of Luyten and others during the New York meeting. The subject is an emotional one and everybody wants to let steam off. I should have liked to have seen the fight."[45] The ringleader of the protestors was Luyten, with Bok, Dutch American astronomer Jan Schilt (then chair of astronomy at Columbia University), and especially Biesbroeck sharing Luyten's views. Harlow Shapley thought the protest "disgusting," but it blew over. "If it were not so sad it would be amusing to find among the most noisy critics men who had sat at their desks throughout the war," Struve remarked.[46]

The Future of German Scientists

On June 30, 1945, Kuiper received a memorandum from a Major Fisher instructing him to make recommendations for the future of German scientists. Kuiper reported that most German scientists showed little moral courage, although the leaders tended to be anti-Nazi. Contributions varied with field—some fields could have made greater contributions but were closed before the war because of Jewish influence.[47] Kuiper felt it was important to allow pure research in the universities to continue after the war under a national agency that reported to the Allies, but that industries should be kept out of the universities. Some scientists did not realize that their research was used in the war effort through companies like Telefunken.

That same day, something happened to harden Kuiper's attitude against the Germans. In a second memorandum to Fisher, Kuiper reported that his initial favorable impressions of German scientists had been changed by conversations with scientists in occupied countries.[48] German scientists took from occupied countries to enrich their institutions, he said. Kuiper reported that Heisenberg's views were typical: Heisenberg argued that democracies are not strong enough to compete with Russia, so either Germany would rule Europe or Russia would. German scientists lacked social responsibility, Kuiper claimed, and he cited their offer to develop the V1 and V2 rockets for the Allies. He also mentioned the concentration camp surgical atrocities, made worse because the German scientists regarded the accounts as exaggerated and regretted only the harm done to Germany's reputation. Kuiper continued, saying that Germans are good at whining and arousing sympathy. By that point Kuiper had seen and heard from a great many people, mostly Germans and Dutch, and seen many scenes of destruction, and was losing his normally objective manner.

Scientists in Holland

Having completed his interview of German scientists in June 1945, Kuiper was sent to interview scientists in Holland.[49] His instructions were detailed. At the observatory in Liège he was to conduct an interview of J. Pauwen for the Navy about photometry. For the remaining Dutch scientists, he was to ascertain status and results of physical and astronomical research in Holland during the war and the fate of certain key scientists: Hendrik Casimir of the Philips Physics Laboratory in Eindhoven, Hans Kramers, Wander Johannes de Haas, Cornelis Jacobus Gorter at the University of Amsterdam, and Hertzsprung at Leiden University. It must have been a melancholy experience to be interviewing his old professors in Leiden. The result of all these interviews was a realization of the true horror of the war and the abuses committed by the Nazis.

Damage to European Observatories

Kuiper was greatly impressed with the heroism of the French astronomers during the occupation and wrote of this in news notes to astronomical journals.[50] In 1947 he was awarded the Janssen Medal of the Astronomical Society of France in recognition of his supportive words. He was equally concerned about the fate of the observatories in Germany and wrote an article on the subject that was published in the *Popular Astronomy*.[51] The Astronomiches Rechen-Institut, Germany's equivalent of U.S. or British National Almanac Offices, now consisted only of the "bare essentials," the library being dispersed. The Heidelberg, Bonn, Göttingen, Breslau, Potsdam, and Babelsberg observatories suffered little or no damage, but it seemed likely the Soviets would remove equipment from Babelsberg Observatory as reparation for some Soviet observatories destroyed by the Germans. The Hamburg-Bergedorf Observatory, the principal German observatory, which had important cataloging work underway, was undamaged, although two hundred incendiaries fell in the grounds. Munich Observatory was severely damaged and Leipzig Observatory was largely destroyed. Kuiper also listed the research and publications that came out of Germany during the war.

Goudsmit's Parents

On April 10, 1945, Goudsmit sent a hand-written note to Kuiper asking him to find out what happened to his parents, who were in Westerbrock camp in the Dutch province of Drente and taken from there in January 1943.[52] The Westerbrock camp was established October 9, 1939, by the Dutch authorities to receive Jews evacuating from Germany. When Holland was overtaken by the Germans on July 1, 1942, the camp became a place for rounding up and organizing weekly shipments of persons to Auschwitz or Sobibor, the selections being made by a committee of Jews. During the war, 103,000 persons were moved east. When the camp was liberated in September 1944, it had 900 prisoners. There is no indication that Kuiper ever replied to Goudsmit, but a biography for Goudsmit published by the NAS says Nazi records showed that Goudsmit's parents were both killed in the concentration camps.

The Max Planck Incident

A noteworthy event during Kuiper's wartime work in Europe was the Max Planck incident. Max Planck is one of the great theoretical physicists, noted especially for conceiving quantum theory, which states that energy is quantized into small packets. The theory explains many phenomena, including how the frequency of radiation is directly proportional to energy, an important concept for understanding spectra. On November 17, 1945, writing from Harvard, Kuiper sent a letter to Hans Clarke at Columbia University at Planck's request, saying that Mr. and Mrs. Planck were well and living his niece in Göttingen.[53] In the second half of May 1945, Kuiper had found them living in one room of a farm house in Rogätz, on the Elbe, thirty kilometers north of Magdeburg, in poor health and without medical care. He evacuated them to Göttingen, where his scientist friends "are enjoying Planck's company." Professor Planck was eighty-seven. A few days later, Clarke replied that he had heard from another source that Planck was living with his niece (Mrs. Seidel) in Göttingen. Clarke gave his warmest thanks and gratitude that Mr. and Mrs. Planck had adequate medical care and were among friends and relatives. While the talk was of physical conditions and health care, the underlying story, as often repeated by Kuiper many years later, was that Kuiper had rescued Planck from the Soviets in a dramatic cross-country dash in a commandeered jeep.

Aftermath

Kuiper published ten scientific papers during the war, five on his search for white dwarfs, two on Beta Lyrae, two on novae (Nova Pupis and Nova Herculis), and the discovery paper of the atmosphere on Titan. Despite all the distractions, he stayed in touch with his astronomical colleagues and tried to attend conferences.

The war was to have profound consequences for science. It brought the rocket to the forefront, which launched the Space Age and the race to the Moon. It brought the nuclear bomb, which ignited the Cold War. It demonstrated the importance of science and technology and that it was worthy of government support. For Kuiper, it demonstrated the rapidly developing field of infrared technology. It had clearly been of major importance to the military with a wide breadth of applications. All three military services saw the value of infrared technology as a means of seeing the otherwise unobserved, and so would astronomy. However, Kuiper was already attuned to the value of infrared spectroscopy through his discovery of the atmosphere on Titan. He found one line in the visible spectrum. He knew more information was available in the infrared spectrum, but the photographic emulsions available at the time were inadequate for recording infrared wavelengths. What he needed was an infrared detector, and the war had told him that such devices, if not already available, were not far off.

The war was over, and after six months serving Alsos as scientific liaison officer, it was time for Kuiper to return to Yerkes. The process did not go smoothly. First, there was uncertainty as to when the military would release him. Originally, Kuiper expected to be back at Williams Bay in late June, but this did not happen; his return was delayed until January 1, 1946.[54] Second, the number of houses at Yerkes Observatory was limited, and the houses were in high demand. During their absence, the Kuiper's home at Williams Bay had been rented to Louis and Mary Henyey.[55] Louis, a graduate of the Chicago department, was appointed assistant professor in 1942. As plans for the return of the Kuipers to Williams Bay began to materialize, the Henyeys were asked to move out, but they could not find new accommodation. In view of the uncertainty over Kuiper's return date, Struve was reluctant to demand their departure. With the stress of the war taking its toll, Sarah wrote to Struve on June 4, 1945, "I feel a little like the man without a country, or rather a family without a home after reading your letter of May 31. It is rather sad after looking forward to returning home to Yerkes after these two years of uncertainty and separation to discover that we have no place to go." Expecting to return during the second half of June, she had already started to make arrangements.[56] Struve eventually assigned the Kuipers a different house and had their furniture moved there.

On December 17, 1945, Kuiper asked Terman for a copy of his final wartime report (known as RRL 91). He also asked for advice on how he might obtain an army search light for use in the infrared astronomy program at McDonald Observatory; he wanted to measure "infrared solar radiation and atmospheric transmission."[57] After so much activity, so much hurt and damage, so many reports and flights across the Atlantic, it is interesting to note Kuiper's words at the end of this period of his life. Kuiper ends his December 17 letter to Terman with thanks for being able to participate in the interesting work at Radio Research Laboratory and associated field work overseas. "This has been a great experience," he wrote. There is no indication that he received the final report or the spotlight.

A Reunion

Fifteen years after the end of the war, the *Chicago Sun-Times* ran a story with the headline, "30 Former Top-Secret U.S. Spies Hold a Reunion Here," by Donald Brucker.[58] The story reported on a reunion held at the Edgewater Hotel the previous day that consisted of a lunch, a business meeting, and a banquet. The reunion included "some of the most unusual agents in the history of U.S. espionage." The story told of how in 1944 the American scientists believed Germany could be developing the atomic bomb and could still win the war. It explained how in November 1943 Lieutenant General Leslie Groves, chief of the Manhattan District atomic bomb project, decided to form a team that would search Germany for the bomb. This team consisted of James B. Fink of Bell Telephone Laboratories; R. A. Fisher,[59] physics professor at Northwestern University; A. Allan Bates, director of Westinghouse Laboratories; and Gerard Kuiper. The article quotes Colonel Pash as saying, "We were told not to fight. We went in with or ahead of the invading armies, we captured our people and documents, and we got out." "We were looking for atom bombs," Groves said. He went on by saying that Alsos was "the most successful spy mission in the war," capturing fourteen German scientists. By spring 1945, Groves could tell President Roosevelt that Germany did not have the bomb.

Kuiper found the newspaper article waiting for him when he returned from an IAU meeting about the Moon that had been held in Leningrad. He was not well pleased—in fact, he went ballistic.[60] "This is a shocking report to someone who believes that scientists must attempt to maintain decent international relationships," he wrote to Goudsmit on December 24, 1960. "My own role, for one, was no different from that of a historian or a sociologist who examines available facts after they have occurred; at no time was I in front of the lines. Publicly to call such men spies is sensational and could be regarded as slander; it obstructs their future work at home and abroad. I certainly hope that this kind of show can be avoided at future reunions." Ever conscious of the need to work with the international scientific community, including

Germans, in the final analysis Kuiper chose to consider his war work simply that of a "historian or a sociologist."

The war was long over. As with most of its active participants, it had a profound effect on Kuiper. It gave him an opportunity to contribute in two major ways: he helped develop and implement the counter-radar measures known as chaff in the United States and windows in the U.K., and he tracked down information, mostly in the form of interviews, that was needed by the military. This is a contribution he continued to provide the U.S. government throughout the Cold War until his death. After his remarkable success in discovering the atmosphere of Titan, he used his European duties to track down instrumentation that would enable him to apply infrared spectroscopy to the study of the planets. But most of all, in seeing the war up close, he witnessed one of the greatest tragedies in our history. He was not without feeling, and he was not immune to the immensity of the events surrounding him. On April 12, 1945, he sent a V-mail message to Moore at Lick Observatory: "Our bombers and our armies have done a job here in Germany. You will know this from the papers, but there is something about being in the middle of it. One of the common sights during this lovely spring: beautiful magnolias or apple trees in blossom between rubble that was once a city. The blasts did not greatly harm the trees while they were dormant last winter."[61]

Kuiper ended with this: "The situation in Holland is dreadful, and I hope relief will not come too late. Thousands have already died."

CHAPTER 6

New Eyes on the Solar System

Infrared Astronomy and Planetary Atmospheres

W ITH THE END OF THE WAR, everything changed. New international politics arose, as did new social programs, new efforts to bring a long-term peace to the world, a Cold War, and new attitudes to science.[1] Recognizing the critical role played by scientists during the war, governments became involved in planning and funding scientific research. James Van Allen exploited wartime technologies by using surplus Navy rockets and balloons, and eventually captured V2 rockets, to measure radiation in space. Kuiper was invited to help manage the U.S. V2 rocket program.[2] Kuiper was on his personal crest of accomplishment caused by the discovery of the atmosphere of Titan. What frustrated him in his otherwise successful discovery was that he could not record the spectrum of Titan's atmosphere in the infrared region of the electromagnetic spectrum, which is rich in molecular lines. The war had shown Kuiper how much progress had been made in infrared technology, particularly in developing a detector for infrared radiation. One of the U.S. experts in the new infrared detectors was Robert Cashman, who worked at Northwestern University, a stone's throw from the University of Chicago. Using Cashman's detector, Kuiper was able to build an infrared spectrometer that he could attach to the 82-inch telescope at McDonald Observatory. He had new eyes with which to view the solar system and was able to make a series of discoveries of fundamental importance.

New Eyes

The visible spectrum was well known when, in 1800, William Herschel discovered that radiation existed beyond the red end of the spectrum. He could detect this

(a) Robert Cashman. L. M. Brown and J. B. Ketterson, "Robert J. Cashman," *PT* 43 (1990): 148. (b) Arthur Adel. Courtesy of the Bentley Historical Library, University of Michigan. (c) Luyten in his later years. Courtesy of University of Minnesota Media Digital Archives.

radiation by placing a thermometer off the red end of the visible spectrum, and he termed it "infrared." Subsequently we have learned that there is radiation beyond the blue-violet end of the spectrum called ultraviolet, and beyond these are other types of radiation with different mechanisms for production and detection. The infrared region of the spectrum is large, and astronomers commonly divide the infrared region into near-infrared (with wavelengths 0.7 to 5 micrometers), mid-infrared (5 to 25 or 40 micrometers), and far-infrared (25 or 40 to 200 or 350 micrometers). The near-infrared radiations are commonly produced by the bending and stretching of molecules—the molecules can emit or absorb radiation depending on their internal energy and the environment. The longer wavelengths are produced by matter depending on its temperature, and this radiation is sometimes referred to as thermal infrared. While Kuiper could produce photographs using light in the visible spectrum and separate the colors by using filters, photographic emulsions were not capable of detecting the less energetic infrared radiations.

Robert Cashman and the Development of the Infrared Detector

In 1933 a professor at the University of Berlin, Edgar W. Kutzscher, first discovered that lead sulfide changed in electrical resistance when exposed to infrared radiation with a wavelength of three micrometers.[3] Soon after, Bernhard Gudden at the University of Prague made sensitive lead sulfide films.[4] Kutzscher continued the work in cooperation with the Electroacustic Company in Kiel, but under wartime

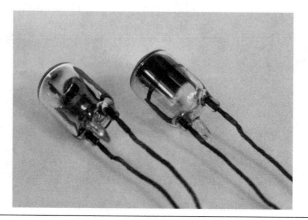

The Cashman infrared detectors. Courtesy of Dale Cruikshank.

conditions the work was performed with great secrecy. During the war he developed a method of chemical deposition that led to the first practical IR detector and a variety of military applications. While the mechanism was not well understood, the ability to make detectors reproducibly from one-micrometer polycrystalline films of lead salts was known, and performance quickly improved as a variety of similar salts were developed.[5] The briefing sheets that Kuiper carried with him in his investigations in wartime Europe listed a series of infrared-detecting systems used by all branches of the German military.[6] Additionally, there was Kiel IV, an airborne infrared system produced by Zeiss in Jena under the direction of Werner K. Weihe, chief of Zeiss's electrotechnical laboratory. It might be that Kuiper met with Kutzscher. When Kutzscher's facilities in Berlin were captured by the Soviets, Kutzcher was able to get to the U.K. where he pursued his work in Southampton with Mullard Limited. He joined Lockheed Aircraft Company in Burbank, California, a few years later as a research scientist.[7]

Cashman had worked on infrared detectors at Northwestern University but set it aside in 1941.[8] Meanwhile, as the briefing sheets show, the military attached a high importance to infrared technologies, so the National Defense Research Committee funded Cashman to resurrect his attempt to develop an infrared detector. Initially Cashman focused on thallous sulfide films, but by the winter of 1944 he produced the first successful lead sulfide detectors and realized they were superior to the thallous cells he had been investigating. By December 1944 he found he could achieve a greater photosensitivity by evaporating the lead sulfide in the presence of oxygen.

Sometime in late 1944 Cashman acquired several captured German lead sulfide cells that were superior to his because of their smaller size. There is no record of how

Cashman acquired the German cells, but it seems likely that Kuiper was involved. No contemporary documents indicate that Kuiper brought back instruments that could detect infrared radiation, although a 1982 letter from Jesse Greenstein to Dale Cruikshank stated that Greenstein "well remembers working with Kuiper on an image intensifier Kuiper had brought back from the war."[9]

Developing the Infrared
Detector for Astronomical Use

On November 23, 1945, Kuiper wrote to Struve about a new venture:

> One of the reasons I want to stay here [at Williams Bay] somewhat longer is the far-infra-red work I have started this week. In my WD [war department] work in Germany I became convinced of the great fields opened by the wartime research, which should enable us to get star spectra to 40,000 or 50,000 A [4 or 5 μm]. Also, it should be possible to get direct photographs up to those wavelengths. Another new field of interest to us is that of the cm and mm frequencies. About the latter I will mention my suggestions after you return here. On the infrared research my plan was, upon return to the U.S., to locate the group in this country charged with working on specific techniques which could be applied to our problem. In the East I secured the proper clearance and introductions, and was lucky in that the group of particular interest to me was located at Northwestern University. My first conference there a few days ago was all I hoped for, and we have set up a small development program that ought to give some important answers in a few weeks. I am anxious to push this thing far enough along before I leave for Texas that a working agreement and a clear program have been made.[10]

A few days earlier, September 28, 1945, Kuiper wrote to Struve that he planned to build an infrared spectrometer at Yerkes Observatory and that Wallace R. Wilson of Northwestern University would come to McDonald Observatory to help with the trials. He mentioned that Cashman was "anxious for success" and that he hoped Struve would present a paper on the spectrometer at the next AAS meeting.[11] On November 13, 1945, he wrote that he was pleased with progress with spectrometer, that he was in frequent contact with Cashman (who was working on improving the lead sulfide cells), and that Wilson was tabulating the wavelengths of interesting molecules.[12] On November 23, 1945, he wrote that they had started to build the spectrometer but six days later was showing impatience at Cashman's progress.[13]

Kuiper's spectrometer would now be called a scanning prism spectrometer. Light from the telescope was focused on a prism by a concave mirror. A second prism

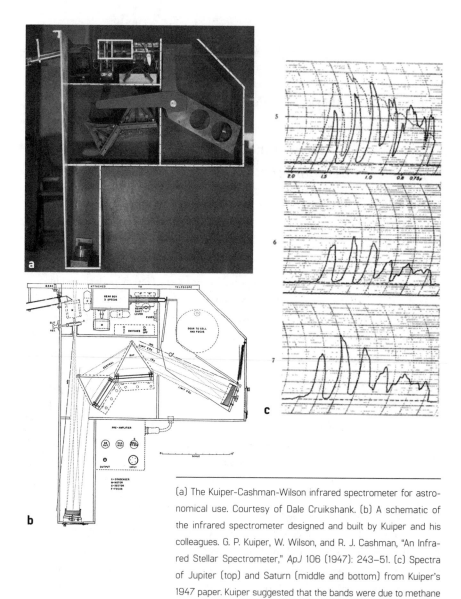

(a) The Kuiper-Cashman-Wilson infrared spectrometer for astronomical use. Courtesy of Dale Cruikshank. (b) A schematic of the infrared spectrometer designed and built by Kuiper and his colleagues. G. P. Kuiper, W. Wilson, and R. J. Cashman, "An Infrared Stellar Spectrometer," *ApJ* 106 (1947): 243–51. (c) Spectra of Jupiter (top) and Saturn (middle and bottom) from Kuiper's 1947 paper. Kuiper suggested that the bands were due to methane and ammonia. G. P. Kuiper, "Infrared Spectra of Planets," *ApJ* 106 (1947): 251–54.

separated the infrared wavelengths from 0.9 to 2.9 micrometers. Another concave mirror reflected the light into the detector. This second mirror was mounted on a pivot and attached to a motor through a system of gears allowing it to scan through the wavelengths. The signal from the detector was then sent to a strip recorder, so as the wavelengths were scanned the signal would trace out the infrared peaks.

By December 18, 1945, the spectrometer was working, and through trials at McDonald Observatory the group had determined the long-wave cutoff, the resolution, and the limiting magnitude, and they had obtained some early results. True to form, on December 28, 1945, Kuiper suggested that it was time to write a press announcement on the new spectrometer and he had drawn up a list of a few hundred objects he would like to look at with the new instrument. Throughout 1946 the group worked on the spectrometer, but by January 13, 1947, Kuiper felt that they were in a race with Mount Wilson to publish their results and, contrary to his earlier comments, asked Struve to not publicize their plans.

The conditions under which Kuiper and his associates built the first infrared spectrometer for astronomical use are described in their 1947 article in the *Astrophysical Journal*.[14] Apart from the importance of its contents, the paper is noteworthy for having multiple coauthors; of the thirty-three papers published by Kuiper in the 1940s, all but three were singly authored by Kuiper—he exhibited a very strong preference for publishing alone.[15]

The paper mentioned the late 1945 meeting between Kuiper, Cashman, and Wilson. Cashman provided the detector. Wilson designed and built the electronics. Kuiper designed the spectrometer. The detectors had been declassified in 1946, and Cashman had presented details at the annual meeting of the American Optical Society and had published a technical paper.[16] The new lead sulfide detector was a thousand times more sensitive than the previously favored techniques using thermocouples and bolometers. The spectrometer was built by Charles Riddel, the observatory's instrument maker, and the optics were built by Carleton Pearson in the optical shop. The steel box consisted of four compartments, one compartment housed concave mirrors (one mirror to disperse the light onto the first of two prisms and one mirror to focus the light on the detector) one compartment to house the two prisms, and one compartment to house the motor that moved the second mirror as the spectrum was scanned. An external compartment housed the preamplifier.[17]

A Navy Sponsor and Arthur Adel

It is not surprising given Kuiper's war experience that he saw a real prospect of support for his infrared astronomy in the military. In January of 1946 he wrote to Walter

Bartky, dean of the University of Chicago's Division of Physical Sciences, for advice on approaching the Navy, and he attached a brief proposal.[18] The proposal is an interesting statement of Kuiper's thinking about infrared astronomy after his wartime experiences. At the time, the lead sulfide cells were still classified, but he was in touch with physicists developing the cells, and application of the cells to astronomy would surely result in pressure to improve sensitivity, which the Navy would appreciate. He planned to place the cells at the focus of a search light mirror and scan the sky, a primitive infrared telescope. Secondly, he could place the cells in the new spectrometer he was building and scan wavelengths using prisms. "Rather than re-write your proposal, in order to make it more attractive to the Navy," replied Bartky, "first talk to the ORI group, and then put it in writing."[19] The ORI was the Navy's Office of Research and Invention. Kuiper did so, and on June 6 was in Washington, D.C., talking to a certain Commander Liddel of ORI. Liddel replied that while the Navy was not interested in astronomy, it was interested in the infrared transmission of the atmosphere, and measurements at Fort Davis would be of value.[20] The proposal, now a "Navy task order," was rewritten to their requirements and carried the title "Determination of the atmospheric transmission between 1 and 1000 μm at 6,800 feet (2073-meter) elevation in dry desert air." Of course, 6,800 feet is the elevation of McDonald Observatory. Coauthoring the proposal was Arthur Adel of the University of Michigan.

Adel was known to have high-order skills with the theory and experimental practice of infrared radiation.[21] He had written a thesis on the topic for his PhD program at the University of Michigan and had shown using both theoretical and experimental techniques that absorption bands in Lowell Observatory astronomer Vesto Slipher's spectra of Jupiter and Saturn were due to harmonics of the fundamental vibrations of the methane and ammonia.[22] He then moved to Lowell Observatory at Flagstaff, and from 1936 until 1942 made a series of important measurements on the composition of the upper atmosphere. During the war he worked on projects for the Navy at the University of Michigan but hoped to return to Lowell afterward. Not only did he return, but for a while it also looked as though he was in line for the directorship when Slipher decided to retire. However, when it was found that he was Jewish, he was told that his work was not suited to an observatory and he was dismissed with a hundred-dollar severance package.[23] The Air Force wanted to continue supporting his infrared research, so he moved his operation to the University of Michigan at Ann Arbor.

Kuiper and Struve learned of Adel's dismissal from Lowell and set about recruiting him to be superintendent at McDonald Observatory and research assistant to Kuiper. Struve and Adel got on well, so he visited the Adels in Ann Arbor to make the final arrangements. Adel recalls in an oral history interview for the American Institute of Physics (AIP), "I remember the ice cream we served Otto Struve. He said, 'Where do you get such wonderful ice cream?' He was a nice man."[24]

On June 26 Adel telephoned Kuiper with the news that he had left Ann Arbor for McDonald Observatory, and Kuiper suggested a meeting with him and Struve in Chicago. Adel declined, so the two men discussed the proposal on the phone. The main issue was the large spectral range required by the Navy and, according to Kuiper's notes on the conversation, Kuiper wanted to subdivide the spectral range. He would focus on wavelengths below twenty-five micrometers and Adel would take responsibility for the rest. Adel wanted both partners to be equally responsible for the whole range. They agreed Kuiper would write a letter summarizing the discussion.[25]

When Adel received Kuiper's summary, he was furious.[26] Perhaps the stress of living with the Lowell rejection made him touchy, but Kuiper was also being insensitive. "Your letter of June 26 reached me today and caused me no little consternation," Adel wrote in a letter three days later. "It was my understanding that your sole purpose in modifying and broadening my application for a grant-in-aid was to improve the chance of securing the funds to continue my work. . . . I can see no reason for applying for a grant in my behalf. . . . Furthermore, the project as you have proposed it burdens me with the responsibility of an almost impossible task." Adel was referring to the range Kuiper was assigning him, twenty-five to one thousand micrometers, and the timescale proposed. "I find this a compromising position and therefore unacceptable." He ends, "Finally, I will under no circumstances undertake any part of this added Navy task unless $10,000 is allocated from the grant to be used at my discretion for the purchase of equipment to reach 25 µm and for the purchase of equipment to go beyond that point." In other words, Kuiper assigned to himself most of the funds and tasks that required little risk, while he expected Adel to perform new, complex tasks with little or no funds and on an unrealistic timescale.

The Adels' move to Fort Davis was a disaster. Adel continues in his AIP oral history, "Catharine and I went to Texas. I went to the McDonald Observatory, but I was in such a state of disarray, I still hadn't got over the shock of being dismissed from Lowell. McDonald was not attractive to Catharine. We saw the quarters where we would be living and so on. We just turned around and left." Adel returned to the University of Michigan for a few years and then moved to Arizona State College (now Northern Arizona University), where he taught mathematics and received a large Air Force grant to establish the Atmospheric Research Observatory for the study of atmospheric ozone. In 1982 the mathematics building at Northern Arizona University was named after him.

A Failed Reconciliation with Luyten

If Kuiper had problems with Adel, they were nothing compared with the ongoing struggles with Luyten. For many years, Struve struggled to bring Kuiper and Luyten into some form of harmony, even trying to get them to collaborate on a paper. In a letter dated February 4, 1941, Struve stressed that he and Kuiper "need to settle collaboration with Luyten for both of us."[27] Then on July 28 he wrote, "I am anxious to keep the new plan operating as smoothly as is possible, and we should try to forestall misunderstanding if we can."[28] Later in the same letter, he wrote, "I want to ask you to be particularly cautious within the next few months in doing everything as stipulated in the agreement with Mr. Luyten. I have had some rather unpleasant correspondence with him in regard to matters which have no relation to your collaboration but I think we must be particularly sensitive at present to anything that might be construed as a lack of living up to the exact meaning of our agreement with him. He is probably watching every step to see whether he can trap us in some misstep." A year later, May 16, 1943, as Kuiper was preparing for a trip to Europe, Struve wrote, "I assume your joint paper with Luyten will be here before I leave."[29]

Kuiper remained cooperative but not enthusiastic, doing as Struve requested. The two exchanged lists of stars they were working on, Kuiper obtained new spectra for Luyten, and Struve arranged for Luyten to spend time at McDonald. On July 12, 1944, Kuiper reviewed a paper of Luyten's that had been submitted to *Astrophysical Journal* and, while finding errors and other problems, recommended publication.[30] Struve was not happy at the lack of references to Kuiper in the paper, which did not bother Kuiper. Nevertheless, "Luyten has his idiosyncrasies, as we know from experience," Struve remarked.[31]

During Kuiper's trips back from his Alsos work he continued to receive pressure from Struve. On March 13, 1945, Struve wrote, "I hope his and your relations will not deteriorate after your establishment of harmony a year ago."[32] Apparently there was a degree of success, for on February 11, 1946, Struve wrote, "Luyten tries to be friendly."[33] However, the master manager never did succeed in getting Kuiper and Luyten to publish together. He tried, often seemingly walking on egg shells, and neither Kuiper nor Luyten refused to cooperate, but it just never happened. Meanwhile Kuiper's observational program at McDonald continued at a feverish pace.

Atmospheres of the Planets

Four days after the paper describing the infrared spectrometer was submitted to the *Astrophysical Journal*, Kuiper submitted a paper, singly authored, on the infrared spectra of the planets. In one of the shortest abstracts in the scientific literature, he wrote, "Infrared spectra of Venus, Jupiter, and Saturn are shown, obtained with a lead sulfide cell."[34]

Kuiper's new passion for infrared astronomy of the planets was beginning to burn strongly, and the first targets were Venus, Jupiter, and Saturn. At the time there was much speculation about the composition of Venus's atmosphere prompted by theoretical chemical and physical arguments. It had become widely assumed that water and carbon dioxide should initially be present, but that they would react to form formaldehyde or formaldehyde polymers, perhaps as solids that resulted in the clouds seen on Venus.[35] Attempts to confirm this theory using ultraviolet spectroscopy failed.

The presence of an atmosphere on Venus was long known.[36] During the transit of Venus across the Sun in 1761 Mikhail Lomonosov noticed that the disc of Venus was blurred, which he attributed to an atmosphere. However, the observation was largely ignored.[37] In the late eighteenth century Johannes Schröter also found evidence for atmosphere on Venus and the movements of clouds.[38] The first good photographs were obtained at Yerkes Observatory in 1928 by Frank E. Ross, who found surprising markings when Venus was photographed in the ultraviolet.[39] In 1932 Adams and Theodore Dunham Jr., working at Mount Wilson Observatory, used new infrared-sensitive photographic plates from Eastman Kodak to find bands in the spectrum of Venus that they assigned to carbon dioxide. In 1947 Kuiper confirmed their finding using his new infrared spectrograph.[40] Then in the 1950s Kuiper at McDonald and Robert S. Richardson at Mount Wilson showed that the clouds of Venus had a banded appearance.[41] Later Charles Boyer and Pierre Guérin observed a Y-shaped feature that suggested a four-day retrograde movement of the atmosphere.[42]

In 1905, Slipher took the spectra of Jupiter and Saturn in the red region of the visible spectrum using newly developed photographic plates. He observed several bands that were not identified until 1931, when Wildt and E. J. Meyer attributed them to ammonia and methane.[43] However, discrepancies remained, and it was Dunham in 1933 that produced a good match when he compared the spectra of the planets with spectra obtained for these gases when light passed two times through a twenty-meter tube filled with these gases at one atmosphere.[44] The main problem in the past had been that the conditions under which gases were measured in the laboratory were not comparable to the conditions under which these gases were observed in planetary atmospheres; mostly the amounts of gases were much too small. Using long path lengths and higher pressures Dunham found evidence for carbon dioxide in Venus's atmosphere and methane and ammonia in the atmospheres of Jupiter and Saturn.

Thus Kuiper ends his 1947 paper with the words, "The interpretation of the present records and similar ones not shown will be treated later in a joint paper by the author and Dr. G. Herzberg, based on absorption measures with long paths."[45] In fact, the collaborative paper never happened, and in late 1947 Herzberg left Yerkes Observatory for a position in Ottawa, Canada. However, probably the biggest impact of Kuiper's 1947 paper was that it showed that his new spectrometer produced results that agreed with the earlier results obtained with photographic means, was more sensitive than photographic means, and extended the spectra further into the infrared.

The Yerkes-McDonald Conference on Planetary Atmospheres and the Redirection of Kuiper's Career

On April 28, 1947, the department consisted of Struve, Chandrasekhar, Kuiper, Morgan, Herzberg, Strömgren, Oort, Kaj Aage Strand, Hiltner, Greenstein, Popper, William Pendry Bidelman, Thornton Page, Paul D. Jose, Paul Ledoux, Edmondson, Lawrence Aller, and Swings when the president, Ernest Colwell, announced that he was appointing Struve as chairman of the Department of Astronomy and Astrophysics for three more years starting July 1.[46] At the same time, Struve announced a plan for "expansion and general reorganization" of the department that, he added, included Yerkes Observatory and personnel of McDonald Observatory of the University of Texas.[47] He would "guide the policies and lay the broader plans." Kuiper would be director of the Yerkes and McDonald Observatories, Chandrasekhar would run the theoretical section, and Morgan would become editor of *Astrophysical Journal*. Hiltner would be deputy director. Struve further made it known—to the university administration if not to the astronomers—that he would continue as "honorary director" of the observatories. One of astronomy's great administrators had just written a recipe for disaster.

The fiftieth anniversary of Yerkes Observatory also fell in 1947, and the astronomers in the observatory set about putting together some form of celebration. Donald E. Osterbrock lays out the details in his book, *Yerkes Observatory, 1892–1950*.[48] Struve published histories of the observatory in *Science* and *Popular Astronomy* that, according to some, were more about promoting the observatory than technical accuracy.[49] Struve and Kuiper both wanted to hold symposia as part of the celebration, but they could not agree on a topic. Struve wanted to focus on stellar topics, and Kuiper wanted to focus on planetary topics. Since the AAS was meeting at Northwestern University in September 1947, both had the idea of piggybacking their symposia onto that. Struve would have his symposium at the beginning of AAS meeting, and Kuiper would have his symposium at Yerkes a few days after the meeting.[50]

Kuiper arranged for the papers in his symposium to be published in a book, *The Atmospheres of the Earth and Planets*, which came out in 1949.[51] Not to be outdone, Struve arranged for the twenty or so papers from his symposium to be published in a book to be edited by J. A. Hynek, *Astrophysics: A Topical Symposium*, which appeared in 1951.[52] Kuiper even contributed an article on the origin of the solar system to Hynek's book.

Kuiper's workshop brought together critical researchers who discussed wartime advances and established a new fraternity of planetary specialists. Whereas most workshop volumes fade in importance after a few years, this book was reprinted in

1952.[53] Certainly there were many challenges that the new field faced and maybe the workshop would address these. Infrared spectrometers were still of poor sensitivity and resolution, the cooling that would improve sensitivity was not yet possible, and interpretation of the spectra was difficult. In the chapter "Survey of Planetary Atmospheres," Kuiper summarized current knowledge of the composition of planetary atmospheres.[54] He distinguished between the atmospheres of the Jovian planets, which were primary, and the atmospheres of the terrestrial planets, which were secondary. His table of atmospheric compositions of the planets has been the basis of considerable theoretical and observational work.

The Atmosphere of Mars and Astrobiology

The following year Kuiper extended the discussion in a paper to *Reports on Progress in Physics*,[55] ideas that were updated by a doubling of the length of his chapter in the 1952 version of his symposium book.

In these papers he mentions that infrared spectroscopy performed at McDonald Observatory in 1947 showed that the atmosphere of Mars was carbon dioxide. Furthermore, the polar caps consisted of water frost and water must be present in small amounts in the atmosphere, although "all spectroscopic tests applied so far have been negative." While the discovery that the atmosphere of Mars is carbon dioxide is usually considered one of Kuiper's most important contributions, in fact the discovery trickled out as a minor topic in several review papers. Nevertheless, the discovery was announced in the illuminated news strip atop the Boston Travelers Building.[56] Kuiper also discussed the pressure on the surface of Mars, reviewing estimates based on cloud heights, the difference in photographic and visual albedo, and the polarization of light reflecting off the surface. Estimates ranged from 0.08 to 0.01 atmospheres, but Kuiper favored a value of 0.01 atmospheres; the currently favored estimate is 0.06 atmospheres.

In the nineteenth century Percival Lowell proposed that Mars's dark regions were due to vegetation because of a perceived seasonal variation. In 1924 Slipher questioned Lowell's idea because his spectra did not match that of chlorophyll.[57] Now Kuiper took up the topic.[58] In the revised version of *The Atmospheres of the Earth and Planets* he wrote,

> Lichens and mosses show a reflection spectrum not very different from the Martian green areas, and they could withstand Martian nights. Certain differences exist however, which show this identification to be too simple: (a) lichens do not show a seasonal cycle, while several of the Martian areas do; (b) some Martian years show a substantial increase in green area, amounting to hundreds of square miles: such rapid changes are

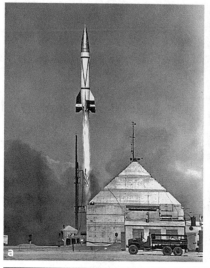

(a) An American launch of the German V2 with a payload that included a monkey called Albert, the first monkey in space. NASA photograph. (b) A photograph of Earth taken from a captured V2 launched by the Naval Research Laboratory. The view is of the Gulf of California and the Pacific Ocean. The photograph appeared as a frontispiece to Kuiper's book *The Atmospheres of the Earth and Planets*. Photographs of Earth from the repurposed V2 were the first to show the curvature of Earth. G. P. Kuiper, *The Atmospheres of the Earth and Planets* (Chicago: University of Chicago Press, 1949).

not observed among lichens; (c) the colour cover on Mars is in the nature of a blanket: the colours are vivid, indicating little space between the elements. Lichens are usually patchy.[59]

A Permanent Shift in Research Direction?

Kuiper's decade or so as a stellar astronomer had been remarkably successful. However, with the discovery of the atmosphere on Titan and the invention of the infrared spectrometer, he was ready to commit himself and his remarkable energy to planetary studies. It was one of the most remarkable redirections of a career in the history of science, and there has been considerable discussion by historians on the cause of Kuiper's shift. Doel suggests that it was because the military might support the new work, that it would be a way to keep Yerkes competitive in the face of better-placed observatories that were growing fast, that there was a global move away from long-term service projects to short-term, high-profile projects.[60] These forces were all at work, but my view is that it was primarily his discovery of an atmosphere on Titan that opened Kuiper's mind to the enormous possibilities of planetary science; it was a vein of gold he was eager to mine.

Following the Yerkes workshop, a stream of planetary papers came from Kuiper's pen covering unidentified absorption features in the spectra of the atmospheres of Uranus and Neptune,[61] the diameter of Neptune,[62] the diameter of Pluto,[63] the pole of rotation of Venus,[64] and the discovery of new moons around Uranus and Neptune.[65] In November 1948 Struve and Chandrasekhar successfully nominated Kuiper, then forty-three years old, to be a member of the AAAS.[66] Two years later he was elected to the NAS. His reputation was made, but considerably more was to follow. As the 1940s ended and the 50s began Kuiper undertook major new efforts on asteroids and the Moon and published landmark papers on the origin of the solar system.[67] While Kuiper's research on stars was at a fever pitch, and well respected by astronomers, he put the topic down and devoted all his energy to the solar system.

CHAPTER 7

Life in Texas, Life on Mars, and New Moons

NEARLY A DECADE AFTER Kuiper's death, Thornton Page, one of the Yerkes-McDonald astronomers from 1946–50, reminisced, "Gerard, like Chandra, was not a member of our Yerkes social group, but he had a great social presence at McDonald, where he made friends with the local ranchers and got several of us invited to cook-outs where we ate roasted lamb, hot green peppers, etc., and drank beer."[1] But the small community of blended astronomers and ranchers was not just set on a busy social life. The ranchers had to make a living and the astronomers were making one of the largest telescopes in the world reap the benefit of technology flowing out of the war. New people and new instruments poured in, improvements were made to the facility, and the observatory was set on a path to being one of the most important centers of astronomy in the world. In the 1940s Kuiper found a fifth moon around Uranus and a second around Neptune. His discoveries placed Kuiper with the likes of Herschel, the genius of observational astronomy. But all was not plain sailing. When Kuiper took over the directorship, Struve would not let go of the reins. His authoritative style did not go down well with the young men he had hired for the astronomy department at the University of Chicago ten to fifteen years earlier. Frictions mounted and eventually led to Struve's departure for Berkeley.

Life in Fort Davis

The Kuipers, Sarah especially, loved life at McDonald Observatory. Sarah said once that the wives of the astronomers and ranchers had much in common. Like the observatory, the ranches are usually far from a town, so children had to be home-educated or arrangements had to be made for a house in town where parents could live when school is open. Homeschooling denies the children the chance to socialize and adapt to a larger society, she thought.[2]

In 1947, the Kuipers' second child, Lucy, was born in Williams Bay. Paul's birth had triggered a serious postpartum depression, but not Lucy's. Sure, there were the usual childhood challenges—chickenpox in the family, then mumps, and other minor afflictions that usually caused delays to the family's travels between Fort Davis and Williams Bay—but nothing more. In fact, there were many births among the young Chicago faculty, and allowing for such interventions was readily accommodated. On the whole, the children's upbringing was very pleasant.[3] "My earliest memories are of my father walking around with me on his shoulders in this sweet little house on Dartmouth Road on the edge of the forest," Lucy (now Sylvia) recalls.[4] She adds, "My father would cut down trees to keep our view of the lake open. He liked to work in the garden, as did my mother." Later she notes, "People came to our house from a great many places; we had such interesting dinner table conversations! When my brother and I were very young, we would eat in the kitchen so as not to be in the way, but when we were older, we would join in. Sometimes, beforehand, Mother would give us reminders about being polite and respecting other people's customs. When we lived in Tucson (we moved there in 1960), I remember sitting around the fireplace with guests and hearing stories." Ewen Whitaker also records how Lucy, when in her teens, would frequently visit her father in the laboratory.[5]

To the children of McDonald and Yerkes Observatories, the 1950s were exciting times. The attitude shown to the children by their parents was very liberal. The children essentially had unlimited run of the grounds and buildings, going in and out of the observatories, running the catwalks, and taking books out of the libraries. Of course, the star catalogs and technical journals held little appeal, but buried among the technical books were all manner of popular books, especially science fiction novels that held a special appeal for Lucy.[6]

Marlyn Krebs was hired as resident astronomer on January 1, 1951, and held the position until September 1957.[7] He succeeded Paul D. Jose, who left in 1950. Krebs was the son of the superintendent at Yerkes and had grown up at the observatory. He had a background making scientific instruments. Morgan was his scout master. The Krebs had two children that went to the local school in Fort Davis, and for twelve years Marlyn held an elected position on the school board.

Carolyn M. Bertins, daughter of astronomer Aden Meinel, was a young child when her father held a position at McDonald Observatory. She describes the nature of childhood at the observatory and her association with the two Krebs boys:

The Krebs boys invited me to play at their ranch. They were showing off their Shetland ponies when 10-year-old Manuel galloped up on a red mustang. He skidded to a stop in a cloud of dust and reared him up like Roy Rogers on TV.

"Squeeze and you'll go faster." Pepper broke into a lope. I caught the rocking horse rhythm. That was the secret. The faster we went, the easier it was. I leaned the reins across his neck and we headed for the Herefords in the meadow southwest of the corrals.

The Krebs boys chorused "Don't run the stock. They'll lose weight." They caught up and we veered west, past a live oak grove and across a lava flow.

The summer I turned nine, the Krebs hosted a barbecue. While the grownups chatted in the shade of live oaks, the Krebs boys and I saddled up. We figured nobody would spot us running livestock. We worked the herd toward the barbecue area. We got to where only one stand of scrub oaks screened us from the party. What the heck. I urged Pepper to a lope, cut out the herd bull and sent him at a dead run toward the sound of chattering adults. As the bull crashed into the scrub I faded back and rejoined the boys.

We made a wide circuit and rode up to the party from the opposite side. My dad shouted, "You kids missed a lot of excitement."

"What happened?" Us little girls can look so innocent.

"A bull came charging out of the forest. Ran right through the crowd!"

The Krebs boys never told on me. It was the best livestock crossing ever, we thought.

About then I vowed that if I ever had a horse of my own, I would let other kids ride it. I went on to fulfill the vow with dozens of children. I supervise them closely. Especially nine-year-old girls.[8]

Another topic that captures some of the sense of life on the mountain and Struve's sensitivity is the issue of dogs. The matter first arose in a telegram from Kuiper to Struve complaining about Page bringing dogs into the observatory.[9] Struve's reply, February 2, 1950, stated he could not contact Page, who has already left, "nor do I know what he would be doing with his dog if he should not be able to take the animal with him. . . . It may be necessary to postpone a firmer action until Page has found someone who would board the dog for him while he is in Texas."[10] Kuiper pushed back, reminding Struve that Jose had to get rid of his dog, and they should be consistent.[11] On February 11, 1950, Struve devoted a one-page letter to the topic. "Elvey kept a number of dogs on the mountain and while they sometimes caused a certain amount of disturbance, I never felt free to suggest the practice be discontinued. I know that people become greatly attached to their dogs and cats and it would be a severe blow to them if they were forbidden to have them."[12] In principle, Struve agreed with Kuiper: there should be no dogs on the mountain, or cats, but enforcing a ban is problematical, especially in the residences. But there was a special reason for not making a point of this matter. He had just announced his departure, and people were "a little

(a) Aden Meinel (left) and Harold Thompson, a Kitt Peak engineer (right), on a field trip in March 1956, accessed September 23, 2018, https://www.noao.edu/news/images/ABM-HT-ostrich.jpg. (b) William Herschel. (c) William Lassell. (d) Kuiper in 1957. Courtesy of Matthew des Tombe. (e) Governor Beauford H. Jester, governor of the state of Texas between January 21, 1947, and July 11, 1949. (f) Tom Gehrels. Courtesy of LPL. (g) Carl Sagan in 1951. Beauford H. Jester; accession ID: CHA 1989.010; courtesy of State Preservation Board, Austin, Texas; original artist: Chandor, Douglas (1897–1953); photographer: Eric Beggs, December 1992, pre-conservation.

disturbed and their nerves may be more sensitive than at other times." Struve wanted to postpone any action until things had settled down. "I most sincerely hope that you will not only withdraw your own objections, if you have any, but prevail upon Jose and possibly Miss Hinds, the departmental secretary, to go easy for the time being."[13] There is no record of any further discussion of dogs on the mountain.

The history of McDonald Observatory and life there during this period has been nicely summarized by David S. Evans and J. Derral Mulholland in their 1986 book, *Big and Bright: A History of the McDonald Observatory*, and much of the following material is taken from there.

There was a flood of visitors to the observatory following the war. They quickly learned the tricks for driving 17 percent gradients, such using stones (chuck stones) to prevent rolling downhill. There were many social gatherings that the Kuipers, Swings, Van Biesbroeck, and the others enjoyed but the Struves tended to avoid because of their dislike of the informality. There was an active social life with locals, potluck suppers, parlor games, and charades. Charades was always popular with Kuiper. At Harvard, in one notable instance, Kuiper stood on a stool, cap on head, when his charade subject was "Dutch Boy Paints," presumably miming painting at the same time.

There was also an ugly side to life on the mountain. As at so many institutions at the time, there was much bigotry, especially concerning the Hispanic workers, who were all called "Mexicans" even though they were U.S. citizens before the telescope was built. Struve had to assure the dean at Austin that Jose was not "Mexican"—his name rhymes with "rose," not "San José." The Hispanic students had a separate schoolhouse. Bertins writes,

> Manuel, the Krebs boys and I used to ride to school in the back of a pickup truck. It was twenty miles to our classrooms in Ft. Davis. The driver always dropped off Manuel at a different school. Only brown-skinned kids ran around on that playground. Why didn't Manuel go to our school? The grownups said kids who speak Spanish have too hard a time in the white kids' school. I didn't understand. Manuel spoke English just fine. The Anglo school was almost empty. My classroom combined first and second grade, and there were only 15 of us.[14]

The Directorship and the Mars Incident

Inevitably there were problems between Kuiper, who considered himself in charge as director, and Struve, who considered himself in charge because he was honorary director. An example is an incident that was triggered by a request from William E. Keys, director of the University of Texas Information Service. The State Parks Board and the governor, Beauford H. Jester, were hosting a group of newspaper and magazine writers at Indian Lodge, a hotel in the state park at Fort Davis, on February 17, 1948.[15] Keys wrote to Kuiper, "Since . . . this is very near the date Mars will be at its closest approach to earth I wonder if it would be possible . . . to make arrangements for the group to visit the Observatory." Being sensitive to the work of the observatory, Keys wanted Kuiper's "opinion and frank appraisal of the extent to which the Observatory staff could cooperate." This would bring the attention of the governor, who was also on the board of regents for the University of Texas, to the Texas-Chicago relationship, and to astronomical research.

On the evening of February 17, T. S. Painter, president of the university; D. K. Woodward, chairman of the board of trustees; G. K. Shearer, of the Texas State Parks Board; Keys; and members of the radio and print press were guests at a 6 p.m. dinner at Indian Lodge. The dinner was followed by speeches by Keys, Woodward, Painter, and Kuiper describing the Texas-Chicago agreement and the international participation by leading astronomers, who all returned to their native countries full of praise for the "fine equipment and excellent climate." At 8 p.m. the group transferred to the observatory on Mount Locke, where they looked through the telescope, took refreshments from Sarah Kuiper and Paul Jose, and gathered in the library for a talk by Kuiper on the various programs of research, in particular those concerning Mars.[16]

The report of the event, presumably written by Kuiper, ends, "Dr. Kuiper concluded his talk by paying tribute to Dr. Struve who was responsible for the creation of the McDonald Observatory as a scientific institution. He pointed out that the current work on Mars is only one of about fifteen programs carried out by staff members, some of which spread over many years. He then invited everyone present to return to the dome to view Saturn and Mars through the telescope." The event went very well. The Texas officials were pleased, and President Painter offered to help get AC power to the observatory.

A few days after the event, Kuiper sent a letter to the Information Service asking them not to emphasize his work and to give due credit to the many other projects being undertaken at the observatory.[17] The letter was a little strange. It is difficult to see how they could do this, given the emphasis on Kuiper's own research as outlined in his report of the event. Besides the request itself, the tone of the request should be noted: "It has come to my attention that there appears to be some danger that the research work on planets which I am carrying on at present at the McDonald Observatory should be unduly stressed to the public as being more important than the many other programs carried out here since early 1939." In other words, Kuiper mostly told them about his work, but he didn't want them to give this impression in their articles.

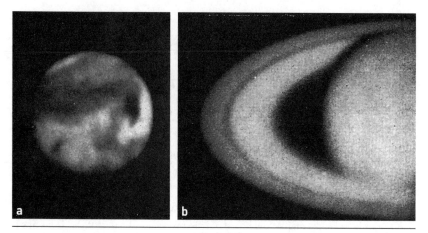

(a) Telescopic view of Saturn from Dollfus's article in *Planets and Satellites*. Dollfus, "Visual and Photographic Studies of Planets at the Pic du Midi" (plate 12, June 18). (b) Telescopic view of Jupiter from Dollfus's article in *Planets and Satellites*. Dollfus, "Visual and Photographic Studies of Planets at the Pic du Midi" (plate 42c).

The last paragraph of the report and the subsequent letter to the Information Service were clearly in response to a letter Kuiper had received from Struve on February 2.[18] Struve wrote, "I have received, indirectly, during the last several days disturbing information about what appears to be a bizarre publicity scheme at McDonald from persons who have been invited there and who are now inquiring about my plans for attending the big celebration. Also complaints have come to me concerning news releases from Texas to the effect that the most important, or even the only important, work at McDonald Observatory is your own present program."

Kuiper responded, "In view of the seriousness of the suspicion you voiced I should appreciate knowing more about this."[19] Struve replied that he expressed no suspicions, simply that he had heard from many about events at McDonald that were inconsistent with "the plan under which we are operating" and that Hutchins, then chancellor of the University of Chicago, should have been invited to underscore the Chicago-Texas relationship.[20]

Kuiper replied that the only "plan" he had was to observe, and he insisted that the event was organized by Texans for Texans, and "I don't see what is wrong with any of this." In truth, he felt like demanding an apology from Struve. He worked fourteen-hour days, more if there were administrative matters that couldn't wait. "I feel this attack is a stab from the back—just when I need my time most."[21]

One more complication was that Struve had arranged for reporters from *Life* magazine to visit the observatory and spend some time with Kuiper. They produced an article titled "Report on Mars." Kuiper had checked the article, but feared "the N.Y. office may mutilate it, of course, and make things even more embarrassing."[22]

Kuiper wrote Chandra a "personal letter" in "confidence" on February 20, 1948, admitting, "[This] has disturbed me greatly, more than almost anything I can remember."[23] In a four-page, handwritten letter, with some phrases underlined and others vigorously crossed out, and with the writing becoming increasingly hurried as Kuiper's emotions fed on themselves, he protested his innocence. There had been no publicity campaign—he had simply responded to a request from Mr. Keys. The only hint he had that anyone was conducting a publicity campaign was that Miss Lowrey of Chicago (presumably of the University of Chicago Information Office) requested to be kept informed of his research progress, but he informed her that she should not expect too much.

Chandra responded with a letter that Kuiper found "long and friendly," pouring oil on troubled waters, and followed up with a letter about teaching arrangements that ending in congratulations on the discovery of Neptune's moon Nereid.[24] Nevertheless, the exchange between Struve and Kuiper was symptomatic of a problem at Chicago that ultimately led to Struve's departure.

Kuiper and Life on Mars

During the February 17, 1948, event Kuiper discussed his work on the green areas of Mars.[25] He reported that they were not higher terrestrial plants but possibly mosses or lichens, and he mentioned the discovery the previous October that the atmosphere is carbon dioxide, saying that it still needed confirmation. He questioned the survival of life on Mars because no ozone layer was present to prevent deadly ultraviolet radiation reaching the surface. Since he expected volcanoes on Mars, he had searched for but failed to find sulfur dioxide. "There is no liquid water on Mars; it never rains and there are no lakes or oceans. Conditions are comparable to those on the Earth at an elevation of 50,000 feet (15 kilometer)," he wrote in his summary of the event.

The *Life* magazine article appeared in the June 28, 1948, issue under the title "Mars in Color," the article being mostly a full-page color image of Mars.[26] "The photograph, made by Dr. G. P. Kuiper with the 82-inch telescope camera at McDonald Observatory in Texas, is one of the few color pictures ever taken of the planet and the first ever to be published," the magazine explained in a single column that also showed Kuiper looking through the 82-inch telescope and a picture of the lead sulfide detector alongside a matchstick to emphasize its small size. The text went on to summarize Kuiper's "recently released findings": Mars has an atmosphere similar to that found 50,000 feet (15-kilometer) above the earth. Its two polar caps are actually thin sheets of snow or

Life magazine for June 28, 1948, and the article in which Kuiper and his infrared detector shared a page with an advertisement for Valvoline oil. Courtesy of *Life* magazine.

frost. Its green patches are vegetation, but of the lowest order, lichens which live by drawing moisture from the air. The orange area is windswept desert. The "canals" are natural valleys. The presence of vegetation could mean that life is just beginning on Mars, but Kuiper thinks otherwise. To him the lowly lichens represent life's last stand on the Red Planet. As we have seen, he later debunked the idea.

A recurring pattern in Kuiper's life is that he would attract the attention of the press, come off looking like the super astronomer he was, and some of his peers would take offense at what appeared to them to be grandstanding and self-promotion. In response to criticism he would, as he did over the February 17 event, protest his innocence and blame the press.

The Cook Telescope

While Kuiper objected to administrative matters taking time away from research, he was accused of micromanaging. On one occasion he had a row with Jose in a parking lot. But, as Evans and Mulholland write, "Whatever his merits and/or frustrations as Director, Gerard Kuiper was leaving everyone else in the shade as an observational astronomer."[27] Maybe this was part of Struve's problem. At the end of 1949, Kuiper resigned the directorship of the Yerkes and McDonald Observatories, leaving Struve to serve out the term in his capacity as honorary director until he left six months later. Kuiper was not alone in his difficulties with Struve; in fact there was a widespread sense that his power was waning. This was never more evident than during an incident with the Cook telescope.

The Cook telescope was loaned to McDonald Observatory by the University of Pennsylvania and set up using a grant from the Research Corporation.[28] The telescope was originally located at Pennsylvania's observatory on private land outside Philadelphia. Tommy Harnett, a long-term night assistant at McDonald Observatory, and Jose, then resident astronomer, took a one-and-a-half-ton pickup truck to Yerkes and then on to Philadelphia to collect the 10.25-inch refractor telescope in August 1949. The telescope came with two cameras with four-inch apertures, and at McDonald it was in heavy use, producing hundreds of plates a month until it was returned to Pennsylvania in April 1956. The telescope was housed in a roll-away building on the south side of the mountain. On September 20, 1949, Page wrote to Struve that he wanted to hire a bulldozer to build a road to the Cook telescope, and that he could do this cheaply because a nearby ranch had one on loan. The cost would exceed the amount available in the discretionary funds, and they needed Struve's permission to proceed.[29] Struve declined. Nine days later Chandrasekhar wrote to Struve, "I have seen the carbon of your letter to Page, advising against the construction of the road to the site below 'H,' which Kuiper and Hiltner have strongly recommended for the Cook instrument. But, as you will see from the minutes of the council meeting of September 27th, we had already decided to go ahead with the construction of the road."[30] The council consisted of the observational astronomers at Chicago who were charged with managing the observatories. Its creation was another indication of the rising mistrust of Struve. Chandra went on to explain that this was not only because they could minimize cost by acting now, but also because they had decided that the "temporary" site for the telescope was to remain because they did not want to move it. The unthinkable had happened. Struve had been defied.

Struve left the University of Chicago in 1950 to become chair of astronomy at UC Berkeley.[31] As suggested by the two incidents described above, things were not going well at Chicago. The bright young faculty, so skillfully recruited by Struve in late 1930s and early 1940s, were now mid-career and came out of the war with increased competence and confidence. Struve's Russian, autocratic style did not go down well, and his repeated threats to resign wore thin. This change in relationship was never clearer than in the exchange between him and the younger faculty over the Cook telescope.

In January 1951 Strömgren assumed directorship of McDonald and Yerkes Observatories and chair of the astronomy department at Chicago.[32] At this point there was no permanent residential astronomer, but whomever was there "acted for the director." Facilities on the mountain improved. In his research activities, Kuiper now had a full head of steam and his next discoveries were ready to be made, this time concerning the moons around Uranus and Neptune.

Uranus, Neptune, and Their Five Large Moons

Herschel discovered a small disk moving among the stars in Gemini on March 13, 1781.[33] Initially there was some uncertainty about whether it was a comet or a planet, but when Finnish Swedish astronomer Anders Johan Lexell, working in Russia, calculated a circular orbit, the new discovery was clearly a planet.[34] By 1783 Herschel concurred. Herschel's proposed name, "Georgium Sidus" (George's Star) was rejected by astronomers outside Britain, and Johann Bode suggested instead "Uranus," pointing out that just as Saturn was the father of Jupiter, Uranus was the father of Saturn. After the existence of the latest planet was confirmed, earlier reports and observations were found in the record by Hipparchos in 128 BCE; by the British Astronomer Royal, John Flamsteed, who observed it at least six times by 1690; and by the French astronomer Pierre Lemonnier, who observed it at least twelve times between 1750 and 1769.[35]

Herschel had been watching the "Georgian planet" for some time when, on January 11, 1787, he saw two points of light that had moved relative to the background stars. Herschel took great pains to confirm his observation until he was ready to report his results to the Royal Society. "The little planets seem to give a dignity to the primary planet," he observed.[36] Herschel may have seen another satellite and suspected that there were several.[37]

William Lassell was an astronomer cast from the same mold as Herschel.[38] He ground his own mirrors, polished them himself, and experimented with mounts, pioneering the use of the equatorial mount for the reflecting telescope so the telescope could track the stars in a single movement.[39] Most of all, he was an amateur, and he was a typical English northerner of the industrial revolution. Born in Bolton, he was educated in Rochdale and apprenticed to a Liverpool merchant until he could go into business himself. Then Lassell made his fortune as a brewer of beer. With his profits he built an observatory at his house in West Derby, a suburb of Liverpool, which he dubbed "Starfield." There he housed a 24-inch reflector telescope.

On November 13, 1851, sixty-four years after Herschel discovered the first two moons of Uranus, Lassell wrote to the editor of the *Monthly Notices of the Royal Astronomical Society* announcing that he had discovered two more, interior to the moons that Herschel had detected.[40] He referred to the four moons as I, II, III, and IV in order of increasing distance from the planet. Like Herschel, he had been searching deliberately for them, "a ten-day painful exercise of patience," he said, being thwarted by bad skies. But now Lassell had found them and been able to determine periods for his two new moons, 2.506 days and 4.150 days, which is consistent with their distance from the planet. "Since I first saw these satellites, I have never looked for them without seeing them," he wrote.[41]

While modern discoverers of asteroids and satellites rush to name them, the astronomers that discovered the first four moons of Uranus were content to number them. It was in a textbook, *Outlines of Astronomy*, published in 1852 by John Herschel, son of William, that the moons of Uranus received their names: Ariel, Umbriel, Titania, and Oberon.[42] Oberon and Titania are the king and queen of the fairies in Shakespeare's *A Midsummer Night's Dream*. Ariel and Umbriel are from Pope's *The Rape of Lock*. Ariel is "an airy tricksy spirit changing shape to serve Prospero, his master," and is also found in Shakespeare's *The Tempest*.

By 1821 Uranus had been tracked for forty years, and it gradually became clear that it was not following precisely the laws of planetary motion. It was as if the planet's movements were being affected by another major planet farther away from the Sun. In 1845 John Couch Adams in Britain and Urbain Le Verrier in France independently predicted the location of a proposed new planet.

On the evening of September 23, 1846, Neptune was discovered by Johann Galle at Berlin Observatory within one degree of Le Verrier's predicted position.[43] Within seventeen days, its first moon was discovered by Lassell with his 24-inch reflector at the Starfield Observatory.[44] He wrote, "One, or perhaps two, luminous points have been seen, which may be satellites; but this will require further scrutiny."[45] Over the next few years, more observations were made until it was possible to calculate a period and an orbit. To everyone's surprise, the moon's orbit was retrograde, opposite to the direction of rotation of the planets and most objects in the solar system. The name "Triton" was first used by Camille Flammarion in his *Astronomie Populaire* in 1880.[46]

As far as the satellites of Uranus and Neptune were concerned, with Lassell's discoveries the matter rested for one hundred years.[47]

Miranda

Kuiper had planned to determine the relative magnitudes of the four known satellites of Uranus when he went to the 82-inch telescope at McDonald Observatory on February 1, 1948. He made a four-minute exposure of the Uranus system using the Cassegrain focus, the location of the eyepiece along the axis of a reflecting telescope.

"The close companion to the planet was noticed at once," wrote Kuiper in his brief discovery paper.[48] It was not until March 1, 1948, that he could return to the subject, make two control plates, and confirm that the new object was not a background star. On March 24 and 25 he exposed eight more plates and was able to calculate the period of $33^h\ 56^m$. The orbit was circular and in the plane of the other satellites.

Until now Kuiper typically worked alone; according to Bok, he was too busy to work with students. However, at this time, Kuiper's work had attracted his first graduate student in planetary astronomy, Daniel E. Harris III. Harris was to complete his PhD work with a dissertation on the satellite system of Uranus in 1949.[49] So in 1948 it seemed appropriate that he would help with Kuiper's observations of the new moon. The two men exposed more plates in October and November of that year.

Kuiper spent half his discovery paper for the first new satellite of Uranus since the days of Herschel and Lassell discussing a name. He reviewed the names of the previous four moons chosen by Herschel and chose Miranda from *The Tempest*, who, Prospero says, is "a little cherub that did preserve me."

On March 8, Chandrasekhar, as chairman of the astronomy department, wrote to Kuiper about changes in teaching arrangements following Strand, Herzberg, and Greenstein leaving the department and Blaauw's pending arrival.[50] He wondered if Kuiper would consider organizing a new course on "problems of the solar system." Chandra added, "Your sensational contributions to this field have made the subject very topical." He ended the letter, "Again with congratulations on your brilliant discovery of a new satellite for Uranus." Kuiper replied,

> Thank you very much for your congratulations on the new satellite of Uranus. In all frankness I must decline accepting your qualification "Brilliant"! In my work on the planets I was carrying out a systematic plan on which I had been working for some time; the results on Mars and the Rings of Saturn are therefore not accidental. But this new satellite most certainly is. The only reason I took plates of the Uranus satellites was to check on their relative magnitudes and to test the relative stability of the Cassegrain focus of the 82-inch for astrometric work. When I found a suspicious object on February 15 I postponed till the last day, March 1, the taking of a check plate because I realized that anything new would mean more work for which I did not have time. Later I borrowed a little of Page's time (two six-minute exposures at the prime focus) to check on distant satellites but none were found brighter than the limit of the plates which must have been about 19th magnitude.[51]

Again Kuiper felt that accidental discoveries do not warrant such applause. Chandra might have replied, "Luck comes to the prepared," but chose silence.

Nereid

The discovery of Nereid, unlike Miranda, was planned. On May 1, 1949, Kuiper reported that the field of Neptune was photographed at the prime focus of the 82-inch telescope in a search for distant satellites.[52] An object was found on two plates of magnitude about 19.5, about 168^m west and 112^m north of Neptune and essentially sharing its motion. Kuiper chose the name "Nereid" because "the Nereids were sea nymphs who, together with the Tritons, were the attendants of Neptune." This was near the end of Kuiper's observing season in Texas, and he did not feel he could extend his stay, so Jose was asked to pursue the work.

Jose was the resident astronomer at McDonald.[53] He was hired in 1945 after the abortive attempt to hire Adel and after Struve assured the Texas regents that Jose was not a "Mexican." A trained astronomer, he had been a high school teacher and was hired as the deputy director of the Steward Observatory in Tucson before coming to McDonald. His time at McDonald eventually came to an end after disputes with Hiltner and he was asked to leave in the summer of 1950.

Jose made two pairs of additional plates during the two dark-moon periods on May 29 and June 18, 1949. The plates were measured and reduced by Van Biesbroeck, who also obtained calibration plates with the Yerkes 24-inch reflector. The positions of the satellite at the three epochs were used by Harris to compute a provisional orbit. The paper describing the orbit of Nereid was authored by Harris and Van Biesbroeck.[54]

The discovery of the satellites entered pop culture. On June 21, 1949, in reporting the discovery of Nereid, the *Chicago Daily Tribune* ran a poem by Carl S. Junge:[55]

Kuiper discovered Moon #30,
three billion miles away.
It is to mankind of no great boon,
this finding of another moon.
A satellite that only he,
the great astronomer, can see.
Heavenly body! Fine, we say,
but not three billion miles away.

The Moons of Uranus and Neptune in the Space Age

The next moon of Uranus was discovered on the flyby of the Voyager 2 spacecraft in January 1986. In fact, ten further inner moons were found. Another satellite, Perdita, was retroactively discovered in 1999 after studying old Voyager photographs. The moons of the planets are of two kinds. Regular moons are those that orbit in the same direction as the planet, are in the plane of the ecliptic, are generally in circular orbits, and are close to the planet. They appear to have formed at the same time as the parent planet by similar mechanisms. Irregular moons are none of these things; they orbit in the opposite direction to the planet's spin, have highly eccentric orbits, and they lie far from the planet. Irregular moons are thought to have been captured by the planet. Uranus was the last giant planet without any known irregular satellites until 1997, when nine distant irregular moons were identified using ground-based telescopes. Two more small inner moons, Cupid and Mab, were discovered using the Hubble Space Telescope in 2003. The moon Margaret was the last Uranian moon discovered. Its details were published in October 2003.[56]

Neptune now has fourteen known moons, by far the largest of which is Triton. The third moon, later named Larissa, was first observed on May 24, 1981. No further moons were found until Voyager 2 flew by Neptune in 1989. Voyager 2 recovered Larissa and discovered five inner moons. In 2002 and 2003 two surveys using large ground-based telescopes found five additional outer moons, bringing the total to thirteen. The fourteenth moon was discovered in July 15, 2013, in images from the Hubble Space Telescope.[57]

The Size of the Planets

Kuiper used the 82-inch telescope at McDonald Observatory four times in 1948 and 1949 to measure the diameter of Neptune.[58] He used a disk meter, a disk he matched to the planet's apparent diameter to obtain an angular diameter. This is easily converted to actual diameter because distance to the planet is known. The device was invented by Henri Camichel of the Pic du Midi Observatory. In July 1949 Kuiper published his result in *Astrophysical Journal*. He found that Neptune had a diameter of 2.044 ± 0.006 seconds of arc compared with the accepted value of 2.440.

After repeated attempts at McDonald, Kuiper finally decided to try to obtain a new value for the diameter of Pluto using the 200-inch telescope at Mount Palomar Observatory in Southern California.[59] Struggling with poor viewing conditions but getting help from Milton Humason, he eventually obtained a value of 0.23 ± 0.01 seconds of arc.

In his 1950 review paper in *Reports on Progress in Physics* Kuiper listed mass, radius, and density for the planets, including Pluto. The data were from the literature, but he substituted his own values for Neptune and Pluto, which were smaller by 10 percent than earlier values. The sizes of the planets, and the discrepancy between Kuiper's values and those of others, were to be an important part of the contentious relationship between Kuiper and Harold Urey in the late 1950s and 1960s.

Pluto and the Law of Distances

As a schoolboy I learned, as I suppose most did, about the Titius-Bode law, commonly known as just Bode's law, which addresses the fact that the planets show a remarkable geometrical relationship in their distances from the Sun. Modern astronomers regard the law as a coincidence with no true scientific justification. Kuiper is noteworthy in being a distinguished twentieth-century astronomer who did not dismiss the law. As usual, his arguments were careful and precise.

The authors of astronomy textbooks have been describing laws relating to planetary distances since 1715, but the credit for the laws seems to have settled on Johann Daniel Titius, a German astronomer and professor in Wittenberg, and Johann Elert Bode, a German astronomer at Berlin Observatory. In 1766 Titius inserted language about the law into Charles Bonnet's *Contemplation of Nature*; six years later Bode included a description of Titius's proposal in his *Anleitung zur Kenntniss des gestirnten Himmels*. The law has been extremely important in the history of astronomy. First it predicted a planet between the orbits of Mars and Jupiter, and not long after this prediction Ceres was discovered. Second, it predicted the location of the next planet after Saturn; when Uranus was discovered, it was near the distance predicted by the law. However, when Neptune and Pluto were found to be far from the positions predicted by Bode's law, the law fell into disfavor. Many systems of planets around other stars have been examined to see if they obey the law, and none do.[60]

It is sometimes surprising where the influences on a career come from. A seminar, a chance meeting, an accidental observation, and any number of spurious events have been known to structure a career. This was true of Kuiper, and in his case it was an invitation to write a book review. While a student in Leiden, Kuiper was asked to review Friederick Nölke's book on the origin of the solar system.[61] The book not only stimulated in Kuiper an interest in the origin of the solar system, but it also left him with a fascination with Bode's law. In the same issue of *Hemel en Dampkring* in which the book review appeared Kuiper published a short article on the law.[62] He began the article by saying that Nölke dismisses the law, but Kuiper thinks this is unreasonable. He described the law and discussed its success, pointing out that it works for all but Neptune and Pluto. There were problems, he said, but it was premature to dismiss the law.

Nineteen years later, Kuiper returned to the topic.[63] This time he found that a better law is found if one includes the mass of the planets. He found that a law can be written $\mu / \Delta^3 = \sim 10^{-4}$ where μ is the mass of a planet relative to the mass of the Sun and Δ is the distance of the planet from the Sun. A more precise form of the equation is $\log \mu = 2.5 \log D^3 - A$, where A is 1.8 or 3.7 depending on whether the planets are of comparable mass. These relationships were derived from simple principles assuming Kuiper's ideas on the formation of the solar system. How can this relationship be explained? The planets formed by break-up of gravitationally bound gaseous rings at a critical density (i.e., $\mu / \Delta^3 = 10^{-1}$ or 10^{-2}), assuming roughly 1 percent of the gases condense into planets and satellites.

The Solar System Book Project

After his success with *The Atmospheres of the Earth and Planets*, published in 1949 with a second edition in 1952, Kuiper developed a taste for assembling books consisting of review papers written by recognized experts. He immediately set out to organize a similar series of books on the solar system. In the 1960s he also produced a series of books on stars and stellar systems. These "editing projects," as Kuiper was fond of calling them, were part of his life from the early 1950s until his death and always featured in his annual reports and his articles about the activities of the Lunar and Planetary Laboratory, which he joined in 1960.[64] There was usually a dedicated editorial office close to his office, and from 1959 onward it was staffed by Barbara Middlehurst.[65] The task was monumental and highly successful. The four books in the *Solar System* series were funded by the Air Force and published by the University of Chicago Press. The first two volumes were edited entirely by Kuiper and the second two jointly by Kuiper and Middlehurst. It is worth considering the chapter topics and the authors because they are a snapshot of where planetary science stood at the time.

Volume 1, *The Sun*, appeared in 1953 and was 745 pages. Kuiper dedicated this volume to his colleague, Bernard Lyot. Shortly before the first volume appeared Lyot died of a heart attack in Cairo while returning from an eclipse expedition. Lyot was a prominent French astronomer best known for inventing the chronograph, a device for observing the Sun's outer atmosphere without waiting for a solar eclipse. He spent his career at Meudon Observatory in Paris. Kuiper relayed Lyot's papers to *Astrophysical Journal* for publication,[66] and Lyot provided Kuiper with equipment to measure the polarization of light by asteroids in return for a double-edged micrometer.[67] Upon hearing of Lyot's death Kuiper immediately set about writing an obituary for *Astrophysical Journal*. Chandrasekhar, then editor of the journal, agreed to publish an obituary but stressed that it should concern Lyot's work and not be "a conventional obituary notice."[68]

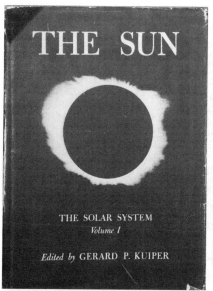

THE SUN

THE SOLAR SYSTEM
Volume I

Edited by GERARD P. KUIPER

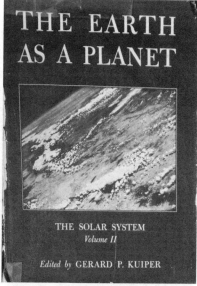

THE EARTH
AS A PLANET

THE SOLAR SYSTEM
Volume II

Edited by GERARD P. KUIPER

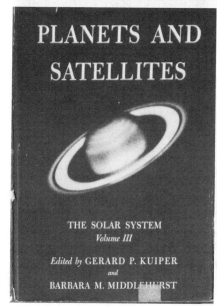

PLANETS AND
SATELLITES

THE SOLAR SYSTEM
Volume III

Edited by GERARD P. KUIPER
and
BARBARA M. MIDDLEHURST

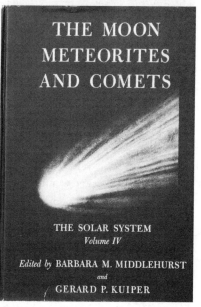

THE MOON
METEORITES
AND COMETS

THE SOLAR SYSTEM
Volume IV

Edited by BARBARA M. MIDDLEHURST
and
GERARD P. KUIPER

The four-volume work *The Solar System*. Courtesy of Dale Cruikshank.

The Sun begins with an introductory chapter by Leo Goldberg and then includes chapters on the Sun as a star (Strömgren), the photosphere (Minnaert), solar lines (Charlotte Moore), chromosphere (van de Hulst), solar activity (Kiepenheuer), radio emissions (Pawsey and Smerd), electrodynamics (Cowling), problems and equipment (thirteen sections by different authors), and finally an appendix on sunspots and instruments.

Volume 2 in the series, *The Earth as a Planet*, appeared in 1954. The preface tells much about Kuiper's view on the evolving field of what we would now call planetary science. He writes that geophysics, geochemistry, and planetary physics come together but are presented in an astronomical setting, which should inform the studies of the planets. There is an emphasis on atmospheres, reflecting his budding success with near-infrared spectroscopy. The two opening chapters concern geophysics, dimensions, and rotation (Spencer Jones) and dynamics of the Earth-Moon system (Jeffries). Then there are chapters on the interior (Bullard); the crust (Tuzo Wilson); oceanography (Sverdrup); geochemistry (Mason); the lower atmosphere (Byers); biochemistry of the atmosphere (Hutchinson); atmospheric absorptions (Goldberg); upper atmosphere (Whipple); emission spectra of twilight, night sky, and aurorae (Chamberlain and Meinel); physics of the upper atmosphere (Bates); dynamics of upper atmosphere (Nicolet); Earth seen from space (Holliday); and albedo, color, and polarization of Earth (Danjon).

There was one critic of *The Earth as a Planet*—Harold Urey. In a letter to David Griggs, a geophysicist at UCLA, he wrote, "It contains some excellent articles as I am sure you will agree. There is one by Tuzo Wilson on the origin of mountains which treats convection in an inadequate manner. I urged Wilson to do more with this but he is so convinced of the correctness of contraction that it would be impossible for him to do an adequate job. I mentioned this to Kuiper but he does not believe that convection is correct and would not secure any one to give an adequate review in the book."[69] In 1955 there was no consensus as to whether the contraction of Earth's crust or convection cells in the mantle were creating the great mountain chains, and the issue was not to be resolved until the discovery of the magnetic bands in oceanic crust in 1963.[70] In fact, Wilson is now known as a strong advocate of the convection-driven theory of plate tectonics.

The next volumes didn't appear until seven years later, after Kuiper's move to Tucson. Volume 3, *Planets and Satellites*, and volume 4, *The Moon, Meteorites, and Comets*, appeared in 1961 and 1963, respectively. In the preface to the third volume Kuiper elaborates on his views on how planetary science was emerging. He saw three phases—we might call them three eras. The first phase was the age of Galileo, Copernicus, Kepler, Newton, etc., who looked at planets and thought of stars as just a backdrop. The second phase was dominated by the work of the likes of Lowell and his

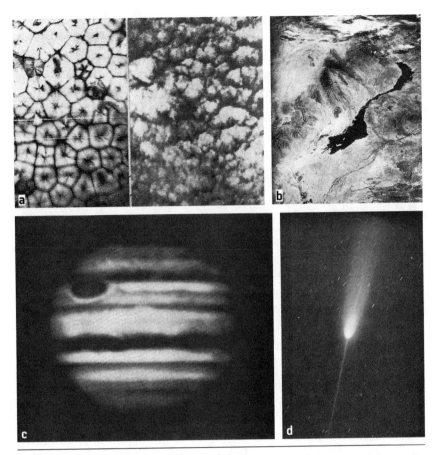

(a) In *The Sun*, Minnaert likened the solar surface to convection cells in the laboratory and alto-cumulus clouds in the atmosphere. M. Minnaert, "The Photosphere," in *The Sun*, ed. G. P. Kuiper, 88–185 (Chicago: University of Chicago Press, 1953). (b) Earth from space, taken from Holliday's chapter in *The Earth as a Planet*. The city in the bottom left is Tularosa, twelve miles north of Alamogordo. The dark structure is a lava flow locally called the Carrizozo Malpais. The photograph was taken at a height of 112 kilometers from a V2 launched from White Sands Proving Ground. C. T. Holliday, "The Earth as Seen from Outside the Atmosphere," in *The Earth as a Planet*, ed. G. P. Kuiper, 713–25 (Chicago: University of Chicago Press, 1953). (c) This photograph of Jupiter (north at the bottom) was referred to as "extraordinary" by Kuiper. It is from Humason's chapter in *Planets and Satellites* and was taken with the 200-inch telescope at the Palomar Observatory. M. L. Humason, "Photographs of Planets with the 200-inch Telescope," in *Planets and Satellites*, ed. G. P. Kuiper and B. M. Middlehurst, 572 (Chicago: University of Chicago Press, 1961). (d) The comet Arend-Roland as it appeared on April 25, 1957, as reproduced in Wurm's chapter in *The Moon, Meteorites, and Comets*. K. Wurm, "The Physics of Comets." In *The Moon, Meteorites, and Comets*, ed. G. P. Kuiper and B. M. Middlehurst, 573–617 (Chicago: University of Chicago Press, 1961).

writings on Mars,[71] while professionals studied stars, nebulae, and galaxies. "Meanwhile those astronomers who occasionally took time off for planetary studies (at the risk of mild scorn from their colleagues) found a rich field for their investigation," he wrote. Kuiper considered that we were now in the third phase of planetary studies, which is best characterized by the use of rockets.

The first chapter in volume 3 (by Fritz and Wexler of the U.S. Weather Bureau) is essentially a series of images of Earth from the first weather satellite, TIROS 1, primitive by today's standards but exciting at the time. Then follows chapters on the search for trans-Neptunian planets (Tombaugh), orbits and masses (Brewer and Clemence), dynamics (Hagihara), interiors (Wildt), lunar photometry (Minnaert), eclipses (Barbier), photometry and colorimetry (Harris), polarization (Dollfus), temperature (Petit), radiometry (Sinton), radio emission (Mayer), radio emission from Jupiter (two chapters by Burke and Gallet), visual images from Pic du Midi (Dollfus) and Palomar (Humason), and color photography of Mars (Finsen). Kuiper wrote the final chapter, which dealt mainly with his survey of satellites from McDonald Observatory. By the time the third volume appeared, Kuiper added a fifth volume to the plans, *Planets and the Interplanetary Medium*, but it never materialized.

Volume 4, *The Moon, Meteorites and Comets*, starts with five chapters about the Moon, including one each from Dai Arthur and Whitaker, and includes descriptions of the recent Soviet images of the far side. Then the meteorite hunter, Harvey Nininger, describes meteorite distributions on Earth, and the leading Soviet meteorite expert, E. L. Krinov, produced two chapters on impact craters and the Tunguska event (the largest impact event in recorded history). Then follows chapters on ancient impact craters (Dietz), impact mechanics (Shoemaker), the physics and chemistry of meteorites (Wood), meteorite ages (Anders), and carbonaceous chondrite meteorites (DuFresne and Anders). The five chapters on comets include chapters by Roemer, Whipple, and Oort and two chapters on meteors, the first by Millman and McKinley and the second by Jacchia.

The four books in *The Solar System* were a success with those who wrote reviews for the journals.[72] Concerning *The Earth as a Planet*, Bok wrote in *Science Magazine*, "All told, this is a first-rate book that deserves to be read by astronomers of all varieties." Of *Planets and Satellites*, John Irwin wrote, also in *Science Magazine*, "I can heartily recommend this volume both as a reference book for the specialist and as a source of information for scientists working in allied fields of study." Also writing about *Planets and Satellites*, John Heard wrote in the *Journal of the Royal Astronomical Society of Canada*, "It does contain a wealth of information, well documented by references," although he did regret the absence of an article by Kuiper on the origin of the planets. When *The Moon, Meteorites and Comets* appeared in 1964 Irwin wrote, again in *Science Magazine*, "The book can be recommended without reservation to all who are interested, professional astronomers as well as amateurs. Its appearance is especially timely now that exploration of the solar system by means of rockets has begun in earnest."

There can be no doubt about the success of these books. When I started a career in meteorite research it was with *The Moon Meteorites and Comets* on my desk, and no book was ever more carefully studied. Certain chapters are still in heavy use as the field returns to subjects neglected since 1963.[73] In his obituary for Kuiper, Morgan wrote, "The four-volume work *The Solar System*, and the nine-volume compendium, *Stars and Stellar Systems*, may turn out to be his greatest and most lasting monument."[74] But the biggest indication of the value of these books is that they triggered one of the best-used and best-known means for supporting research in the planetary sciences. The University of Arizona Press still publishes such books—multiauthored collections of review papers on specific subjects—to this day. There are now more than thirty titles, covering every planet, the planetary satellites, protostars, the Sun, comets, and asteroids. For the asteroids, there are separate books dealing with them as objects of scientific interest, as in situ resources, and as impact hazards. Like so much that Kuiper did, in publishing these review books he left a big legacy. He showed the way.

Kuiper and Students

If books of review papers have become one of the staples in planetary science research, so have students. Students are the main conduit through which most university professors perform their research. We are now encouraged to think of graduate students as apprentices, learning the art of research by doing it alongside their professors. But this was not Kuiper's style, at least not during the McDonald years before and after the war. He preferred to work alone or bring in just a few technical experts when needed. Thus most of his papers were single-authored. Some commentators have said that he did not have time for students, who required training. Despite this, Kuiper did work with several students, both his own and those of other faculty. Sometimes he could be very passionate about his students.

Albert Shatzel is a case in point. On the November 10, 1949, Kuiper wrote a two-page, handwritten letter marked "personal" to Chandrasekhar.[75] Shatzel had been "under fire" and unjustly reprimanded by Struve. Struve had been provided erroneous information from Page to the effect that Shatzel had turned up at McDonald Observatory without warning. In fact, Kuiper argued, Shatzel had informed Page about the Texas trip and had called from Chicago just prior to departure. Apparently Dorothy Hinds, in a bad mood, had written a critical letter to Page that she later retracted. There was a second matter: Shatzel had been missing Chandra's lectures because of a teaching assignment given to him by Page that involved 300–350 students. It was impossible to move such a large class. "I dislike both cases," wrote Kuiper. "A man should not be accused of something he is not guilty of and be put in the dog house for reasons that are not even told to him. While such 'justice' is all too common today, we should not be guilty of it." Kuiper went on to say that he had worked with Shatzel and found him an exceedingly competent observer, able to do anything, and that he was very cultured. "We make a mistake to discriminate against him because his background is a little different. He does possibly injure our vanity, at times, by having his own views, not always correct or well founded. But we ought to be generous enough not to let that affect our verdict. He needs a lot of training in certain scholastic fields; but he is independent and might do very good work in his later years, if we give him a chance."[76] It is hard to imagine a more powerful defense offered by a professor of a student.

But it was to no avail. The next day Struve wrote to Kuiper to reprimand him for interfering. "I am afraid it was quite premature to discuss with Shatzel the question of next year's Office of Naval Research contract or any other form of appointment for him. I trust that no further commitments will be made to him."[77] On December 6, 1949, Shatzel wrote to Struve, "In view of impossibility of further employment, and the opportunity for immediate employment elsewhere at a substantial salary increase, it would seem propitious that I arrange to resign at the end of this month, or as soon thereafter as possible."[78] Shatzel was working with Kuiper on asteroid photometry, but when his results appeared in *Astrophysical Journal* a few years later, he gave his position as assistant director of Chicago's Adler Planetarium.[79]

Harris was busy publishing a log of cometary observations, first in *Astronomical Journal* and then in *Popular Astronomy*, which he took over from Van Biesbroeck.[80] He was working on theoretical problems with Chandrasekhar at the time.[81] However, after Kuiper discovered Nereid he persuaded Harris to work with Van Biesbroeck to determine its orbit, which they published in *Astronomical Journal* in 1949.[82] This became the basis of a PhD thesis, and Harris became Kuiper's first successful graduate in planetary science in that same year.[83] After graduation Harris found a position at the Warner and Swasey Observatory at Case Western Reserve University in Cleveland. Kuiper thought very highly of Harris and persuaded him to come back to Yerkes and McDonald Observatories. After a few years of being supported by various grants, Harris was given a faculty appointment. However, after a negative tenure decision in 1956, supposedly based on a weak publication history and poor relations with Chandrasekhar, he moved to Northwestern University. He died young in 1964.

Tom Gehrels was a student from Leiden who had worked for the Dutch resistance as teenager. He joined Kuiper in 1952 after getting a BSc in physics. Gehrels has written an autobiography, *On the Glassy Sea*, in which he describes how, when he was a graduate student, Kuiper presented him with a large number of plates and told him to determine the magnitude of the asteroids imaged.[84] From an asteroid's magnitude, of course, an approximate estimate of size can be made if the distance of the asteroid from Earth is known.[85] No further help coming, Gehrels obtained a photometer from Bob Weitbrecht, a noted deaf scientist at Chicago who taught electronics, and learned procedures from Nancy Roman, a research associate with Morgan. In the process of understanding asteroid magnitudes Gehrels discovered the now-famous opposition effect: objects illuminated directly are much brighter than objects illuminated obliquely. In 1956 Gehrels became Kuiper's second PhD graduate, with a thesis on the determination of asteroid magnitudes and their phase relationships. He then went to Indiana University on a five-year research assistantship to continue these studies with Frank Edmondson and his research assistant, Delores Owings. In 1960 he joined Kuiper at the University of Arizona, performing a research program in polarimetry of Venus and bright stars. Polarimetry is the study of the plane in which light waves vibrate using lightweight telescopes that were carried on balloons. Following Kuiper's example, Gehrels also earned a reputation as the editor of a series of books of collected review papers. He stayed at Arizona until his death in 2011.

Kuiper's third PhD student to graduate, and last while at the University of Chicago, was Carl Sagan. Little needs to be said about Sagan, whose rise to fame as a popularizer of astronomy and life in the universe is well known. Several book-length biographies have been written about him.[86] Sagan obtained a BS and MS in physics at the University of Chicago, and then a PhD working with Kuiper with a thesis titled simply *Physical Studies of Planets*, which he defended in 1960. In 1958, the two worked on the Air Force's classified Project A119, the secret plan to detonate a nuclear warhead on the Moon.[87] After a postdoctoral position at UC Berkeley, Sagan obtained a faculty position at Harvard. Failing to get tenure, he moved to Cornell University, where he spent the remainder of his career.

For a scientist as successful as Kuiper, and having spent so long on the faculty at the University of Chicago with such abundant resources, the number of research students he graduated is remarkably small. It is probably Gehrels who has thought most about this. In a letter to Dale Cruikshank dated September 29, 1980, he wrote, "I learned a lot from him how not to do things and he was a fascinating person to observe. However, were Carl Sagan and I really his students? My dissertation topic was chosen against his wish. I hardly ever saw him and was in his office only a few times in all three years. Nancy Roman, Adriaan Blaauw, Kees van Houten and Bob Weitbrecht were my real advisors. I carefully avoided him also during LPL years for that was the only way to survive independently."[88] That said, two of the three students Kuiper graduated at Chicago became important people in the field, and the third was well on his way when his death brought a premature end to his career.

CHAPTER 8

Origin of the Solar System

W E NOW KNOW THAT THERE ARE millions, probably billions, of planets orbiting the stars in our galaxy. This has been suspected for some time, but in the last decade or so planets around other stars have been found in spades.[1] Kuiper suspected this; so did Struve. This gives new emphasis to the question of how systems of stars and planets formed. It is an old question. It is fundamental to our sense of who we are and how we relate to the universe. Thus every culture has its origin myths.[2] So when, on October 5, 1949, Kuiper wrote to Struve that he was making progress on his theory for the origin of the solar system, he was addressing an ancient topic.[3] In tackling this topic, Kuiper laid the foundations of modern theories for the formation of the planets and proposed a new category of solar system objects: the Kuiper Belt Objects.

Descartes's and Kant's Views on the Origin of the Solar System

The origin of the Sun and planets has always played a major role in our cultures, from the numerous creation myths of various civilizations; the views of the ancients, who did not know the Sun was at the center of our solar system; to current views derived from modern astronomy. The great French philosopher René Descartes wrote *Treatise on the World and Light* between 1629 and 1633.[4] However, witnessing the fate of Galileo Galilei, who died under house arrest for his views,[5] publication was deferred until 1677, twenty-seven years after Descartes's death. The book describes his philosophy and vision for the world and includes three chapters on the formation of the Sun, stars, planets, comets, and Earth and the Moon. Descartes rejected the ancient approach to understanding the world, which relied on sensations, and replaced it with a more modern mechanical approach. The influential German philosopher Immanuel Kant laid out his views on the origin of the solar system in *Universal Natural History and Theory of Heaven*, published in 1755.[6] According to Kant, our solar system is merely a smaller version of the fixed star systems. Both Descartes and Kant imagined that the Sun was engulfed in a swirling mass of gas and dust that ultimately condensed into the planets.

Later Views

When you list the properties of the planets, their distances from the Sun, their size, their orbital periods, and their spins, you are left with a giant puzzle, especially if you have in mind the nebula-type theories of Descartes and Kant. This puzzle is called the angular momentum problem. Why is 99 percent of the mass of the solar system in the Sun, whereas 99 percent of the angular momentum is in the planets? It is as if while the nebula collapsed into the Sun and planets, a mighty arm reached out from the Sun to whip up the planets into rapid movement while it slowed down the Sun. No one could think of what that whip could be, so the nineteenth-century astronomers abandoned nebular ideas and decided to focus on mechanisms that would transfer angular momentum.[7]

The most popular theory was probably that of Jeans, who, in 1917, proposed that a star had passed close to the Sun and pulled out a filament of material that fragmented and condensed into the planets.[8] It was an elaboration of a previous theory proposed by Thomas Chrowder Chamberlin and Forest Ray Moulton of the University of Chicago in 1905.[9] The relative motion of the two stars was the means whereby momentum was transferred. The theory had opponents. Some suggested that the event was statistically unlikely, and others argued that the filament would be reabsorbed by the Sun. Other astronomers suggested that the Sun and planets came from different clouds— one cloud rotated slowly and produced the Sun, and this encountered another rapidly rotating cloud that produced the planets.[10]

Edgeworth, Weizsäcker, and the Origin of the Solar System

The return to the popularity of nebular theories was probably started by Lieutenant-Colonel Kenneth Essex Edgeworth, of British Army's Royal Engineers, who described his vision for the evolution of the solar system in the *Journal of the British Astronomical Association* in 1943.[11] Besides being an army officer, Edgeworth was an engineer, an economist, and, let's say, an independent theoretical astronomer.[12] He envisioned a rotating gaseous cloud from which local condensations would form that would grow into the planets. He then went through each object in the solar system and explained how it fit into the scenario until he reached the outer solar system. Then he remarked that the cloud must have extended beyond Pluto, where the density of the nebula (he used the term "opacity") was too low for planet formation but where numerous small, loosely conglomerated objects would form that would ultimately become the comets.

Also in 1943 Carl von Weizsäcker, then at the University of Strasburg, published a major paper in the journal *Zeitschrift für Astrophysik* describing the origin of the solar system in terms of a number of vortices in the flattened solar nebula.[13] There were, argued Weizsäcker, convection cells of increasing size as one proceeds outward from the Sun, all circling like clockwork. In the spaces between these convection cells were tiny sites at which material could get trapped and form into planets and their satellites. Kuiper was aware of the paper and, while at Harvard during the war, wrote a letter to Chandrasekhar requesting that he have a print copy made from the astronomy department's microfiche and sent to Russell who was "much interested,"[14] While it quickly became a well-known theory, it was immediately criticized. It required a high degree of order, but the solar nebula was expected to be highly turbulent. There was also no way to make the satellites with their diverse properties in such a highly organized system.

Weizsäcker was a protégé of Heisenberg, one of the principal founders of quantum mechanics as well as the leader of Nazi Germany's (unsuccessful) effort to build a nuclear bomb during World War II.[15] A raid on Heisenberg's laboratory convinced the Allies that Germany was far from producing a nuclear bomb. After the war he claimed that Germany did not build the bomb because the scientists did not want to, and he went on to become a major advocate for world peace.

Weizsäcker was born into a family of highly successful politicians. He was educated in physics, mathematics, and astronomy at Berlin, Göttingen, and Leipzig, getting his PhD with Friedrich Hund, the well-known German molecular physicist. He also worked with Heisenberg and Bohr. Weizsäcker visited the Kuipers in Yerkes in February 1950. No doubt the origin of the planets was discussed, but uppermost in everyone's minds was the fate of von Weizsäcker's father, Ernst von Weizsäcker.[16] Ernst was a prominent member of the Nazi government, and after the war he was sentenced by a U.S. military tribunal to seven years imprisonment for war crimes. The Kuipers were outraged. Gerard published a letter in the *London Times* from his hotel at 14 Henrietta Street, Covent Garden, London,[17] that appeared in December 1949.[18] He had interviewed Carl during the later stages of the war and was aware of his visits

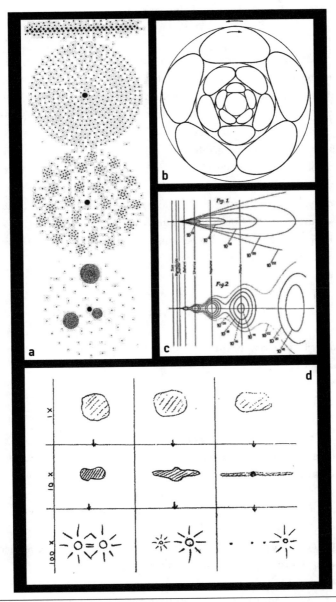

(a) Edgeworth's theory for the origin of the solar system. K. E. Edgeworth, "The Evolution of Our Planetary System," *Journal of the British Astronomical Association* 53 (1943): 181–88. Courtesy of the British Astronomical Association (http://www.britastro.org/). (b) Carl von Weizsäcker's theory for the origin of the solar system. C. F. von Weizsäcker, "Über die Entstehung des Planetensystems," *ZfA* 22 (1943): 319–55. (c) Berlage's theory for the origin of the solar system. H. P. Berlage, "Spontaneous Development of a Gaseous Disc Revolving Round the Sun into Rings and Planets I," *Proceedings of the Royal Academy Amsterdam* 43 (1940): 534–40. (d) Kuiper's theory for the origin of the solar system. G. P. Kuiper, "Formation of the Planets, Part III," *Journal of the Royal Astronomical Society of Canada* 50 (1956): 158–76.

to Bohr in Copenhagen to stress their anti-Nazi feelings and develop a mechanism for publishing their scientific papers. Gerard had ample chance to explore the family's wartime activities. Sarah spoke for both of them when she wrote, "They [the Weizsäckers] are among the finest people we have known."[19] She pleaded to Chandrasekhar to speak up for him and, in an attempt to play on his birthright, likened the Weizsäckers to Jawaharlal Nehru, who needed help in making the world understand his plight. "He is completely innocent," she pleaded. Carl von Weizsäcker brought a letter from his father to help people understand what Ernst had been going through. In the end, Ernst's sentence was commuted to five years, and he was released after three, probably not so much because of the Kuipers' intervention but because of Winston Churchill's.[20] Despite the distractions, Kuiper was fully aware of Weizsäcker's theory for the origin of the solar system and its many critics.

Kuiper on the Origin of the Planets

Kuiper described his view of the origin of the solar system in his 1971 Kepler Gold Medal discourse:

> I felt that I had come to understand the problem of double-star origin, at least in outline; that it was identical to the general process of star formation, from slightly turbulent prestellar clouds upon contraction, with conservation of angular momentum. It followed that the solar system was no more than an "unsuccessful" double star with the companion mass spread out radially into a disk that in time developed the planets. The mass partition [between the primary and companion masses] would be random mass fractions of the total, a result I had derived empirically from a statistical study of double-star ratios. Thus, planetary systems clearly had to originate as the low-mass extremity of the almost universal process of double star formation. A basis had thus been found for estimating the frequency of planetary systems in our galaxy.[21]

Kuiper first submitted a paper to Struve for inclusion in Hynek's *Astrophysics: A Topical Symposium* on January 31, 1950.[22] The work was apparently completed in October 1949, when Kuiper wrote to Struve,

> I have worked very hard on this matter during the last six or seven weeks and have shown that the Weizsäcker theory is definitely wrong. His primary vortices violate the laws of turbulence (laminar flow in a medium with Reynolds number 1014) and his roller-bearing eddies are not dense, cool and semi-permanent regions where planets can condense, but are instead hot and short lived affairs with a density no greater than that of

the surrounding medium. This result was established quantitatively. Instead a new theory of segregation of the proto-planets has been worked out, somewhat along the lines of my paper in ApJ of last March. The history of the solar nebula has been determined quantitatively, including the rotation of the planets is explained as well as the systems of regular satellites. There is little doubt in my mind that the problem has been solved now in its essentials.[23]

Struve's reply is interesting: "I have not yet read this paper but I am sure it represents an important addition to the work of Russell, von Weizsäcker, and others who have discussed this problem in the past. I may say that I have always regarded the discussion of broad problems of this kind as constituting the most important scientific occupation of an astronomer and I am especially glad to see that your long preoccupation with the important details of the solar system has now led to this comprehensive study in cosmogony."[24]

Kuiper also presented his ideas on the origin of the planets on October 15, 1950, to the NAS in Washington, D.C., who published them in the *Proceedings* the following year.[25] After summarizing the properties of the nine planets, thirty satellites, asteroids, meteorites, comets, and meteors, Kuiper again discussed the analogy with binary stars. He also believed that stars must always form in multiple systems. The present size, distribution, and composition of the planets suggest a mass for the nebula of 0.01 solar masses. Weizsäcker had shown that turbulent dissipation would cause the solar nebula to collapse with a half-life of 10^7 years. Initially, collapse would be even faster (comparable to the orbital period of rotation around Sun), but then (after 10^{4-5} years) would slow at a rate governed by the rate at which heat could be lost. Thus the nebula reached its maximum density after 10^{4-5} years. When the gases reacted to form various molecules they would become supersaturated and form solids, which would gravitationally accrete to form numerous small bodies. Weizsäcker had argued that these would collect at the vortices of his convection cells. Chandrasekhar and Dirk Ter Haar questioned whether these would be stable,[26] and Kuiper argued that random distributed aggregates would grow if they could avoid tidal disruption. The key was to grow quickly while being sufficiently separated from other growing aggregates. This determined the spacing of the planets throughout the solar system. This idea has been so widely accepted that it has become part of the IAU's definition of a planet.[27]

What is interesting is that while astronomers had turned away from nebular theories because of the angular momentum problem, Kuiper embraced the theory and presented it in a way that was convincing. Yet he had not solved the angular momentum problem any more than Descartes or Kant had. The slow rotation of the Sun is still not explained. However, observed Kuiper, the Sun is a slow rotator like all G stars, so this problem is not part of a theory for the origin of the solar system but a problem for the origin of the G stars. In other words, he kicked the problem into someone else's backyard.

The Abundance of Stars with Planets

An outcome of Kuiper's belief that the formation of the solar system was a special case of binary star formation caused him to propose that stars with planetary systems were common, a finding that has been confirmed only very recently.[28] His argument was also laid out in his article in Hynek's book.[29] He started by pointing out that the average distance between the stars in a binary system was very similar to the distance between the Sun and Jupiter. Second, he knew the distribution of relative masses in a binary and how many had the Sun-Jupiter ratio. He suggested that in these cases, instead of a second star, a planetary system often formed. Thus the ratio of stars with planets was one in a hundred to one in a thousand. He repeated the conclusion in his article in *Proceedings of the National Academy of Sciences*,[30] and in later years he increased the fraction even more, ultimately believing that half the stars had a planetary system. As with the Kuiper Belt, this was another example of where his rigorous line of reasoning led to predictions decades ahead of their time.

Accusations of Plagiarism

After the appearance of Kuiper's papers, Willem Luyten, still no friend of Kuiper's, wrote to prominent astronomers arguing that Kuiper's papers on the origin of the solar system had been plagiarized from an unpublished paper by H. P. Berlage. Luyten argued that Kuiper had found Berlage's paper when going through the filing cabinets of the *Astrophysical Journal*, at that time edited by Struve. To back up his allegation, Luyten recalled an incident where Kuiper had asked an author for copies of a figure from a paper that had been submitted to the journal and not yet published.

Berlage was a Dutch meteorologist at the Meteorological and Geophysical Service in Batavia (now Jakarta), Indonesia. He put the abbreviation "Jr." after his name in to differentiate himself from his father, a distinguished architect. Between 1930 and 1940 Berlage published eight papers,[31] starting with the effect of the Sun's electric field on the gases around it and ending with a full-fledged theory for the origin of the planets and their satellites from a viscous disc surrounding the Sun. These papers were published in the *Proceedings of the Royal Amsterdam Academy of Sciences* and the *Annals of the Bosscha Observatory*. Berlage was imprisoned by the Japanese during World War II, and when he was liberated in 1945, he was found in poor health and carrying a large manuscript on the origin of the solar system. After consulting with

Luyten, Berlage sent his manuscript to Struve, then editor of the *Astrophysical Journal*. Struve returned it, saying that the article needed considerable shortening.[32] A few years later, Berlage returned to the Netherlands and in 1968 published a summary of his theory in a book, *The Origin of the Solar System*.[33]

Like Kuiper, Berlage proposed that the Sun was surrounded by a large viscous disc. In his case, turbulence caused the formation and accumulation of solids that were spread into rings as the system rotated. Eventually these accumulated into planets. It was a theory that was essentially similar to Kuiper's but proposed more than a decade earlier. Others have repeated the allegation that Kuiper plagiarized Berlage but tempered it with the comment that he may have not realized that he was doing so.[34] In fact Struve frequently consulted the Yerkes astronomers on the disposition of papers submitted to the *Astrophysical Journal*, and it is likely that he sought Kuiper's opinion before returning Berlage's paper.

However, the vindication for Kuiper comes from Berlage himself. As pointed out by Cruikshank,[35] Berlage attributes to Kuiper many fundamental aspects of the evolving theory for the origin of the planets in his book, and he does so without arguing about priority: random variations in local density led to condensations of solids, solar radiation pressure drove 90 percent of the mass from the solar system, the planets formed rapidly (10^{4-5} years), planets were surrounded by "sediment rings" of material that would later collide with the planet—this is especially important for the Moon, whose majors features were produced this way—and satellites at large distances from their planets could be lost by gravitational perturbation by the Sun and other planets. Berlage ascribes all these attributes of his model to Kuiper.

The Kuiper Belt and Pluto

The outer region of the solar nebula, Kuiper wrote, must have had a surface density too low for planetary accretion,[36] It was also cold, 5–10 K when the solar nebula existed and about 40 K thereafter. Condensation products, like the ices of water, ammonia, and methane, must have formed, and the flakes must have aggregated into large objects, up to one kilometer in size or more. These condensations can account for the comets and their size, number, and composition, he wrote. He had proposed what has now become known as the Kuiper Belt.

Kuiper saw a special role for Pluto. Moving through this belt of objects he predicted to be at 38–50 AU, Pluto would cause the comets to be scattered and find their way into the inner solar system. But there is another importance to attach to Pluto: it is, in fact, the first Kuiper Belt Object.[37]

Pluto, Tombaugh, and Another Controversy

Lowell made many predictions as to the location of a planet beyond Neptune. With each prediction, systematic photographic searches were made. They all failed. About twenty years before Kuiper presented his theory to the NAS, a new search began at the Lowell Observatory with a better telescope and a patient observer, Clyde Tombaugh, who used the observatory's 13-inch refracting telescope to search for what Lowell called "Planet X." Lowell concluded that a seven-Earth-mass planet beyond Neptune was responsible for small unexplained discrepancies in the calculated motion of Uranus. With a large-format camera attached to the telescope, he imaged the same section of sky several nights apart and compared the images in a blink comparator. On February 18, 1930, Tombaugh found a faint object in Gemini six degrees from one of Lowell's two last estimated positions in 1915.[38] It was found to have an orbit beyond that of Neptune and therefore was not an asteroid. The ninth "planet" was given the name "Pluto" at the suggestion of an eleven-year-old English schoolgirl, Venetia Burney.

Over the next decade or so, astronomers began to challenge whether this was a new planet or a captured comet or lost satellite. In 1932 UC Berkeley's Armin Leuschner questioned whether it was massive enough to significantly perturb Uranus.[39] Two years later, the Japanese astronomer Issei Yamamoto, professor of astronomy at Kyoto University and director of the Kwasan Observatory in Kyoto, suggested that Pluto was an escaped satellite of Neptune.[40] R. A. Lyttleton of Cambridge made the same suggestion two years later.[41] Twenty or so years later, Kuiper presented his lengthy discussions of his solar-system theory at the National Science Foundation (NSF) in Washington and the Royal Astronomical Society of Canada, arguing that Pluto was a captured comet.[42] After the paper was published in three parts in the Canadian Society's journal and *Proceedings of the New York Academy of Sciences*, Luyten fired off a letter to *Science* that began, "The recent announcement of G. P. Kuiper—with the usual fanfare of a sensational magazine, radio, and TV accompaniment that we have come to expect of him—that Pluto might not be an original planet, strikes most astronomers somewhat humorously, coming, as it does, nearly 20 years after the original suggestion."[43] One can only imagine Kuiper's private response, but in a public response published in *Science* Kuiper says he was not responsible for the headlines and, in any case, he had cited the earlier work, which was quite a different from his proposal.[44] Luyten was wrong in this instance, and his biographer for the NAS archives describes how Luyten could be aggressive toward other scientists, but his letter to *Science* is indicative of a real problem that Kuiper had: he repeatedly attracted such attacks.

In 1992 another object was found in the Pluto region, 1992 QB$_1$. Since then, the number of known Kuiper Belt Objects has increased to more than a thousand. In

(a) Clyde Tombaugh around 1940. "Pluto: First, Best Look at a Mysterious World," CAA: Cuyahoga Astronomical Association, July 14, 2015, https://cuyastro.org/2015/07/14/pluto-first-best-look-at-a-mysterious-world/. (b) The blink comparator at Lowell Observatory being used by Tombaugh in the way he did to discover Pluto. "Blink Comparator at Lowell Observatory Used in the Discovery of Pluto," taken by Pretzelpaws, August 16, 2005, https://commons.wikimedia.org/wiki/File:Lowell_blink_comparator .jpg, licensed under GDFL 1.3. (c) Kuiper Belt object 1992 QB1 (circled). NASA photograph.

2006 the IAU designated Pluto and four similar-sized objects in the solar system (Ceres, Eris, Makemake, and Haumea) as dwarf planets, believing them to be fundamentally different from the true planets.[45]

The Kuiper Belt, the Edgeworth Belt, or the Edgeworth-Kuiper Belt: Priority vs. Influence.

The Kuiper Belt is occasionally referred to as the Edgeworth-Kuiper Belt. If you type "Kuiper Belt" into a web search engine, there will be more than a thousand hits. If you type "Edgeworth Belt," there will be just a few hits. History has given to Kuiper the honor of having this belt of objects named after him. However, there is no doubt that

Edgeworth has priority over the idea that there are a large number of objects in a belt beyond the orbit of Uranus.

In his 1943 article on the origin of the solar system,[46] Edgeworth's exact words were the following:

> It is not to be supposed that the cloud of scattered material which ultimately condensed to form the solar system was bounded by the present orbit of the planet Pluto; it is evident that it must have extended to much greater distances. It must also be supposed that the opacity of the cloud diminished at greater and greater distances from the Sun. Since the formation of a single large planet is only possible when the opacity is not too low, it is evident that the condensations formed in this outer region would be unable to coalesce; they simply retained their individuality and condensed upon themselves. It may be inferred that the outer region of the solar system, beyond the orbits of the planets, is occupied by a very large number of comparatively small bodies.

Eight years later, when Kuiper wrote his paper for the *Proceedings of the National Academy of Sciences*,[47] he used the following words:

> The outermost region of the solar nebula, from 38 to 50 astr. Units (i.e., just outside proto-Neptune), must have had a surface density below the limit set by equation (7). The temperature must have been about 5–10°K. when the solar nebula was still in existence (before the proto-planets were full grown), and about 40°K. thereafter. Condensation products (ices of H_2O, NH_3, CH_4 etc.) must have formed, and the flakes must have slowly collected and formed larger aggregates, estimated to range up to 1 km. or more in size. The total condensable mass is about 1029 g., but not all of this could be collected. These condensations appear to account for comets, in size,[13] number[13] and composition.[14]

Equation 7 in Kuiper's text describes how the surface density of the nebula drops with distance from the Sun and is analogous to Edgeworth's use of opacity. References 13 and 14 refer to papers by Oort and Whipple.[48]

In other words, Edgeworth and Kuiper used the same reasoning to arrive at the same conclusion. Edgeworth clearly has priority on the suggestion that there is a belt of comet-like objects beyond Neptune, and there is a case for calling it the Edgeworth Belt. But the name is not determined by priority—as much as many people would like it to be—but by established usage. Kuiper was a much better-known astronomer than Edgeworth and much better networked into the community of astronomers that would work on these objects. Kuiper did not cite Edgeworth's paper, and it is possible he did not know about it. But Kuiper, who tended so often to publish alone, was not prone to citing earlier work. He tackled a problem, applied his own style of back-to-basics science, and with great confidence and thoroughness presented his deliberations. If an earlier paper had not influenced him, however relevant and whatever claims to priority, he did not see the need to mention it. This was a trait that was, in some part, to set Luyten and Urey against him, and it has possibly played a part in other scientists' treatment of him too. But the main responsibility rests with the many authors that adopted and used the term "Kuiper Belt." Kuiper never proposed this or any other name for the objects. In science, much like life in general (excluding patents and copyright), influence trumps priority every time.

The theory for the origin of the solar system that Kuiper put together has stood the test of time. It is a good theory, and not only because it sent Kuiper in new directions of research, but also because it has served well the several generations of scientists that have followed. So ingrained is Kuiper's view of the origin of the solar system that it has become part of the IAU's definition of a planet: a planet is an object that has "cleared the neighborhood."[49] Now we see the theory being tested and used to help understand the formation of planets around other stars.

CHAPTER 9

The Urey Affair

T WAS PROBABLY THE ORIGIN of the solar system, the Moon in particular, that was at the heart of the conflict between Harold Urey and Gerard Kuiper. Their conflict has become a cause célèbre in the history of planetary science. Whitaker, who based his talk for the fiftieth anniversary of the LPL on the conflict, called it the "Clash of the Titans."[1] The historian of science Ronald Doel used it as a case study to illustrate several major themes in twentieth-century planetary astronomy in his 1996 book, *Solar System Astronomy in America*.[2] Sagan, a student of Kuiper's who went on to become a household name as a popularizer of astronomy, called it the "maddeningly counterproductive feud."[3] Whether you consider it a major step in the evolution of planetary science as a recognized discipline, a sign of the upheavals that the subject had to endure during its birth, a truly damaging event for a new research field struggling to exist, or just an irritating distraction, it pitted two vastly different views of the Moon against each other at the time when America was readying itself to land humans on that very object. The arguments the two men pounded out were to thread through the struggle to the Moon and were only to be settled when spacecraft landed on the Moon and performed chemical analysis of the surface. When the crew of Apollo 11, Neil Armstrong, Buzz Aldrin, and Michael Collins, handed over the moon rocks to the scientists at the Manned Spacecraft Center in Houston, they placed the final nail in the coffin of Urey's ideas.

A Moon Rediscovered

We can trace our efforts to understand the nature of the lunar surface to the observations of Galileo, who saw the light mountains and dark maria through the first astronomical telescope. Four hundred years and many astronomers later, in a paper titled "On the Face of the Moon," geologist Grove Karl Gilbert argued that those mountains were volcanic.[4] However, our present story begins with Ralph Baldwin.[5] Baldwin was educated as an astronomer and received his PhD from the University of Michigan in 1937. His thesis was on the spectroscopic study of novae. From 1935 to 1942 he taught astronomy at various universities and at the Adler Planetarium in Chicago. Photographs lining the walls at the planetarium sparked his interest in lunar craters, and he gradually developed the idea that the surface of the Moon was covered with meteorite impact craters rather than volcanoes. He published his ideas in *Popular Astronomy* after they had been rejected by the major research journals.[6] In 1941, Baldwin gave a talk at Yerkes Observatory at Struve's invitation, but this led to disappointment. Despite the preeminence of the audience, not one person was convinced. Philip C. Keenan, Greenstein, Morgan, Struve, Van Biesbroeck, and Adel were listed by Baldwin as being present during what he described as a "fiasco." He later recalled, "Not one of these great men had his opinion changed. Early astronomers such as Robert Hooke and J. H. Schröter have said that the Moon's craters were volcanic. Therefore they were volcanic." The war came, and Baldwin was hired as a physicist by Johns Hopkins University's Applied Physics Laboratory where, like Van Allen, he worked on the proximity fuse, one of the secret weapons that won the war. After the war, he returned to working for the family-owned Oliver Machinery Company, which made woodworking tools. In his spare time he wrote *The Face of the Moon*, which was published in 1949.[7] At his publisher's request, this was enlarged and published again as *The Measure of the Moon* in 1963.[8] At a time when there was still controversy about the origin of some of the most spectacular impact features on Earth, Baldwin had now fully laid out his arguments that the craters covering the Moon were of impact origin. First, the numerous valleys and ridges surrounding Mare Imbrium followed great circles that intercepted each other in the middle of Imbrium. Imbrium was the site of a giant explosion, too big by many orders of magnitude to have been a volcanic explosion. He found the same at Crisium, Humorum, and Nectaris. Thus he concluded that all the craters were caused by impact, not volcanism.

There was one person who was convinced, and it is not clear whether he was present when Baldwin gave his talk at Yerkes—one must assume not. It was Gerard Kuiper. Kuiper was spending a considerable amount of time looking at the Moon through the 82-inch telescope at McDonald Observatory. He had a binocular eyepiece fitted to the telescope to avoid the eye strain that would follow from looking at such a large, bright object for so many hours. The result was a paper published in the *Proceedings of the National Academy of Sciences* for 1954 that showed a strong influence by Baldwin's 1949 book.[9]

(a) Harold Urey. NASA photograph. (b) Ralph Baldwin. Courtesy of AIP Emilio Segrè Visual Archives.

Kuiper clearly regarded his paper as incomplete with many outstanding questions, and he clearly thought the Moon was ancient, perhaps 5 billion years old. Such an ancient surface, unlike Earth's, could teach us much about the early evolution of the solar system. A year later he launched a major new lunar initiative that threw him into the heart of the effort to land humans on the Moon. His efforts also triggered one of the most important feuds in the history of astronomy, the Kuiper-Urey feud over the nature of the Moon.

Harold Urey

Harold Clayton Urey received his PhD from UC Berkeley in 1923, studying under that giant of physical chemistry, Gilbert N. Lewis.[10] After periods at the Niels Bohr Institute in Copenhagen and Johns Hopkins University in Baltimore, Urey was hired as an associate professor at Columbia University in 1931, where he began work with the separation of isotopes. This work resulted in the discovery of deuterium, for which he received the Nobel Prize just three years later. It must have been a bittersweet moment for his mentor Lewis, who had been nominated for the prize thirty-four times and never received it.

During World War II, Urey headed the Columbia University group that developed gas diffusion methods for isotope separation as part of the national effort to develop a nuclear bomb.[11] Urey then moved to the University of Chicago and became interested in the formation and history of the solar system and the origin of life, publishing the classic Miller-Urey experiment by which life precursor molecules are made in the laboratory from simple molecules by a flash of electricity.[12] The experiment is now the starting point for most scientists thinking about the origin of life. In 1958 Urey moved to the new University of California, San Diego (UC San Diego), where he helped establish their school of chemistry.

Urey had a career every bit as successful as Kuiper's, and for anyone interested in applying the principles of chemistry to problems in planetary science, he was the master. His book, *The Planets*, published in 1952, was the universal manual in showing how to do this.[13] Disciples would be mining the book for decades, trying to explain every nuance of meteorite and planetary composition using his techniques. The very origin of life could be addressed, so it was thought, using his discussions of carbon chemistry. Important in Urey's thinking were the meteorites whose age, composition, and textures suggested that they were the building blocks of the planets.

Urey's Theory—The Moon as a Meteorite

In the late nineteenth and early twentieth centuries, Russell and Adolf Erik Nordenskiöld showed that the most common meteorite types have essentially the composition of the Sun's photosphere.[14] All the elements in meteorites—or to be more precise, the meteorites called chondrites—are present in the same proportions as they are in the Sun, except for the elements that were in a gaseous state when the meteorites formed (hydrogen, helium, noble gases, etc.). Urey ran with this idea, postulating that the planets formed from chondrite-like material through a series of chemical reactions. Thus a well-known problem of the 1940s, why the planets have various iron-silicate proportions that differed from the Sun's if they shared a common origin, was resolved. Urey saw the Moon as a product of the same process that produced Earth and meteorites—the whole solar system in fact.[15] The Moon formed from the same materials by the accretion of solar system solids at low temperatures. It was never molten because it never had enough radioactive elements to produce any significant melting.

Harold Urey also liked Baldwin's impact conclusions. The Moon has been bombarded by planetesimals soon after its formation, many in around one hundred kilometers in size. This left a surface full of basins and craters, and the bottoms of these basins and craters were once filled with pools of material melted by impact. Sure there were scars, what might be called valleys, but Urey thought they were simply grooves cut by flying meteorites made of metallic iron and nickel, just like the iron meteorites.

That famous bulge on the Moon's equatorial regions, described by the prominent English geophysicist Sir Harold Jeffreys and pointing toward Earth,[16] was the effect of Earth's gravity pulling the once plastic mantle and crust of the Moon. Thus the Moon, Urey said, was essentially a large meteorite, a meteorite of approximately solar composition, very much like the most Sun-like meteorites, called carbonaceous chondrites. However, the surface of this massive "meteorite" showed a history dominated by the effects of impact. It was covered with impact basins, impact craters, and impact melts.

Kuiper on the Origin of the Lunar Surface Features

Kuiper and Urey shared time together at the Rancho Santa Fe Conference organized by the NAS in 1950.[17] Kuiper gave Urey a prepublication copy of his paper on the origin of the solar system. Urey explained to Kuiper why many of his views on the Moon were in error and how he had figured out a view of the Moon that most colleagues accepted. Four years after the Rancho Santa Fe meeting, Kuiper published his views on the surface of the Moon in the *Proceedings of the National Academy of Sciences*.[18] There are advocates of both volcanism and impact as the major surface process, he pointed out, with the same observations being used by both camps. "It seemed to the writer that an independent study, based not on published records but on new visual observations with a large telescope, might resolve some of the ambiguities." In his first few lines, Kuiper was blowing off the whole of Urey's arguments, replacing them with observations made with one of the largest telescopes in the world. The words "independent" and "large telescope" must have been especially hard for Urey to read. "It will appear that the impact hypothesis is confirmed as the chief cause of lunar topography." This was something both men could agree on. "But the source of the 'planetesimals' is found to be neither meteorites nor small masses of the type that formed the moon by accretion.... Furthermore, evidence is presented that the moon was nearly completely melted by its own radioactivity, some 0.5 to 1 billion years after its formation, and that the maria were formed during this epoch."

Kuiper goes on to discuss the dynamics of the Moon and the clues they provide to its internal structure. In 1693 the Italian astronomer Giovanni Domenico Cassini described the Moon's movements, and from these one could estimate its moments of inertia, which reflects the extent to which an object resists rotational acceleration.[19] A uniform homogenous sphere has a coefficient of inertia of 0.4. Earth, with its structure of core, mantle, and crust, has a coefficient of inertia of 0.334. The best value for the Moon at the time was 0.43, but the experimental uncertainty on this value was so large (±0.09) that it could be as small as 0.36, consistent with the Moon having a core like Earth's. Kuiper thought that radioactive elements could melt hundred-kilometer

(a) The full Moon showing the dark maria and the light highlands. "Primordial Light: The Moon," accessed May 27, 2017, http://primordial-light.com/moon.html. Courtesy of David Ilig. (b) The lunar surface from the *Photographic Lunar Atlas.* G. P. Kuiper, *Photographic Lunar Atlas based on photographs taken at the Mount Wilson, Lick, Pic du Midi, McDonald and Yerkes Observatories, Edition A* (Chicago: University of Chicago Press, 1960). (c) Rilles on the floor of the lunar crater Gassendi, from Apollo 16. NASA photograph. (d) The Alpine Valley on the Moon. Alpine Valley is now considered to be a graben, not an impact feature, which is an elongated block displaced downward between two faults. Courtesy of Jean-Luc Dauvergne, ©S2P/OMP/OPM/Dauvergne/Colas/Rousset, accessed May 27, 2017, http://astrosurf.com /eternity/temp/lune_2010-09-29_06-09-07-s1.jpg.

asteroids because there were iron meteorites, and these clearly were the cores of melted asteroids. Therefore, radioactive elements could melt the moon, with its diameter of 3,474 kilometers.

In the telescopic images Kuiper saw three kinds of surface of differing age; the oldest he called the "premelting surfaces" (essentially the uplands) and the intermediate were the "maximum-melting" surfaces, the maria. The third kind of surface were the "post melting" surfaces, those showing evidence for events that occurred after maria formation, such as impacts. These demarcations are still recognized today, with some modification. The pre-Nectarian system is the oldest, the Nectarian and Imbrium systems are the intermediate, and the Eratosthenian and Copernican systems are the youngest. Kuiper speculated that the tektites found strewn in large fields in Australia, Europe, the Ivory Coast, and the eastern United States were impact ejecta from the Moon, mainly on the basis of their young age and bubbles with low-pressure gases, but he stressed the need for trajectory calculations "with the aid of a fast electronic computer."[20]

In getting to the details of the lunar surface, Kuiper discussed rilles and ridges, "always puzzling," the maria and continents, lava flows and grooves caused by flying objects, volcanism triggered by impact, and craters that are explosive and caused by impact, not volcanism: "volcanism adds mass, impact removes it." The central peaks were of special interest to Kuiper, who suggested they were caused by lava upswelling through weaknesses in the crust caused by the impact. They were only present in craters that were formed more or less contemporaneously with the maria because they required the presence of subsurface lava, which explains their presence in some craters and not others. The famous bulge that points toward Earth was due to the accretion of material onto an already solidified crust.

Urey's Response

Urey's response, "Some Criticisms of 'On the Origin of the Lunar Surface Features' by G. P. Kuiper," appeared in the July 15, 1955, issue of the *Proceedings of the National Academy of Sciences*.[21] It was an angry piece of writing, referring not just to Kuiper's published article but to communications they had at the Rancho Santa Fe meeting.

Kuiper has made observations of the Moon with the 82-inch at McDonald Observatory, says Urey, "but, so far as I can deduce from his discussion, he has not observed anything markedly different from what has been previously observed." This must have been a body blow to an observer proudly using one of the largest telescopes in the world to look at the Moon more thoroughly than ever before. Urey goes on by pointing out that in the fall of 1953 he showed Kuiper his latest calculations, which indicated that the Moon did not have enough radioactive elements to cause melting. Urey claims that Kuiper is also wrong about the cause of the bulge. Kuiper's theory would place the bulge in the wrong place; the bulge would be pointing ninety degrees from Earth, instead of straight at it. "Kuiper's very brief discussion of this subject is at least internally inconsistent."

If the Moon was completely molten, the resulting expansion and cracking would mean the flow of lavas onto the surface. "All these effects are ignored in Kuiper's presentation." Also, the Moon would still be hot and volcanically active. "This necessitates some consideration of possible processes for making large astronomical objects of markedly different chemical composition, which Kuiper has not done. . . . No consistent explanation for the rigidity of the moon as indicated by its irregular shape has been offered by Kuiper, for the moon could hardly retain an irregular shape if its interior were melted." Kuiper had written that melting by radioactivity was near-surface, yet solidification would start at the center. If the moon were even 15 percent chondritic in its uranium, potassium, and thorium, it would be molten out to 0.8 of the Moon's radius, but how would it be possible to make a Moon so chemically different from meteorites? Urey also claims that tektites are terrestrial, Alpine Valley is a gouge made by a nickel-iron object, and that Kuiper's talk of "squeeze ups" are something Urey is "wholly unable to entertain."

"It would be a thankless task to review adequately this paper in all details. Features of the moon that have been observed and recorded during the last century are again described as though they were newly observed and then are followed by a discussion of the completely melted moon theory." Urey goes on,

> On page 1104 there is a long paragraph listing facts well known for many years, and then at the bottom of the page is the statement: "It is at first somewhat surprising that the mountain ranges around Mare Imbrium should have formed during the period of maximum melting, at which time the crust must have been fairly thin and possibly was not able to support the extra load." It is certainly surprising to this reader. Then came a discussion of the Alpine Valley that still leaves me at a complete loss as to how these massive mountains were supported.
>
> . . . it is my belief today, as it has been for some five years, that very probably the moon was accumulated at low temperatures from a primitive dust cloud of solar composition with the iron in oxidized states and that the concentrations of radioactive substances within the moon are sufficiently low that melting has never occurred. In fact, I believe that present temperatures are so low that the interior of the moon has a high strength and that such low temperatures require the moon to have been formed at low temperatures and never to have been melted at any time.

Kuiper's Response to Urey

Kuiper's response was typical Kuiper.[22] Even the title, "The Lunar Surface—Further Comments," is typical Kuiper: no reference to Urey, no mention of rebuttal or defense. In the face of a vicious attack it is business as usual—gracious, polite, almost a plea for innocence, and calculated to stoke Urey's fury. Kuiper explains that he was aware of previous work and knew others held different opinions, but this was a preliminary work to develop "a working hypothesis that had been built up in the course of visual studies made with a large telescope." Furthermore, "an effort was made to give sufficient detail to indicate how the new conclusions were reached to avoid controversial discussion." In other words, in the politest way, Kuiper is denying Urey any authority in the subject and hanging his conclusions on the second largest telescope in the world.

Kuiper then summarizes the differences between the two men. He cites evidence that impacts cannot do what Urey wants and lists new explanations for previously unexplained observations that his theory makes possible. "These data seem very strong, if not incontrovertible," he concludes. Kuiper admits that the degree of melting of the interior is unclear: "In other words, while it probably implies that the entire interior went through a stage of melting and solidification, there is no need whatever for assuming that the entire moon was liquid simultaneously. This would not be expected. Perhaps this point was not made sufficiently explicit in my article." On the subject of Alpine Valley, Kuiper writes, "The explanation of this valley in terms of flying fragments would never have been made if its proponent had studied it visually through a large telescope—or, in fact, had studied published descriptions of it by Elger (1895) and others."[23] He goes on to discuss the figure of the Moon and why the Moon's "bulge" faces Earth and not its leading edge as it orbited Earth: "I was well aware that this difficult problem was not fully solved, and I so stated." About expansion and contraction of a melting core, Urey is not correct; on the sediment ring, Urey's discussion is misleading; and on tektites, "The last paragraph is not a fair statement of my position."

Kuiper ends his response with the statement, "I find nothing of importance in my article that is appreciably affected by Urey's remarks, with the possible exception of the precise timing of the Mare Imbrium event. I believe that my article contains a synthesis of lunar observations that is essentially new; but it is natural to expect that refinements can be made whenever new data become available. I am convinced that the best way to progress is to get the very highest resolving power in future surface studies." He not only sticks to his guns, he also makes an opening gambit for his next major project, namely, mapping the Moon.

Many of Kuiper's ideas were to be refuted over the next few decades (for example, the cause of the Earth-aligned bulge and the lunar origin of tektites), but the fundamental conclusion that the Maria were lava flows and that there were three epochs in lunar history, were to become fundamental tenets in our understanding of the Moon. Urey's arguments that the Moon should have a bulk composition similar to the commonest meteorites were based on sound scientific reasoning. When it was found that this was not the case, profound challenges arose, which resulted in the current best theory, as extraordinary as it is, that the Moon was made from the ejecta of a Mars-size body colliding with Earth.[24] Incidentally, one of the astronomers to propose the giant-impact hypothesis for the origin of the Moon was one of Kuiper's students, William Hartmann.

Two Fundamentally Different Views

Two leading scientists of the day had profound disagreements stemming from differences in their backgrounds. Urey took every argument Kuiper made and criticized it in detail. Nothing survived intact. It was a hatchet job. The bottom line is that Kuiper saw evidence for volcanism, mostly in the form of the Maria, while Urey argued that radioactive elements are not present in the Moon in sufficient abundance to cause melting. For Urey, the Moon was, essentially, a huge chondrite meteorite. What Kuiper thought were lavas, Urey thought were impact melts, even entire maria. The theory being promoted by Kuiper is exactly the opposite of Urey's carefully honed and highly argued views. The simple rules of chemistry, as any undergraduate student would know, clearly indicated a low-temperature origin for the Moon. Kuiper goes on: "They were not, as has been supposed, primarily the result of melting by the impacts themselves." Urey's book and his several major papers had been brushed aside. Indeed, Urey's stature as the foremost authority in this subject was ignored. Is it any wonder Urey was angry? The two men were coming at the problem of the nature of the Moon from very different directions. To Kuiper, the origin of solar system was a special case of binary star formation, where the Sun, the planets, and the Moon were formed at high temperatures. Kuiper accepted just about everything in Baldwin's 1949 book, *Face of the Moon*.[25] So did Urey, in fact. The surface may be heavily cratered, but Kuiper saw a massive role for volcanism in sculpting the surface of the Moon or, as he called it, melting.

The Feud

The correspondence of the day, kept in the special collections of UC San Diego, tells the rest of the story. It is a remarkable story of two brilliant minds and two men behaving badly. It also illustrates the problems facing planetary science as it emerged from planetary astronomy.

The relationship between Kuiper and Urey began cordially enough. They were quickly on a first-name basis with a mutual desire to cooperate. But the signs of problems were there from the start. On December 1, 1949, Kuiper wrote to Urey, "The physico-chemical approach promises to be a powerful one but, of course, I am not able to evaluate this subject critically,"[26] and on June 13, 1951, Urey wrote to Kuiper, "Of course my approach to this problem is of necessity of a distinctly non-mathematical kind. I try to find bits of evidence rather than deduce things from fundamental principles."[27] The two men were admitting that they were finding it difficult to cross into each other's disciplines. Planetary science was going to be a different field from planetary astronomy and would require a new kind of researcher, one that is comfortable not just with astronomical concepts and methods but also those of geology, chemistry, physics, and even the principles of engineering necessary to send spacecraft out into the solar system.

Despite claiming little expertise in astronomy and mathematics, Urey began scrutinizing Kuiper's data tables, looking for the information he needed for his calculations—in particular he needed to know the densities of the planets. Scouring the tables, he found that Kuiper's planetary radii were systematically smaller than earlier values and wrote to Kuiper about this on November 13, 1951.[28] Kuiper explained that he had superior telescopes, superior viewing conditions, and better methods,[29] but Urey would not accept this and Kuiper must have been irritated at Urey's frequent return to this topic.

As relations began their inevitable slide, Urey began to complain about the lack of citations to his work in Kuiper's. Kuiper laid out his attitude to citations in some detail in a letter to Urey dated January 23, 1952.[30] He started out, "I need to tell you if I failed to give credit where credit is due this has been in spite of great efforts on my part to be accurate." He goes on, "so many papers in geophysics and related matters appear written as if no one worked on the subject before." The way Aston's work is ignored "has been a real surprise to me."[31] "If I have credited Suess with something you did before then I suppose this happened because the argument became part of my thinking as a result of discussion with Suess and because, after this, I failed to encounter the point again in your writings."[32] He continues,

> One is quite sure of certain ideas, because they appear rational and mutually consistent; one hears other facts or ideas, but they may not fit in and are therefore put aside. Later the same idea may click and be integrated with the theoretical picture one is developing. One may then only belatedly discover that this idea is not new. While I made a real effort to be historically correct and stress the development of ideas (something in which I have evidently not fully succeeded), it might have been better to limit discussion to the ideas themselves, leaving it to the writers of the next decade to apportion credit and give references. The greatest compliment, I suppose, an idea ever gets is that everyone immediately adopts it (and forgets its source).

This paragraph really captures Kuiper's attitude to citing the work of others and explains why so many people found him so exasperating. He never deviated from this method of working, and Urey, like Luyten, could never let go. Both continually repeated their criticism that their work was not sufficiently cited and became increasingly frustrated, while Kuiper continued without change. Many colleagues accepted this quality of Kuiper's, preferring to focus on his catalog of important accomplishments, but nevertheless resented it. Urey and Luyten would no doubt have believed that the selective memory and unusual views on citation were a deliberate way of manipulating the community and sending them unpleasant messages, that in effect he was taunting them, but the viciousness of their attacks were not helping anyone. Eventually, Urey's attacks on Kuiper spilled over to others. British geophysicist Edward Bullard complained of this when he received such a letter from Urey,[33] and Urey confided to Page, "One of the worst things about this sort of thing is the warping of the personalities of the people involved. This, I am sure, has happened to me to some extent. One cannot go on for years trying to protect oneself until almost everything one does has something of this objective in it without becoming affected by it. However, someone should stand up to this and I just will not back down. This is the way I am made, I guess."[34]

On June 3, 1954, Kuiper circulated among members of the IAU Commission 16 (a committee of astronomers interested in the solar system) asking for them to help him in assembling a summary of work done in the physics of planets and satellites and proposing to change the name of the commission from "Commission pour les Observations Physiques des Planetes et des Satellites" to the "Commission pour les Etudes Physiques des Planetes et des Satellites" (my emphasis), reflecting, one might suggest, part of the growth of planetary science out of planetary astronomy.[35] Kuiper added Urey to the mailing lists even though Urey was not a member of the commission. Urey replied that he knew nothing about the commission and was not a member.[36] How much acrimony could have been avoided if he had responded as requested!

When Urey received Kuiper's annual report of progress for the IAU he thought that it did not adequately mention his work. Kuiper had referred to every paper he had published in the last few years, but nothing Urey had published. Urey submitted two supplements to Kuiper's report, one on the age of the solar system and one on the origin of the solar system. Kuiper accepted the first and added the material to his report, but he rejected the other, arguing that "no useful purpose is served by a short report on this subject."[37] He went on to add, in a letter to the IAU, "The situation may be different three years from now, when out of the many contradictory claims and views an organized and more or less generally accepted pattern may have emerged." Urey sent the remaining supplement to the IAU and to members of Commission 16, but to no effect. "It has been customary for many years that the President of a Commission is the only person responsible for the Draft Report of his commission," wrote Pieter Oosterhoff on behalf of the IAU.[38] The IAU approved Kuiper's changes to the Commission and endorsed the new lunar atlas project he had proposed in 1954.

Claiming success, since at least one of his supplements had been incorporated into Commission 16 report, Urey continued his attacks on Kuiper. On February 9, 1955, Urey wrote to Alan Waterman of NSF about the nine-volume work on stellar systems being compiled by Kuiper.[39] The AAAS and the AAS had endorsed Kuiper's proposal but also requested comments from the community on funding this massive undertaking. Urey began by mentioning Kuiper's four-volume series *The Solar System*,[40] which he thought "beautifully produced and very convenient," and *Atmospheres of the Earth and Planets*,[41] about which he wrote, "They are valuable and useful books."

However, Urey points out, he was breaking his contract with the University of Chicago Press by not writing his chapter for the *Solar System* series because he considered Kuiper unsuitable to edit such important volumes. "Kuiper in his writings minimizes the work of other people in his field by referring to minor things they do with pin-point references, and then includes their important work at other places without references, so that the reader infers that it is his own work. Subsequently he refers to the subject by referring to his own paper."[42]

But Kuiper's problem was not just unfair in citing of the work of others, but also in neglecting subjects he did not agree with, such as mechanisms for mountain building on Earth. Kuiper accepted the idea of Tuzo Wilson, for instance, who thought that the major mountain ranges were due to contraction of Earth's crust.[43] However, there were other views, such as they were the result of convection currents in the mantle. This was an idea that Urey favored, and he complained to Kuiper that it should be better represented in volume 2 of the *Solar System* series, *The Earth as a Planet*. Kuiper persuaded Wilson to mention this in his chapter, but Urey was not satisfied and wrote a long letter to David Griggs, a professor at UCLA and advocate of the convection theory, suggesting he launch a campaign against Kuiper on the subject of mountain building to parallel his campaign against Kuiper's views on the origin of the Moon.[44]

Urey also wrote to the chancellor of the University of Chicago, Lawrence Kimpton, making these complaints and trying to persuade him that the University of Chicago Press should either not publish any more books by Kuiper or should appoint a strong editorial board or associate editor to monitor him.[45] The chancellor was sympathetic, but did not intervene.[46]

The attacks continued into 1957. This time Kuiper wrote an article on his views on the origin of the planets for *Vistas in Astronomy*.[47] Urey wrote a scathing attack and sent it to *The Observatory*, since *Vistas* did not publish letters from readers, with a long explanation of his grievances in an April 17 cover note.[48] The reply came from John Guy Porter, the editor, on June 4, "I would wish to express once more our regrets that we are unable to agree to publishing it," referring to Urey's letter.[49]

As the 1960s approached, the attacks continued and took an even uglier turn. In January 1960 Kuiper was negotiating with the University of Arizona for a position at that university, where the chairman of the astronomy department and director of the Steward Observatory was Edwin F. Carpenter. On January 8, 1960, Carpenter wrote to Kuiper mentioning the new Kitt Peak Observatory and the University of Arizona's decision to hire five new astronomers.[50] He was astonished to learn that Kuiper was leaving Yerkes and contemplating a move to Arizona. "His current studies, however, are so far afield from the conventional astronomy in which we have been engaged, and indeed so peripheral with respect to the basic needs of graduate students, that I have some doubt of his conformation to the new graduate program, and both my colleagues and I lack the acquaintance with his current fields for adequate appraisal." Carpenter had also heard rumors of problems at Chicago with which Urey would be familiar, since he was there at the time. These factors persuaded Carpenter to request an appraisal from Urey.

On January 19, 1960, Urey replied.[51] "I would not be in a good position to recommend Kuiper," he began. People who work with him "invariability have difficulty with him." Urey claimed that "I would not wish to be on the staff of the same university with him. I came to California for a variety of reasons, but certainly one of them was the desire to be farther away from Kuiper than I could be at Chicago." On the subject of his research, Urey wrote that "Kuiper's work in recent years, it seems to me, has not been very interesting. . . . No observations on the moon at the present time by conventional methods are worth tying up a great telescope for." Then the body blow: "As time has gone by, I have drifted from the position of a person who greatly admired Kuiper to one who believes that his work has been consistently second-rate." And it is not just in his research that Kuiper fails. "He does not take a really genuine pleasure in the success and advancement of his students and young people when it results in some professional sacrifice on his part."

Six days later Carpenter must have felt the rug pulled from under his feet. "The University seems to be overwhelmingly fascinated [by Kuiper], and hired him before the arrival of your letter."[52] External appraisals or departmental recommendations had made no difference. But Carpenter made it clear that he had not been convinced by Urey's negative letter. He was not sure that the department would have voted against the hire, and he had had friendly relations with Kuiper since the 1930s.

Urey immediately wrote to Albert Whitfield, director of Lick Observatory, and Aden Meinel, director of Kitt Peak Observatory, suggesting that they oppose the hiring, but to no avail.[53] "I would like to assure you that I do not anticipate difficulties," replied Meinel, stressing that the observatory and the university were separate institutions and Kuiper's interests overlapped very little with the observatory's.[54]

Having spent a decade attacking Kuiper at every opportunity, and Kuiper refusing to be affected by these attacks, sticking to his policies of citation that upset so many, Urey started to attack Kuiper's students. On November 2, 1962, Urey wrote to

Hartmann about a paper he had just published describing the spectacular concentric structures Kuiper and Hartmann reported on the moon.[55] The best known is Mare Orientale. Because this structure lies on the eastern edge of the moon, it is difficult to see from earth. The technique Hartmann used was to take the best photographs of the moon taken from Earth and project them onto a large white sphere. Then Hartmann could move round the globe and photograph the image as if observed from overhead while being ninety degrees from earth. True to form, Kuiper and Hartmann implied, if not specifically stated, that this technique was new and, also true to form, Urey took offense. He pointed out to Hartmann that he described the method in a two-part article he published in *Sky and Telescope* in January and February 1956 and should have been credited.[56] He then goes on to say, "I wonder if it is not your policy to refer to papers published in the serious scientific literature by well-known scientists," and he cites prior work of his that appeared in *Sky and Telescope* seven years previously. Gilbert Fielder and Kurd von Bülow had also written on this topic, "Here again you should have referred to the previous literature and given your opinion in regard to these questions. . . . As an older man to a young man, it seems to me that it would be well not to engage in practices of this kind. It is a poor way to start a career."

It was not only the inference that the technique was new that Urey objected to. Urey also questioned the reality of the concentric structures. "I find it very difficult to convince myself that these concentric circles have any real meaning. . . . I wonder if such prejudices have not worked themselves into the description of the Mare Orientale."

Hartmann replied that he stuck by his work—Fielder and von Bülow were talking about very different features, but he welcomed the comments about references.[57] A long reply followed, continuing the reprimands and emphasizing Urey's views that the moon formed cold.[58]

Hartmann was not the only victim. A day before writing to Hartmann, Urey wrote to another of Kuiper's group, L. Harold Spradley, also "in the spirit of an older man giving advice to a younger one."[59] Spradley was on detail to LPL from the Air Force's Aeronautical Chart and Information Center (ACIC). Spradley had written in "Contributions of the Lunar and Planetary Laboratory" that foreshortening can be removed by projecting lunar images onto a white sphere and photographing from the side.[60] Urey again referred to his *Sky and Telescope* publication and reprimanded the younger man for not giving him prior credit for the method. "I wish to emphasize however that failure to refer to other people's work in an obvious way is a mistake on your part." In a postscript Urey explains that his paper has been reprinted many times and "it seems very doubtful to me that you overlooked it."

As is often the case in such heated exchanges concerning priority, the truth is that neither Urey nor Kuiper were the inventors of the projection technique for rectifying

(a) Student William Hartmann taking a photograph for the Rectified Lunar Atlas. A telescopic image is projected onto a white sphere and a photograph is taken from the side to reveal foreshortened structures on the limb. Urey argued that he proposed the technique but was never given due credit. "Multiring Impact Basins on the Moon," Planetary Science Institute, last updated January 18, 2011, http://www .psi.edu/epo/multiring_impact_basins/multiring_impact_basins.html. (b) Mare Oriental on the eastern limb of the Moon detected by Hartmann using the projection technique. W. K. Hartmann and G. P. Kuiper, "No. 12 Concentric Structures Surrounding Lunar Basins," *COMM* 1 (1962): 51–66 (plus 77 plates), figure 12.31. Courtesy of the LPL. (c) Mare Orientale on the eastern limb of the Moon as photographed from an altitude of 2,773 kilometers by Lunar Orbiter 4 in 1967. NASA photograph.

lunar images. In February 1935 Frederick Eugene Wright, optical scientist and geophysicist at the Carnegie Institution of Washington, published an article titled "On the Surface Features of the Moon" in *The Scientific Monthly* (now known as *Science*) in which he described the technique. History has a way of having the last word.

Epilogue

Perhaps the most remarkable letter among the collection at UC San Diego is the final one in the Kuiper folder. Written to Sarah on December 28, 1973, four days after Gerard's death and eleven years after the group reprimands.[61] Urey wrote, "I am so very sorry to hear of the death of Gerard. He was a good scientist and man, and I hope that you will have many things to do in the future to help you bear up under this enormous disappointment. Frieda and I send our condolences to you and to your family." After over twenty years of the most vicious and persistent attacks, apparently death had brought peace to the troubled Urey.

Much has been written about the famous dispute between Urey and Kuiper over the nature of the Moon. As remarked at the outset of this chapter, some think it part of the growing pains that were experienced by the new field of planetary science. Planetary science now required the skills of many disciplines—it was not pure astronomy where observations at the telescope were the bread and butter and every one focused on data they obtained from observations and spectra. Neither was it pure geology, since these were alien worlds that had so far to be studied without actual rock samples, except for the meteorites whose study was also experiencing growing pains as geologists had to hand over some of the field to chemists and astronomers. Moon rocks were in the future. Both Kuiper and Urey were using the best techniques in their armory to study the same objects, and their views could not have been more different. It would take many years and many students of both Urey and Kuiper to learn to work together as planetary scientists. Many years after the feud, spacecraft landed on the Moon and discovered that the surface was made of basalt, a volcanic rock. Kuiper was right. A few years later, the astronauts of Apollo 11 gave rocks from the Tranquility landing site to the scientists who confirmed that the rocks were basalt and found that they were much younger than the chondrite meteorites. Long after its formation, there had been volcanism on the Moon, just as Kuiper had said.

Personalities were surely a problem. Urey could not let go, and Kuiper would not change his ways. In some senses, the Urey-Kuiper affair was an irritating distraction. If both men could have put personal feelings aside, then maybe both would have had a more peaceful life and planetary science would have had a gentler birth.

But it cannot be denied that they both men contributed enormously to the growth of planetary science. They led the way, demonstrating the existence of a new subject, with new problems and methods, requiring new thinking patterns. Chemists had an example of how to think about the planets, satellites, asteroids, and meteorites using their techniques. Astronomers were learning, with more difficulty than the chemists, that planets were not moribund or to be the sole province of the amateurs, science fiction writers, or even cranks. They would have to give way to laboratory measurements, terrestrial models that could help and hinder, and the application of unfamiliar geophysics. By their contributions, and the excitement they generated through their work, they attracted many people to the field, surely the biggest accomplishment of all.

CHAPTER 10

Asteroids and Life and Death on Earth

KUIPER'S THEORY FOR the origin of the solar system was successful in many ways, one being that it suggested new avenues of research. One new avenue was the asteroids, fragments of which come to Earth as meteorites. There are few more dramatic ways we are made aware of our solar system environment than by the fall of a meteorite to Earth—every time a meteorite falls, even today, it makes the headlines. Alexander Humboldt first connected meteorites with asteroids, writing that meteorites are "the smallest of all asteroids."[1] The hunt for asteroids had been going on for more than a century when Kuiper became interested in these rocks, but when he jumped in he began an effort that continues today.[2] He was primarily interested in their physical properties—their size, shape, spin rate, and colors—and with obtaining data for a statistically significant number of asteroids. In 1950 he started a systematic search for asteroids using the Cook telescope, and he and his collaborators found all asteroids brighter than magnitude 16.5. The observatory Kuiper established now leads the effort to find new asteroids, especially those that come close to Earth. However, now it is a matter of life and death. The Chelyabinsk meteorite was eighteen meters (about sixty feet) in size when it entered the atmosphere and exploded over Russia's Chelyabinsk Oblast, and its arrival sent more than twelve hundred people to the hospital.[3] The ten-kilometer asteroid that hit Earth 65 million years ago destroyed species, most notably the dinosaurs.[4] Therefore the hunt for asteroids is of more than scientific interest.

Chixulub, Tunguska, and Chelyabinsk

Some of the impacts of objects from space have been so dramatic, so spectacular, and so well described that everyone knows about them. Such are the impacts of the Chicxulub, Tunguska, and Chelyabinsk objects. In fact, scientists are excited about an

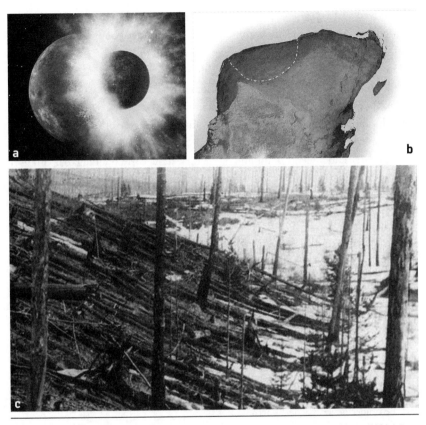

(a) Artist's painting of a major impact of the sort thought to have created the Moon. NASA image. (b) The Yucatan peninsula false color image to reflect slope and height of the topography. Subtle but clearly visible is a semicircular trough truncated by the northern coastline that reflects the rim of the crater produced by the impact of the 10-kilometer object that caused the major biological extinction 65 70 million years ago. This is indicated by the arrows. NASA image PIA3379 with author's annotations. (c) Scorched and fallen trees that resulted from the impact of an object at Tunguska.

even more dramatic impact, but it is perhaps less known to the public. It is the impact that made the Moon.

When Kuiper wrote his article about the Moon in 1959, he argued that it formed by the same mechanism as the planets: the condensation of a ring of nebular gas and dust surrounding Earth. "The Earth and moon formed in this protoplanet as a binary planet," he wrote.[5] However, the Moon has long been known to be an anomalous satellite, and his binary-planet idea did not survive the test of time. The Moon's size, dynamics, density, and (most of all) composition are hard to reconcile with co-condensation. Other theories, like capture by Earth or fission from Earth, have also not survived scrutiny. However, there is now widespread acceptance of the idea that the Moon was made by the impact of Earth with a Mars-sized body about 4.5 billion years ago, an idea championed by Hartmann, a student of Kuiper's.[6]

On a much smaller scale, but still highly significant, is the Chicxulub impact 65 million years ago. This impact of a 10-kilometer (6.2-mile) asteroid near the coast of Mexico changed the environment of Earth sufficiently to cause the extinction of all of Earth's animals too large to burrow underground, most notably the dinosaurs that had occupied Earth for more than 200 million years.[7] The tell-tale signal of such an impact is large enrichments of iridium in the rock strata corresponding to 65 million years ago.[8] Iridium (and similar metals) are very rare on Earth but relatively abundant in meteorites. We now know that there have been many major extinction events, and asteroid impacts are a possible cause of several.[9]

On an even smaller scale was the Tunguska event of June 30, 1908.[10] It has become almost legendary, and it has led to some highly imaginative explanations. One suggestion is that an alien spacecraft crashed into Earth. Another is that antimatter collided with Earth. One reason for all the speculation is that while reports of all the major impact effects have been documented, including seismic signals picked up all over the world and a forest of burned and fallen trees covering two thousand square kilometers, there is no crater. The consensus of the scientific community is now that the Tunguska object was a thirty-meter (about one hundred-foot) asteroid with stony meteorite composition that exploded in the atmosphere near Russia's Yeniseysk Governorate (now Krasnoyarsk Krai). The airburst caused the spectacular effects on the surface, and the object never reached Earth.[11]

The Chelyabinsk event of February 15, 2013, was caused by an eighteen-meter (about sixty-foot) object. In this case meteorite fragments were recovered, and we know exactly what the object was like. What was notable about the fall of the Chelyabinsk meteorite is not just the fall phenomena, the fireball, the sounds, the pressure waves, the pits in the ground, and the recovered meteorites, but that twelve hundred people were sent to the hospital mostly because of cuts from shattered window glass.[12] An even smaller event was the meteorite that fell on November 30, 1954, in Sylacauga, Alabama. This meteorite crashed through a roof and gave Ann Hodges a very large bruise on her hip.[13] On August 14, 1992, in Mbale, Uganda, a small meteorite fragment is supposed to have struck a young boy in the head,[14] and there are somewhat ambiguous reports of an ancient Chinese meteorite that killed five or six people.[15]

Meteorwrong

Kuiper once thought that he had found a meteorite, but it turned out to be a mistake. The story is told by Cruikshank and Whitaker in the August 2010 issue of *Meteorite*

Magazine.[16] Apparently, in July 1960 Kuiper was called by a farmer who found a large rock in a cornfield where nothing was present a few weeks earlier. Kuiper drove out to the farm, gave the farmer one or two hundred dollars, and returned the stone to Yerkes Observatory in the trunk of his car. He then called the several meteorite researchers located at the University of Chicago, who agreed to drive to the observatory to take a look, but before they could arrive Whitaker found fossils in the stone and the two astronomers concluded it was a block of limestone. Cruikshank, a student of Kuiper's at the time, notes that the incident was too embarrassing for his students to ever mention in Kuiper's hearing.

Asteroids and Techniques

The detection of asteroids had a long history when Kuiper became interested in them. When Giuseppe Piazzi at the Palermo Astronomical Observatory discovered 1 Ceres in 1801,[17] it was widely heralded as the "missing" planet between Mars and Jupiter predicted by Bode's law. However, three more objects were quickly found, 2 Pallas by Heinrich Wilhelm Olbers in 1802, 3 Juno by Karl Ludwig Harding in 1804, and 4 Vesta by Olbers in 1807.[18] It was then presumed that these four asteroids were the fragments of the missing planet that had exploded. For forty years these were the known asteroids. Then, starting in 1845 with the discovery of 5 Astraea by Karl Ludwig Hencke, a long list of astronomers, each beating the record number of discoveries of their predecessors, increased the number of known asteroids to thousands before Kuiper became interested in the subject in the 1950s.

Techniques developed steadily from 1800 to 1950. Visual observations were augmented by photography in the mid to late nineteenth century. The plates produced were then examined using the blink comparator to detect moving objects like asteroids.[19] The blink comparator became obsolete when digital methods of performing the same function became available, but it was indispensable for almost a century. Electronic methods for measuring light intensity (or "photometry") came along in the early twentieth century and eventually became automated when digital techniques for analyzing the data became available.[20]

A Statistical Sample

Kuiper came out of his study of the origin of the solar system with many ideas, and over the course of the next few decades he turned his attention to almost every solar-system object. His views on the early behavior of the solar system were highly honed, albeit somewhat speculative at times, as he admitted. His approach is exemplified by a passage in his article on satellites, comets, and interplanetary material published in the *Proceedings of the New York Academy of Sciences* the same year as his paper on the origin of the solar system.[21] Using the simplest concepts and basic equations, he wrote, "For the remainder of this paper these general considerations suffice, without detailed numerical analysis." He then launched forth into a detailed description of the process of planet formation. He argued that planets formed by condensation and accretion in a cooling solar nebula, and that large satellites were formed the same way while small satellites were essentially shed from the outer fringes of the protoplanetary cloud that formed the planet. Everything was governed by gravity and gravitational instability.

Concurrent with his article on satellites, comets, and interplanetary material was a paper on asteroids in which he argued that the original asteroid belt also formed by direct condensation and accretion. However, the early solids were prevented from accreting into a planet by Jupiter, whose mighty gravity kept separating the objects we now refer to as asteroids. The problem is the regularity with which Jupiter and the asteroids interacted; astronomers refer to it as "orbital resonance." After condensation and accretion into, for example, one-hundred-kilometer objects, they collided and produced even more fragments.[22] However, Kuiper argued, any detailed understanding of the formation and history of the asteroid belt was limited by the number of asteroids known. He stressed the need for a statistically significant number of observed and characterized asteroids. In the early 1950s, following a suggestion he made in 1949, he set forth to study these objects, creating a program that involved many junior colleagues, students, and temporary visitors to his laboratory. The work was funded by a grant from NSF and resulted in a string of papers between 1954 and 1960, when Tom Gehrels took over the project. It is interesting to note the coming and going of various authors, which reflects the dynamics of the group during this period.

First out of the gate were two 1954 papers Kuiper coauthored with Ingrid Groeneveld.[23] Born in Berlin in 1921, Groeneveld was at Yerkes on leave of absence from the Heidelberg-Königstuhl State Observatory. The first of their two papers concerned observations with the McDonald 10-inch Cook telescope between November 1949 and April 1953.[24] With a field of twenty by twenty-four degrees, it was perfect for asteroid hunting. Over twenty-seven days, fourteen light curves—plots of light intensity against time—were obtained for nine asteroids.[25] From the light curves the asteroid's rotation rate could be determined, and sometimes an indication of shape was possible. The second 1954 paper with Groeneveld concerned observations made in January 1954 using the 82-inch telescope, with Harris and Shatzel assisting.[26] The paper compared the color of the asteroids with those of with stars and satellites.

In 1949 Shatzel obtained data for 44 Nysa with help from Hiltner and Groeneveld. They found a light curve that gave a period of $6^h 25^m$ and suggested an angular appearance. As described earlier, Shatzel was dropped from the graduate program at the

(a) Kuiper with his suspected meteorite. Courtesy of Dale Cruikshank. (b) Ingrid Groeneveld (left) with Sarah Kuiper. Courtesy of Matthew des Tombe. (c) Albert V. Shatzel (right) and F. Wagner Schlesinger, Alder's director, (left) show part of the Zeiss projector of the Adler Planetarium to a group of boy scouts. Adler Planetarium photograph. (d) Ingrid van Houten-Groeneveld in 2005 with the cabinet housing the collection of plates on which she and her husband located asteroids using the blink comparator. Research Institute Leiden Observatory, *Annual Report 2005* (Leiden, The Netherlands: University of Leiden, 2005), 6. Courtesy of Leiden Observatory.

University of Chicago at Hiltner and Struve's instigation and over the emotional protest from Kuiper. By the time Shatzel's singly authored paper appeared in print in 1954, he was associate director of the Adler Planetarium.[27] He resigned a few years later in a dispute over whether the planetarium should perform research or focus exclusively on public education.[28]

The fourth paper in the series was also singly authored, this time by Imam I. Ahmad, a visiting astronomy student from Egypt.[29] The observations were made between January and July 1953 using the 82-inch reflector at McDonald and 12-inch refractor at Yerkes and published the following year. Light curves enabled periods to be determined for 1 Ceres (9^h 4.7^m), 6 Hebe (7^h 16.5^m), and 22 Kalliope (4^h 8.8^m), the period for 22 Kalliope being the shortest yet known. For 8 Flora the light curve was flat, suggesting a spherical shape, so a period could not be determined. The fifth paper was singly authored by Gehrels and concerned the light curve and phase curve for 20 Massalia.[30] The brightness of a planet or asteroid's surface depends not only on its size and distance from the Sun, but also the angle between the Sun, asteroid, and telescope. Objects are brightest when that angle is zero and dim as the angle increases. The dependence of magnitude on the angle depends on the nature of the surface—rougher surfaces have greater drop in magnitude with phase angle.

Gehrels was Kuiper's second graduate student, after Harris. He started in 1952, and for his thesis project Kuiper gave Gehrels more than twelve hundred plates and told him to determine the magnitude of the asteroids. The project turned out to be a lot more complicated than it at first appeared, and it was in its execution that Gehrels discovered the phenomenon described above, which he called the "opposition effect." The results were published in the sixth paper, which came out in 1957.[31] The survey of magnitudes included about three hundred asteroids that had previously shown considerable disagreement in the literature.

In July 1953 an exchange of letters between Jan Oort and Kuiper resulted in Cornelis J. van Houten (usually called Kees van Houten) being invited to Yerkes Observatory to help with the asteroid observations.[32] Oort described Van Houten as "a candidate I can recommend very highly. He is extremely quiet and modest, and he is very shy, so much so the we have wondered what would become of him if he were to leave the observatory. A year at Yerkes would be a fine thing for him." However, Oort was adamant that Van Houten finish his present work on variable stars so he could get his doctorate. Kuiper's decision was to hire him three-quarter time at $2,700 per year (equivalent to $24,000 today) and to suggest he come straight to McDonald in January 1954, where he could also obtain plates for his own work.

Van Houten was born in The Hague in 1920 and died in 2002. He spent his career at Leiden except for the time he was at Yerkes (1954–56).[33] He obtained his first degree at 1940, but his studies were interrupted by the war. He succeeded in obtaining his PhD in 1961. After his two years at Yerkes he returned to Leiden Observatory with Ingrid Groenevald. Two years later Kees and Ingrid were married.

The seventh paper in the asteroid series was jointly authored by Groenevald and Van Houten.[34] It reported light curves for nine asteroids, three of which were new, using data gathered with the 82-inch telescope between December 1955 and January

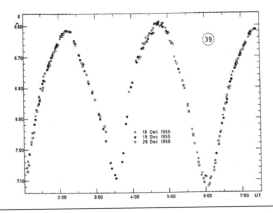

The light curve of Laetitia shows that the asteroid is rotating every 5 hours, 52.2 minutes, the shortest rotation period known to the Van Houten's when their paper was published. The vertical axis is magnitude and the horizontal axis is time. I. Van Houten-Groeneveld and C. J. Van Houten, "Photometric Studies of Asteroids VII," ApJ 127 (1958), 253–73.

1956. They confirmed 15 Eunomia was retrograde, and they also studied 321 Florentina (which was found to have a period as short as $2^h 52^m$) and eight other asteroids.[35] That was the last asteroid paper in the series to appear while Kuiper was at Yerkes, but the group published two more papers after his move to Tucson.[36]

The main instrument for Kuiper's efforts to discover new asteroids was initially the Cook telescope, which could observe asteroids down to a magnitude of 16.5.[37] Kuiper's survey of asteroids with the 10-inch Cook telescope probably enabled the first statistical study of asteroid size distribution. Between 1950 and 1952 the group took 1,094 pairs of eight-by-ten-inch plates, each subtending forty degrees, going twice round the ecliptic, plus another 149 to fill in details. They located all asteroids down to a magnitude of 16.5. In a 138-page article, the group systematically listed the new and missing asteroids, documented their orbits, recorded magnitudes, compared their visual estimates of magnitudes with photometric estimates, and plotted size distributions as a function of distance from the Sun.[38] The paper appeared in 1958 with the disarmingly simple title, "Survey of Asteroids."[39]

As early as 1950, Kuiper had in mind using a superior telescope to push the limit of detection down to a magnitude of 20.5. On March 25, 1950, he wrote to Chandra to say he had obtained permission from Ira Bowen and Edwin Hubble at the Palomar Observatory to use the plates obtained with their Schmidt Telescope for a statistical study of faint asteroids to complement bright-asteroid work with the Cook telescope.[40] In 1960, after his move to Tucson, Kuiper handed this work over to Gehrels, by then a professor at the University of Arizona. Gehrels took plates at the 48-inch

Schmidt telescope at Palomar Observatory and sent them to Groenevald and Van Houten in Leiden, who used the blink comparator to locate the asteroids.[41] They discovered several thousand, which they published in a supplement to *Astronomy and Astrophysics*.[42] In 2005 the famous collection of plates was transferred from Leiden Observatory to the Astronomical Calculation Institute in Heidelberg, where it fills a large fraction of an eight-by-four-foot steel cabinet.[43]

On November 3, 1970, Van Houten wrote to Kuiper, "We would like to name asteroid (1776) 'Kuiper.' Since you initiated the McDonald Survey and the Palomer-Leiden Survey we consider it most appropriate that a minor planet be named after you. Please let us know whether you agree with this proposal." On November 9 Kuiper replied, "Your suggestion came as a very pleasant surprise. I deeply appreciate your proposal and I shall be honored to accept it."[44]

An Impacted Earth

With the exploration of the solar system by spacecraft, and the observation that almost every solar system object has been heavily impacted by asteroids (and presumably comets), it became clear that impact was an important, widespread process in the history of the solar system.[45] The iridium enrichment at the Cretaceous–Paleogene geological boundary—formerly known as the Cretaceous-Tertiary (K-T) boundary—which corresponds to 65 million years ago, convinced most scientists that an impact had caused a major biological extinction on Earth.[46] Then in 1994 the world's astronomers watched as comet Shoemaker-Levy 9 impacted Jupiter.[47] As a result, the U.S. government established a national program to detect threatening asteroids. The Spaceguard Program was proposed in 1992 and funded by NASA starting in 1998. The program provided funds to the Air Force's LINEAR (Lincoln Near-Earth Asteroid Research) project operating at Lincoln Laboratory's site on the White Sands Missile Range near Socorro, New Mexico. Starting in 2005, it also funded the Catalina Sky Survey at the University of Arizona using the refurbished 61-inch telescope Kuiper built in 1961.[48] The target was to detect 90 percent of the near-Earth asteroids larger than one kilometer. The number of known near-Earth asteroids is now approaching twenty thousand.

CHAPTER 11

Mountain Man

Observing the Heavens

K UIPER WAS NEVER FAR from telescopes. He was first and foremost an observational astronomer. Some of the most informative photographs of the man were taken of him standing by a telescope. At Fort Davis he helped in the final stages of construction of McDonald Observatory and bringing the 82-inch telescope into service. There is a beautiful model of the observatory in its visitor's center that includes small reproductions of people working around the building. Standing in the opening of the dome is a prominent figure with a striking resemblance to Gerard Kuiper. Kuiper had a lifelong passion for bigger and better telescopes and better locations for those telescopes. In March 1953 he sent a graph to Chandrasekhar, on which he plotted telescope aperture against time.[1] This graph showed that a 400- or 500-inch telescope should be possible by about 1985. Sure enough, in the 1990s telescopes in this size range started to appear. The Keck Observatory on Hawaii's Mauna Kea houses two of them; they came into operation in 1993 and 1996. With a little exaggeration, astronomers could argue that the observatory sites in Chile and Hawaii, two of the most important observatory sites in the world, would not be what they are today if not for Kuiper. Airborne astronomy would not be as important as it is today without Kuiper. The story of Kuiper and his telescopes is a story of success, controversy, and bitter disappointment.

Memorandum on Large Telescopes

A confidential memorandum, more like a seven-page report with a one-page supplement, lays out in detail Kuiper's attitudes and thoughts about large telescopes. Kuiper

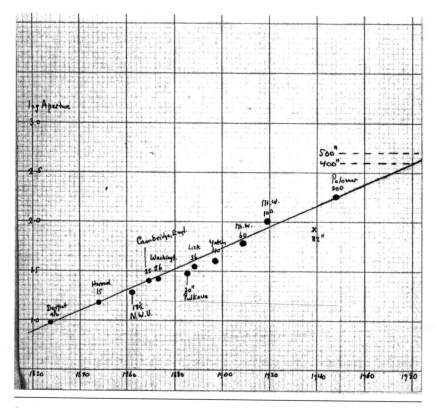

Graph produced by Kuiper in 1953 describing the growth in telescope aperture with time predicting 400-inch telescopes by 1980. The graph was probably originally attached to the memorandum UChSC box 19, folder 21, Kuiper to Chandrasekhar, March 26, 1953. Courtesy of Special Collections Research Center, University of Chicago Library.

sent it to Chandrasekhar on March 26, 1953, in his capacity as chairman of the astronomy department, for discussion with Walter Bartky, former astronomy professor and now dean of the Division of Physical Sciences.[2] In his response, Chandrasekhar writes, "I doubt if anyone else in our group has such clear ideas on the subject as you have set forth."[3]

Kuiper writes that a 400- or 500-inch telescope is technically feasible, but the question is whether telescope size advance in conservative steps, exhausting the capabilities of each telescope before advancing to the next. The physics community does not think so, believes Kuiper, who advocates a bold push of technology to the limit. This would be reasonable if a very good case can be made. Big apertures mean greater light-gathering power and higher resolution. He suggests that stars would appear as disks, asteroids would appear as nonspheres,[4] planetary satellites would come within range of "adequate study," and better images could be obtained of the Moon and planets' surfaces. Several types of program would be possible: visible spectroscopy and radial velocities; infrared astronomy; radiation measurements; photoelectric and polarization measurements; visual work on close double stars, stellar disks, planets, and satellites; direct photography with and without filters on special projects. Meinel's studies, cited by Kuiper, suggest that optimum for infrared work would be a light-collector-type telescope, which has lowest construction cost, lowest operating cost, and allows conventional methods for observing.

Kuiper writes that Prescott, Arizona, or McDonald Observatory are the best locations for a new big telescope, and if it is placed at an existing observatory site, where it can share accommodation and housekeeping resources, only three people are needed to operate the observatory—a superintendent, an optician, and a technical assistant—with an annual budget of $50,000, exclusive of research costs. Today this would be about $500,000. Such a large telescope could only be used by the most experienced observers, of which there are only one or two. It would be managed by Meinel, the Chicago astronomers, and men like Bowen, C. Donald Shane (director of Lick Observatory), and Lyman Spitzer (director of Princeton University Observatory), with Oort representing foreign observers. Funds could be sought from the Ford Foundation, if in Arizona, or Texas benefactors, if in Texas. The supplemental note favors Arizona over Texas because of high winds and atmospheric pollution at McDonald Observatory.

In his cover note that accompanies the memorandum Kuiper adds a handwritten postscript suggesting that Struve not be informed of this matter but that Roger Putnam, a "powerful man" and major benefactor for the Lowell Observatory, could help address the management issues.[5] He ends, "I hope that you will be able to prevent the project from being killed." Nothing more is heard of Kuiper's 500-inch telescope, so presumably it was, in fact, "killed." However, this memorandum succinctly describes the factors Kuiper cares about as he enters into a phase of major observatory building.

Criteria for an Excellent Observatory

Kuiper was driven by science. He wanted better infrared spectra because it would provide access to improved information on the nature of the atmospheres of the planets and the Sun. However, he constantly fought against the technical limitations of his instruments and the water in Earth's atmosphere, which produced a great many strong absorption lines. For the instruments, he kept his ears to the ground and waited for new developments, quickly assimilating new ideas and recruiting the best people to develop and exploit those new ideas. An example of his reasoning and the nature of his exchanges with NASA's William E "Bill" Brunk are given in a letter dated October 30, 1968: "Since we are succeeding reasonably well with this approach at the Catalina Observatory, I have become increasingly convinced of the need of going still higher which will definitely get us over the hump. I have considered both Mt. Lemmon, 8 miles from the Catalina Observatory, 9160 ft (2800 meter) elevation, the site of a radar base run by the Air Defense Command; and Mt. Agassiz, 12,000 ft (3650 meter), north of Flagstaff."[6] Presumably Brunk knew that the Catalina site is at 8,450 feet (2,600 meters). Kuiper goes on to explain to Brunk how he will try the infrared solar spectrometer and a Johnson-type 5-foot telescope at Mount Lemmon, north of Tucson.[7] "Ultimately, we shall need an IR telescope larger than 60-inches aperture and I have discussed with President Harvill of the University of Arizona steps that may be taken to acquire such an instrument during the next few years. It will probably be necessary to approach more than one potential source of support for such an instrument."

To combat the interference from atmospheric water, as determined from a hygrometer or water bands in the infrared spectrum, there was one option: to get as high as possible to minimize the amount of water-bearing air between the telescope and space. Kuiper summarized the situation in a letter to Brunk dated June 24, 1967.[8] The choices were to go to mountain tops, use aircraft, use balloons, and, finally, use spacecraft. He used all of these. However, the cost and complexity increased along this series, so he spent most of his energy pursuing better ground sites. Thus when it came to siting large telescopes, subjective or intuitive judgments were no longer adequate. The usual criteria beyond access and political stability were the latitude (which determined what part of the sky was visible), number of nights that were clear enough to use highly sensitive light-detecting equipment, the humidity and elevation (as discussed above), wind speeds (since wind can affect the structures of the telescope and affect seeing), and, most of all, the astronomical seeing.

Mount Wilson Observatory and Palomar Observatory

When MacDonald Observatory came online in 1933, it had the second largest reflecting telescope in the world. The largest was the 100-inch Hooker telescope at Mount Wilson Observatory in the San Gabriel Mountains near Pasadena, northeast of Los Angeles. Prior to the Hooker telescope, the largest was the 60-inch, also at Mount Wilson, which became operative in 1905. At 5,710 feet (1,740 meters), Mount Wilson is above the unsteady, polluted air of the Los Angeles Basin. The Hooker telescope remained the largest telescope in the world until 1949 when the Hale 200-inch (508-centimeter) reflecting telescope was constructed at the Palomar Observatory.

The Hale telescope, built by the California Institute of Technology (Caltech) with a $6 million grant from the Rockefeller Foundation, was the largest until two years after Kuiper's death in 1973. The observatory is in San Diego County in the Palomar Mountain Range at an altitude of 5,617 feet (1,712 meters). While Kuiper made use of both Mount Wilson and Palomar for his research, these were both established facilities when in the 1950s Kuiper began to take an interest in the creation of bigger facilities at better locations.

The Mountaintops of the Andes

According to Victor Blanco, who has written the history of the Cerro Tololo Inter-American Observatory, in 1960 there were only ten astronomical observatories in the southern hemisphere, compared to the eighty-eight in the northern. The southern observatories had only 10 percent of the light-collecting power of the northern observatories.[9] Interest by northern astronomers in making observations in the southern hemisphere prompted several expeditions. In Africa, the Royal Observatory, Cape of Good Hope, had been founded in 1820. Herschel carried out his famous survey of the southern stars and nebulae from the Cape between 1834 and 1838. In 1849 a small observatory was established on Cerro Santa Lucia in Santiago, which eventually became Chile's National Astronomical Observatory. Then in 1903 astronomers from Lick Observatory erected a 36-inch (92-centimeter) reflecting telescope on Cerro San Cristobal, Chile, which took ten thousand stellar spectra over the next twenty-five years. The telescope is still in use by the Universidad Católica de Chile. Finally, a Harvard College Observatory expedition to Chuquicamata, Chile, during 1922 and 1923 obtained spectra of stars in the Magellanic Clouds, which remain standard reference material today.

In 1959 Professor Federico Rutllant of the University of Chile approached Kuiper and Hiltner about establishing a cooperative observatory in Chile.[10] Rutllant's idea was to establish a bigger facility at the site of the existing observatory in Santiago. Kuiper received funds from the Air Force for a site survey with the promise of support for a 40-inch (102-centimeter) telescope.

Jurgen Stock had been hired by Kuiper to be resident astronomer at McDonald Observatory, but before he could settle in Kuiper persuaded him to go to Chile to run the site survey.[11] Stock obtained his PhD in Hamburg in 1951 under the supervision of Otto Heckmann, who later became the first director general of the European Southern Observatory. After some years in Cleveland, and one year at Boyden Observatory, South Africa, Stock moved to McDonald. As soon as he got to Santiago, he saw that it was no place for an observatory. The Santiago observatory had good skies when it was established, but by that point it was engulfed by the city. With Kuiper's and air force approval, Stock began looking for an alternative site outside the city. He considered Cerro Robles, where the Soviets later put their Maksutov telescope. But feeling he could do better, he kept looking farther north, testing sites as he went using a small telescope and spectrometer until he reached Cerro Tololo (which has an elevation of 7,241 feet, or 2,207 meters), near La Serena, and Cerro La Peineta (9,570 feet, or 2,916 meters), near Copiapo Peineta, a long ridge whose name means "comb." Spending nearly three years on horseback, on mountain trails unsuitable for vehicles, Stock checked almost a dozen mountains. He made notes every day of atmospheric conditions, astronomical observations, problems with the mules, the progress in the construction, some shelter on the mountains, the need for a support team to bring food and water, the conduct of the local people, and so on.[12] Each time Stock was back in Vicuña, his base of operation, he produced a lengthy description of his activities and observations that were circulated as the "Stock Reports" to interested members of the astronomical community.

The agreement under which Stock's site surveys were conducted involved the three universities: Chile, Chicago, and Texas. Texas was still a player because the Chicago and Texas astronomy departments were still acting as a joint department. When Kuiper left Chicago to go to Arizona, the University of Chicago was left in a predicament.[13] They had major a project whose principal investigator had left the university.

At the AAS meeting in August 1959 in Toronto, Kuiper approached Shane,[14] director of Lick Observatory and president of the Association of Universities for Research in Astronomy (AURA), explaining the situation and asking them to take over managing the project.[15] AURA was founded October 10, 1957, with the encouragement of NSF, by a group of seven U.S. universities: California, Chicago, Harvard, Indiana, Michigan, Ohio State, and Wisconsin. The original task was to manage the National Optical Astronomy Observatory at Kitt Peak. A stumbling block to AURA taking over the construction of an observatory in Chile was that the University of Texas would not agree to terminate the present agreement with the University of Chicago unless they had a role in the new arrangement. But the University of Texas did not qualify for membership of AURA, which required that the department had four astronomers, a PhD program, and an observatory. The University of Texas had an observatory and one astronomer. For many decades the home for astronomy in the University of Texas was the department of mathematics and astronomy, which was formed in 1925, a year before the McDonald bequest. The first astronomer hired was recommended to President Painter by Struve. Frank Edmonds, a recent student of Struve's, joined the Texas faculty in 1952.

The problem of Texas's participation in the management of the Chile observatory was overcome in discussions between AURA; William Morgan, who was acting chairman of the Chicago department after Kuiper left; and Edmonds for the University of

Texas. The solution was that Texas would become a member of AURA through its affiliation with Chicago but without voting rights.[16] On the AURA letterhead the Universities of Chicago and Texas were on the same line in a vertical list of members. The dean of the Texas graduate school, Gordon Whaley, was appointed as an administrative consultant to the AURA board and attended all the board meetings, including executive sessions. The University of Texas had its foothold in the southern hemisphere.

Two sites emerged as favorites, Tololo and Morado, which is sixty miles (one hundred kilometers) southeast of Santiago. Tololo was finally chosen because the site had superior seeing and was downwind from the city of Vicuña. Another factor favoring Tololo was the view. If Stock was going to live on the mountain, so the rumor went, he wanted to have a good view. Decisions concerning the location and design of the observatory buildings would take that view into account.

While the Chile site survey was spending Air Force funds, NSF agreed to contribute to the costs of a telescope. AURA decided that a 40-inch telescope was too small for a major new observatory and combined the funds from the two sources to work toward a 60-inch telescope. Kuiper did retain some interest in the Chile telescope after moving to Arizona, and Alika Herring performed some site tests between February 19 and March 6, 1963.

Herring was Kuiper's expert at determining the quality of potential observatory sites.[17] He was born in Hawaii, graduated from the University of Oklahoma in 1932, and worked as a mailman and musician during the Great Depression. In the aftermath of Pearl Harbor, he enlisted in the navy, where he was a sonarman. After the war, he worked for the Cave Optical Company, making telescope mirrors in his spare time, until in 1961 he was hired by the LPL. Herring was an avid amateur archaeologist with a deep interest in Native American culture and artifacts, and he had an important personal collection.

There was some suspicion, in light of Kuiper's continued interest in the project, that he was maneuvering to have greater control of the AURA facilities in Chile and Kitt Peak. This is borne out by a letter Kuiper sent to Rutlland on July 23, 1960. In response to a letter from Rutlland expressing concern about the project upon hearing of Kuiper's resignation from the University of Chicago,[18] Kuiper replied, "My natural inclination is to encourage you to take whatever steps you deem appropriate to bring about my continued association with the Chile Observatory."[19] Nothing seems to have come of this, but a month later Kuiper sent what amounted to a plea to Stock to hire his son Paul, then nineteen years old.[20] "We dreamed of the possibility of Paul helping you," wrote Kuiper, who went on to remark that he had in mind a summer appointment, but Rutlland favored a longer period and suggested that period should extend to January, 1962. He ended, "If you can use him it would be a wonderful experience for Paul and his parents would be very grateful."

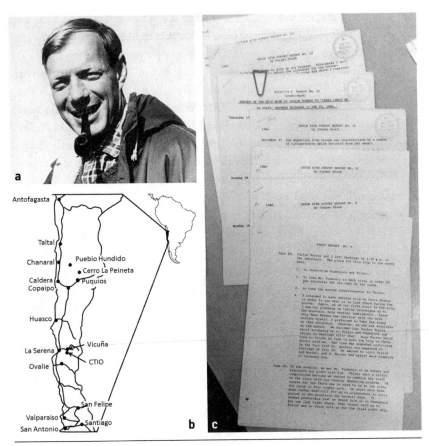

(a) Jurgen Stock. "Jurgen Stock," Cerro Tololo Inter-American Observatory, accessed May 29, 2017, http://www.ctio.noao.edu/noao/content/jurgen-stock. Courtesy of University of Arizona Libraries Special Collections. (b) Map of Chile from Santiago to Antofagasta locating the places mentioned in the "Stock Reports." Author image. (c) The "Stock Reports." Jurgen Stock sent regular reports back to Kuiper (and then Shane when AURA took over responsibility for the development of the Chile observatory) describing his progress in selecting and developing a site for a southern observatory. UASC box 50, file 9, reports from Jurgen Stock describing his progress in selecting a site for the southern observatory. Courtesy of University of Arizona Libraries Special Collections.

Things went well for four or five months, but then there was trouble. The problem was that a husband-and-wife team, William and Sheri Mathias, had been hired to the project. Nothing had gone well for them.[21] They complained to Nicholas Mayall, who was then president of AURA. "I receive the second highest pay on the expedition, yet I have less responsibility than the hired Chileans," William Mathias wrote. He was

(d) In the foreground is Cerro Tololo, near La Serena, and in the distance is the Cerro La Peineta range, near Copiapo Peineta. UASC, box 25, file 8, photographs of CTIO. Courtesy of University of Arizona Libraries Special Collections. (e) Cerro Tololo Inter-American Observatory under construction. UASC, box 25, file 8, photographs of CTIO. Courtesy of University of Arizona Libraries Special Collections.

hired as an astronomer but found himself having little to do other than type Stock's reports. He concluded that he must have been hired to secure the translation services of his Spanish-speaking wife. Thus "she is underpaid for her work, and I am overpaid for mine." Attempts to resolve this with Stock had made things worse, and Stock was telling Sheri and Paul that William's work was unsatisfactory. Worse yet, a report was circulated that amounted to a major assault on Stock. Attached to the report were the signatures of all three: William Mathias, Sheri Mathias, and Paul Kuiper.[22] The report included a statement of resignation, but in a handwritten addendum Paul excluded himself from the resignation. Independently, Paul sent a long letter, which included all the complaints about Stock, to his parents.

Given Kuiper's long struggle to get the southern observatory underway—the total commitment, the endless hours consumed—it is not surprising that Kuiper found it difficult to let go, and he had written to Rutlland to see if he could retain a role. The letter from Paul describing the problems with Stock were more than Kuiper could bare, and he immediately dashed off letters to everyone in authority enclosing a redacted version of Paul's letter. Having received no reply from Mayall and Shane, he wrote to NSF, repeating the accusations against Stock.[23] Kuiper added his own criticisms, complaining that the quality Stock's reports had deteriorated and were now preoccupied with trivia. Further, Stock was not cooperating with the Chilean authorities, was failing to use appropriate fiscal authorizations, and was failing to follow Kuiper's procedures that were calculated to nurture international collaboration. Instead, Stock was favoring a "go-it-alone" policy "like that of a colonial power." Not content with complaining to the authorities at AURA and NSF, Kuiper finally wrote to Whaley that "from Dr. Stock's own reports, (or lack of them), and his personal behavior, I would consider his continued stay in Chile undesirable and very harmful to the project."[24]

The end of 1961 saw an end to Paul's period in Chile and brought peace to Kuiper's mind. On January 11, 1962, Kuiper wrote to W. B. Harrell, vice president of the University of Chicago, that Shane had resolved the difficulties and that the Chile operation "was now running well again."[25] Kuiper may have made overtures to have a hand in running the Chile project after he left Chicago, and Paul may have aggravated Kuiper's position, but AURA remained in firm control. When later asked about Kuiper's role in creating the Cerro Tololo Inter-American Observatory, Shane wrote, "I think that among Gerard's accomplishments this should be counted among the more important ones. It was he who developed the initiative."[26]

Whatever forces attached Kuiper emotionally to the Chile project, they did not last forever. A year or two later, Kuiper found a new cause to devote his considerable energies to. He decided that Herring's site testing at Stock's preferred Chilean sites had been disappointing. On March 3, 1963, he wrote to NASA headquarters advising against investing in the Chile observatory and urging them to support, instead, an observatory on the top of Mount Haleakala on the Hawaiian island of Maui.[27]

The Mountaintops of Hawaii

The story of how Kuiper was responsible for creating the most important observing site on Earth, Mauna Kea on the Big Island of Hawaii, was told in some detail by M. Mitchell Waldrop in an article in *Science*[28] and by John Jefferies in an essay, "Astronomy in Hawai'i 1964–1970," on the Mauna Kea Observatories website.[29] The story has more than a touch of melancholy for Kuiper. It is often the case with NASA that one group will have a good idea, spend considerable time and effort developing and advocating the idea, and then the agency will fund another group to execute the work. That is what happened in the case of the first observatory at Mauna Kea, and in the words of the NASA official involved in awarding the contract to the University of Hawaii, "We went out on a limb."[30]

In the fall of 1962 Kuiper decided to pursue Hawaii as a possible site for major telescopes. Unlike the Sierras and Andes on the mainland, the volcanoes of Hawaii do little to disrupt the prevailing weather patterns, and their summits are high enough to penetrate the dry upper air. First Kuiper considered Haleakala on Maui, whose summit is 3,055 meters (10,000 feet).[31] It is not as high as Mauna Kea at 4,205 meters (14,000 feet), but it had the necessary logistics already in place. It was the site of the University of Hawaii's Mees Solar Observatory, so it had a paved road and a power line. However, after prolonged examination of the site, Kuiper found it unsuitable. Cold air and fog collected in the crater and sometimes billowed up over the summit,

(a) M. Kuiper, J. Akiyama, J. Texereau, G. Fielder, and an unknown person on Mauna Kea, October 12, 1964, in a photograph by A. Herring. Courtesy of the LPL. (b) Kuiper on a cinder cone, Kilauea Volcano, Hawaii, in the 1960s. Courtesy of Dale Cruikshank. (c) N. U. Mayall and Kuiper taken by Dale Cruikshank on Kilauea Volcano the day following the dedication ceremony for the Mauna Kea observatory, June 27, 1970. Courtesy of Dale Cruikshank.

producing turbulent air or fog. But, as Kuiper is quoted as saying, the last thing he saw as the fog closed in on Haleakala was Mauna Kea, standing clear in the sunlight.[32]

It was Howard Ellis, a scientist at the weather station on Mauna Loa, who suggested to Mitsuo Akiyama, manager of the Hawaii Chamber of Commerce, that Mauna Kea would make a fine site for astronomical telescopes. Akiyama wrote to many prominent astronomers, but only Kuiper responded.[33] Kuiper studied maps of Mauna Kea, flew around the summit and saw cinder cones sitting on a lava plateau but, most important, saw no crater to capture warm air as at Haleakala. He consulted scientists at the university, Akiyama of the Hilo, Hawaii, Chamber of Commerce, and Hawaii's governor, John Burns. Burns immediately saw the economic potential for the chronically depressed Big Island and gave Kuiper $50,000 ($400,000 in 2018 dollars) to build a road between the existing road at Hale Pohaku, an old ranger station at an elevation of 9,200 feet (2,800 meters) on the south flank, and the summit. Always hands on, Kuiper sat beside the bulldozer driver to direct and plan the road as they climbed to the test site. The Tucson team erected a small wooden dome, put up a sign that read "Mauna Kea Station, Lunar and Planetary Laboratory," and Herring set up his 12.5-inch reflector with which to check the seeing. The site testing lasted for more than a year, and it finally became clear that, as Kuiper once remarked, "The mountaintop is probably the best site in the world, I repeat in the world, from which to study the moon, the planets, and the stars."

In the 1960s, under the inspired leadership of James Webb, NASA decided to spend about one-third of its budget on science, the remainder going on its single-minded drive to place a man of the Moon.[34] Kuiper asked NASA to fund an observatory with a 61-inch telescope for solar-system studies, to be operated by LPL. In 1963 astronomers interested in the planets found it difficult to get telescope time in the face of competition from stellar astronomers, and NASA concluded that it had to build telescopes for planetary science. It was already funding a good deal of ground-based observation of the moon and planets in support of its flight programs. It had recently funded a 107-inch instrument for McDonald Observatory,[35] and Kuiper just completed a 61-inch telescope at Steward Observatory in Arizona's Catalina Mountains.[36] Kuiper submitted his proposal to NASA, but the agency, while receptive, delayed in giving a response. Then, with Kuiper's proposal in hand, they requested a proposal from the astronomers at Harvard University.

There is considerable speculation about why NASA solicited a competing proposal. Some have suggested that Kuiper was getting too influential, that NASA officials saw him as an empire builder. However, there is nothing unusual about NASA selecting on the basis of management decisions rather than technical merit. When questioned about this, Brunk, NASA's director for planetary astronomy at the time, said only that NASA did not want to go with a single proposal.[37]

Word spread, especially among the four solar astronomers at the University of Hawaii led by John Jefferies, a thirty-nine-year-old theorist specializing in the interpretation of solar spectra and who was on sabbatical at Hawaii from his permanent position at the University of Colorado.[38] When Jefferies and university officials complained to NASA, they were initially told that NASA would choose between the two heavyweights, the University of Arizona and Harvard University, but NASA relented and agreed to receive a proposal from the University of Hawaii.

To everyone's surprise, the proposal from the University of Hawaii was selected. NASA would provide $3 million for the design, development, construction, and installation of an 84-inch telescope, larger and more sophisticated than the one Arizona or Harvard proposed. The state of Hawaii, acting through the university, would put up $2.5 million for the building ($24 million today), the telescope piers and dome, the site preparation, and the power line. This was the customary arrangement NASA made for supporting astronomical observatories. The university also promised to build up its astronomy program.

"It was," said Brunk, "the best proposal, especially considering the University of Hawaii's proximity to the mountain."[39] It was also the biggest risk NASA had ever taken. The contract was signed on July 1, 1965. We now know that NASA had created one of the most important centers for astronomy in the world.

It is not difficult to guess Kuiper's reaction. It is said that he was furious. Kuiper maintained that Jefferies was not competent to build a night-time telescope and that the University of Hawaii was in way over its head.[40] For many years he would claim that John Jefferies had stolen the Mauna Kea observatory from him. But all of this has been questioned by Whitaker, one of Kuiper's closest associates, who claims that Kuiper never spoke against NASA's decision.[41] The telescope was dedicated in June 1970 and began full-time operation that November. Kuiper was invited to the dedication ceremony, and he came, reconciled to his defeat. During the dedication ceremony, Kuiper examined the observatory and the telescope and was said to have been very impressed. Apparently, if Jefferies did not know what he was doing, he had friends who did.

Mexico

While hunting for observatory sites in mainland America, Kuiper would typically rent small planes that were piloted by his friend, John Summer, and in later years his student, Godfrey Sill. He would have the pilot fly so low that he could not only see potential sites, but also feel the wind flow, air currents, and turbulence. He flew over the southwestern region of the United States and Mexico; one trip was over Mount Graham and the Pinacates, while another was down to the southern tip of Baja

California, over Cabo San Lucas, where he and his team met with local astronomers for site testing. They took a small telescope for the purpose.[42] Another was Mount Tequila, near Guadalajara.

Kuiper spent several years searching for a better location for the National Astronomical Observatory for the University of Mexico. The first observatory for Mexico was installed on the balcony of a castle in Mexico City in 1878. In 1929 it was placed under control of the National Autonomous University of Mexico and moved to the western outskirts of the city, near the terminal of one of the metro lines.[43] Soon this location was not suitable, and the observatory moved to Tonantzintla, Puebla State, in southern Mexico, but by 1967 air pollution and light pollution were again problems.[44]

Some of Mexico's highest mountains are in Baja California. Kuiper made several flights over the peninsula, and on at least one occasion took Sarah, Lucy, and her fiancé along for the ride.[45] Eventually he found a site on the summit of the Sierra San Pedro Mártir. Reminiscent of Kuiper's bulldozed road to the top of Mauna Kea, the superintendent of LPL observatories, Arnold Evans, used a bulldozer to carve an eight-mile road to the summit in "nine very long days without respite." Herring was again involved in site testing, using his 12.5-inch telescope. On one occasion an unexpected snowstorm marooned him on the mountain, and he had to be rescued by helicopter.[46]

In the late 1960s the University of Arizona loaned a 33-inch telescope to the observatory—although Kuiper later refers to it as belonging to the University of Mexico—for direct imaging, optical spectroscopy, and photometry.[47] In 1970 a 60-inch telescope with a metal mirror was built by Harold Johnson at LPL and installed with their assistance in the observatory, again for direct imaging, optical spectroscopy, and photometry—and, adds Kuiper, infrared astronomy. The observatory still functions, and after Kuiper's death an 83-inch telescope, the largest in Mexico, was added to the site.

The Observatories That Never Were

Kuiper considered several other sites for major observatories.[48] These were Agassiz Peak in the San Francisco Mountains (1963), Pikes Peak in Colorado (1968–69), Mount Shasta in California (1970), Mount Logan in Canada's Yukon Territory (1970–71), and White Mountain in California (1950, 1970–73). For a variety of reasons, including local apathy and presumably the hostility of some of these environments, they never materialized.

Kuiper invested particular energy in proposing Pikes Peak as a superior observing site, at one time believing it to be superior even to Mauna Kea.[49] Pikes Peak had been seen as a favorable place for an observatory since the nineteenth century and in 1889 was visited for the purpose by American brother astronomers affiliated with Harvard College Observatory, William H. Pickering and Edward C. Pickering. However, it never quite made the cut, and eventually another site William surveyed, Mount Wilson, ended up playing a major role in astronomy at the time.[50] Throughout 1969 Kuiper sent a string of letters to Brunk documenting his efforts at locating an observatory on Pikes Peak. They started with optimism and grew increasingly impatient as the year progressed. The problem was the apathy of the local astronomers, something he found especially difficult to handle.

Observatories in the Sky

Earth's atmosphere not only presents a problem for astronomers interested in infrared wavelengths, but also makes observations in the ultraviolet impossible. At ultraviolet wavelengths it is not the water that absorbs the radiation, but ozone and oxygen. During the 1950s and 60s there was considerable interest in performing experiments from balloons that could reach higher than thirty-five kilometers, above most of these interfering species. The February 1967 issue of *Applied Optics* was dedicated to the topic. In the issue Gehrels described how he had organized observations from a high-altitude balloon in December 1965 and May 1966. Two telescope-gondola systems were provided by the National Center for Atmospheric Research, one with two 3-inch reflectors and one with a 28-inch Cassegrain reflector and two vidicon (video camera tube) cameras.[51] The instrument, which he called the "polariscope," had command and telemetry by radio link and a star tracker to stabilize the instrument platform. Gehrels reported polarization measurements for the moon at 290 micrometers and of interstellar medium at 282 and 220 micrometers. Kuiper considered the work "excellent," but complicated to execute.[52]

A couple of years after Kuiper submitted his Hawaii proposal to NASA, he decided that another means of getting above the water in Earth's atmosphere was to fly telescopes on aircraft. The first attempt was in April 1965, when Kuiper and Pierre St. Amand of the Naval Ordnance Test Station at China Lake, California, flew a military A-3B Skywarrior jet that could accommodate reflecting telescopes of three-to-four-inch aperture.[53] After the arrangements had been made, the flights were taken over by Frank Low and Carl Gillespie of Rice University, who flew three flights in 1966. Low and Gillespie made measurements of the infrared radiation being emitted by Jupiter and Saturn. An important finding was that these planets were emitting twice the energy they were receiving from the Sun.[54] Apparently, the planets are still undergoing gravitational collapse and their gravitational energy converted to radiation. This gives Jupiter and Saturn a star-like property—they are emitting radiation

a

b

c

(a) A Navy Skywarrior in flight in late 1950s. U.S. Navy photograph. (b) The airborne observatory Galileo. Courtesy of Dale Cruickshank. (c) Dedication of the Kuiper Airborne Observatory on May 21, 1975. Left to right, Hans Mark (director of Ames Research Center), Sarah Kuiper, and Carl Sagan. Kassander is seen at extreme right, exiting the field of view. Courtesy of Dale Cruickshank.

(d) Kuiper observing on the NASA Convair 990 (Galileo) aircraft. Courtesy of Dale Cruickshank. (e) The NASA Convair 990 (Galileo) with solar spectrometer in place in July 1968. In the background are A. B. Thompson, G. P. Kuiper, and G. T. Sill. Courtesy of Dale Cruickshank.

rather than just reflecting sunlight. Tragically, the flights were suspended on March 15, 1967, when the Skywarrior aircraft stalled during a routine military mission, resulting in the loss of the entire crew.

The Kuiper-Cashman lead sulfide detector had a spectral range of about 0.75 to 2.5 micrometers. Several similar sulfides have infrared-detecting capabilities, and some are in use today. Indium antimonide can detect infrared radiation to about 5 micrometers, depending on temperature. It is used for the Infrared Array Detector in the Spitzer Space Telescope. Detecting longer wavelengths required a totally different approach, and this was provided by Low when he invented the gallium-doped germanium detector. This detector has a range of 1 to 1,200 micrometers. Low made great use of his solid-state detector, placing it on a string of robotic telescopes. It will be placed on the next of NASA's Great Space Observatories, the James Webb Space Telescope due to be launched in 2019.[55]

However, in the late 1960s Kuiper was already planning for better facilities for airborne astronomy. In December 1966 he made a proposal to NASA headquarters and NASA Ames to use a NASA Convair 990 Coronado for such flights, this time flying mostly out of NASA Ames Research Center's Moffett Field.[56] The proposal to use the aircraft was successful, and funds were forthcoming from the University of Arizona's Space Sciences Committee in Tucson to build a suitable instrument.[57] In April, May, and June of 1967 eight flights were made, the larger plane (named "Galileo") having windows that would allow a 12-inch reflector to be mounted in the cabin. For observing the Moon and Mars, flights out of Moffett Field flew down the coast and out over Pacific Ocean, allowing five-to-six-hour flights at 40,000 feet (12,200 meters).[58] One flight on May 14, to observe Venus, required a route that encircled the Great Lakes and was therefore flown out of Dulles Airport in Washington, D.C.[59] A specially designed high-resolution spectrometer for airborne infrared observations was designed for the flights.[60] In the next few years Kuiper and his group published results from the airborne astronomy for composition of the atmospheres of Venus, Mars, Jupiter, and Saturn.[61] In 1967 and 1968 Kuiper conducted a program to make an atlas of the infrared solar spectrum at high resolution above most of the atmosphere. The atlas was published in ten articles in the LPL's *Communications of the Lunar and Planetary Laboratory*.[62]

The Convair 990 was destroyed in a crash in 1973. On April 12 it was replaced with another Convair 990, dubbed "Galileo 2."[63] After Kuiper's death, the plane was reassigned to testing landing gear for the Space Shuttle, and a Lockheed C-141A Starlifter became NASA's airborne observatory. This much larger aircraft could carry a 1-meter telescope, and it flew between 1974 and 1995.

On May 21, 1975, representatives of NASA and the University of Arizona met for a ceremony to dedicate the next airborne observatory to Kuiper; it was to be known as the Gerard P. Kuiper Airborne Observatory. The observatory housed a 36-inch (92-centimeter) Cassegrain reflector telescope and a spectrometer with a range of one to five hundred micrometers. Richard Kassander, vice president for research, introduced the speakers; John Schaeffer, president of the university, greeted the attendants; Hans Mark, director of the Ames Research Center, performed the dedication; and Carl Sagan gave an address. The participants then ate a buffet supper and toured the observatory. Attached to the plane was a plaque that read, "Our expanded knowledge of the Solar System owes much to Dr. Kuiper's innovative applications of modern technology. This memorial honors the distinction with which he promoted and practiced airborne astronomy. The Gerard P. Kuiper Airborne Observatory adds a new dimension to the strong, Earth-based national astronomy program that he influenced not only by the many scientific achievements of his career but also by his special emphasis on the role of the individual university scientist in support of the United States space effort."[64]

CHAPTER 12

Eyes on the Moon

THE AIRBORNE STUDY OF the solar photosphere's composition and the planets' atmospheres by infrared spectroscopy was one of two major research efforts Kuiper had underway as the 1960s unfolded. The other effort concerned the Moon. With the invention of rocketry and the world rushing into the space age, the Moon was the top target. The United States and Soviet Union squared off for what many feared would be the next world war, which would be a nuclear war. One of the prospects that faced a nervous world was the Moon becoming a strategic base for spy telescopes or even weapons. The U.S. Air Force knew this. They wanted to be prepared to establish a military center of operations on the Moon should it ever become necessary, and first they needed maps. Kuiper's thoughts were already there. He had the idea of collecting the best lunar photographs and rationalizing lunar nomenclature. In his book on the history of planetary science in the twentieth century, Ronald Doel paints a colorful picture of a hot, nervous Kuiper, bulging briefcase under his arm, arriving in Dublin for the ninth General Assembly of the IAU in 1955.[1] Kuiper's mission was to solicit the help of the international astronomical community for his next venture. His proposal drew very little attention; however, it did attract the help of Whitaker, who was able to bring enough other people into the project. Eventually, Kuiper and his associates found themselves mapping the Moon on the eve of the Apollo Moon landings.

Mapping the Moon

The history of lunar cartography has been nicely described by Whitaker in his book, *Mapping and Naming the Moon*.[2] Early cultures saw the mixture of light and dark regions on the Moon in many ways—for example, as a rabbit, a dog, a bush and a little man, and as a face—but William Gilbert in the last half of the sixteenth century first produced a map we would consider modern. It featured twelve regions of seas and islands, half of which we now recognize as the various maria. Thomas Harriot made the first telescopic map in 1609, a year before Galileo published the *Sidereal Messenger* containing many now-famous sketches of the Moon. After Galileo there followed many attempts to describe the Moon through sketches at the telescope, culminating in Pierre Gassendi's unpublished map and Claud Mellan's engravings (1635–37).

Whitaker credits the birth of selenography (the mapping of the Moon) to Michael Florent van Langren in 1645, who named 325 features. Although many of Van Langren's names moved to other features, 168 survive today. Two years later Johannes Hevelius produced his famous images of the Moon, which included the effects of libration, seeing slightly more than half the Moon because of the "wobble" of its orbit. Whitaker discussed the accuracy of the Hevelius maps, describing some as "quite remarkable," while others have mysterious mistakes. Maps continued to gradually improve, and by the end of the seventeenth century Giovanni Domenico Cassini produced maps that were widely reproduced. Nevertheless, it was Hevelius' maps in his *Selenographia* that remained the most popular.

The eighteenth century saw huge improvements in telescopes and lunar maps. Tobias Mayer produced a small-scale map, and John Russell made a pastel painting and a small globe. Johann Hieronymus Schröter planned to produce a full-scale map but soon decided it was too ambitious—he instead published studies of specific features and titled his book *Selenotopographische Fragmente* (*Selenotopographical Fragments*).[3] Whitaker suggests that lunar cartography came of age with Wilhelm Lohrman, who divided the face of the Moon into twenty-five squares and produced detailed drawings of each region in something of a precursor to the Kuiper photographs 130 years later. Lohrman immediately found he had competitors in Wilhelm Beer and Johann Mädler, who were producing similar maps and introducing new names.

The Victorian era saw continued improvement to lunar maps by using photography as an aid to visual observations. The period also saw the production of plaster models by James Nasmyth in the 1870s that were often photographed for use in popular books.

By the end of the nineteenth century, there was considerable confusion over the names of lunar features. Under the auspices of the Royal Astronomical Society, Mary Blagg attempted to remedy this by publishing the *Collated List of Lunar Formations* in 1913. After the interruption by World War I, the effort was assumed by the IAU, which was founded in 1919.[4] The IAU is composed of thirty-three "commissions," and Commission 16 was tasked with arbitrating on lunar nomenclature. When the IAU published *Named Lunar Formations* by Blagg and her colleagues in 1935, with fuzzy

(a) Ewen Whitaker with H. P. Wilkins at the Greenwich Observatory in the mid-1950s. Courtesy of the British Astronomical Association (http://www.britastro.org/). (b) Alan Lenham. R. J. McKim, "Obituary of Alan Pennell Lenham 1930–1996," *Journal of the British Astronomical Association* 107 (1997): 50. Courtesy of the British Astronomical Association, (http://www.britastro.org/). (c) Ewen Whitaker with Dai Arthur. "David 'Dai' Arthur (Lunar Scientist)," accessed December 8, 2016, https://the-moon.wikispaces .com/D.W.G.+Arthur#David%20%22Dai%22%20Arthur. Courtesy of the LPL.

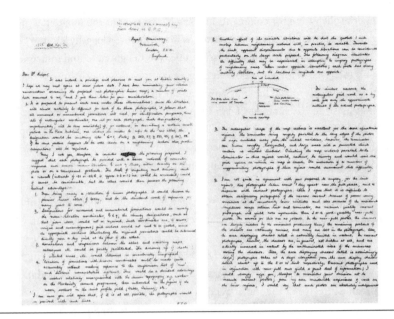

Letter from Ewen Whitaker to Kuiper, following up on his proposal to produce an atlas of the Moon. The letter is reproduced in Whitaker, "Clash of the Titans (or Why and How LPL was Born)." Courtesy of the LPL.

pencil drawings and uncertain, handwritten labels, it had been the result of a thirty-year effort. It was described as a stop-gap measure by the authors, but at least it had the blessing of the IAU. From then on, names would have to be approved by the IAU, which in the 1940s and 50s was not slow to decline proposals.

Kuiper's Lunar Mapping Program

In 1955 Commission 16 reported, "The Commission recommends to the Union that at the present—and particularly pending the completion of the proposed photographic map of the Moon—no official recognition shall be given to additional lunar nomenclature." In 1955 the chairman of Commission 16 was Kuiper, the person proposing the new photographic map. At the ninth General Assembly of the IAU, Kuiper distributed two memoranda to members of the Commission. One dealt with some current lunar nomenclature problems, the other, "Considerations on a New Photographic Lunar Map," emphasized the need for large-scale, high-resolution photographs of the lunar surface.

Kuiper's solicitation for comments and suggestions about the lunar atlas elicited one reply. Several months after the meeting, Whitaker wrote a lengthy letter to Kuiper discussing Kuiper's plans for a lunar atlas.[5] Kuiper was delighted to find someone who shared his interests. Whitaker was invited for a two-or-three-month visit to select and enlarge the best McDonald lunar negatives and help with editorial work on the atlas. Whitaker's visit in the summer of 1957 must have been a success because on January 30, 1958, Kuiper wrote to Whitaker inviting him to Yerkes for a one-year appointment to work for the lunar-atlas project. The salary he offered was twice Whitaker's salary in the U.K. Although the initial appointment was for a year, it most likely would be extended indefinitely.

During World War II Whitaker was engaged in ultraviolet spectroscopy of material used in the manufacture of pipes that were being strung under the English Channel to provide gasoline to the allied armies in continental Europe.[6] After the war he applied his experience with ultraviolet spectroscopy to stellar spectra at the Royal Greenwich Observatory. In 1951 he became interested in lunar studies, and in a few years became director of the Lunar Section of the British Astronomical Association (BAA). In 1954 he published the first accurate chart of the lunar South Pole. Whitaker was excited to receive Kuiper's invitation and readily accepted. The Astronomer Royal, Richard Woolley, who is best known for his disbelief in space flight and rocketry, was uncooperative, having no affection for the Moon or Kuiper. Kuiper had to spend considerable time sorting details and persuading Woolley to give Whitaker leave, and in the end Whitaker had to surrender a year's worth of leave to visit the United States. Whitaker arrived at Yerkes in October 1957.

Also involved in the work was another British amateur lunar observer, Alan Lenham. Lenham had written articles for the *Journal of the British Astronomical Association* on Mare Humorum and other topics of lunar topography.[7] He was offered a position at Yerkes and McDonald Observatories by Kuiper in 1956 and eagerly set to work making lunar observations with the 82-inch telescope.

When Kuiper, Lenham, and Whitaker were discussing lunar matters, Whitaker mentioned another member of the BAA Lunar Section, David Arthur, commonly called "Dai."[8] Arthur was a photogrammetrist for the U.K. Ordnance Survey, making measurements from photographs to determine exact positions for government map making, but he had a strong interest in the Moon.[9] He self-published five issues of *Contributions to Selenography*, which were based on measurements made on glass plates obtained from Mount Wilson and Lick Observatories. The first issue was a list of 1,400 crater diameters measured by James Young. The second gave precise positions for 490 features in Mare Imbrium. The third listed 300 features in Oceanus Procellarum. The fourth gave formulas for converting the x-y positions observed on a lunar plate to their positions at zero libration. The fifth listed 580 features around the impact crater Copernicus. It is probable that Kuiper came to hear about these from a 1956 article by Joseph Ashbrook in *Sky and Telescope*.[10] These studies demonstrated exactly the skills Kuiper needed for his lunar atlas project. In January 30, 1958, Kuiper wrote to Arthur inviting him to Yerkes with Whitaker for a one-year appointment to work for the lunar atlas project. Arthur played a leading role in lunar work at the University of Arizona for nine years, after which he went to the U.S. Geographical Survey's astrogeology branch in Flagstaff.

In April 1957 Kuiper received a $5,750 ($57,500 in 2018 dollars) grant from NSF to finance the Lunar Atlas Investigations Research Project; these funds paid for Whitaker's trial visit. Then on April 1, 1958, he received a one-year, $39,000 ($390,000 in 2018 dollars) grant from the Air Force. The budget, now in the University of Arizona archives, shows the scale of the operations.[11] Funds were provided for four research associates, a photographic specialist, a secretary, three graduate students, three photographic assistants, and five months' support for Kuiper.

Photographic Lunar Atlas

At the time of the Dublin meeting, Kuiper was making large prints from original high-resolution negatives. The few good negatives that Yerkes had were printed by Joe Tapscott, the observatory's photographer. For the roughly 130 Lick negatives and more than 500 from Mount Wilson, Kuiper would select the best and make prints. Kuiper and Lenham then produced 300 negatives using the 82-inch telescope at McDonald, and 30 more were obtained from the Pic du Midi Observatory. Since each negative would translate to six pages in the atlas, and each page was fourteen by seventeen inches, the full moon would be one hundred inches in diameter. The operation was expensive. In spring 1957 Kuiper went to Caltech to select the best Mount Wilson glass-plate negatives and then to Lick Observatory, where over the next two months their photographer, Rulon E. Watson, made about six hundred enlargements.

By the time Whitaker and Arthur arrived at Yerkes, the work, titled *Photographic Lunar Atlas*, was getting close to ready for the printers. Tapscott was assigned the task of traveling between Yerkes and the printer in Chicago to deliver the prints and ensure the quality of the reproductions. Kuiper generated publicity for the atlas, so by the time it went into production it was well anticipated. There were two versions. A loose-leaf boxed version was published by the University of Chicago Press for civilian use,[12] and a bound version was published by the Air Force for government use.[13] Each was large (eighteen by twenty-two inches) and heavy (twenty-two pounds), and the civilian version cost thirty dollars. They now sell for about a thousand dollars.

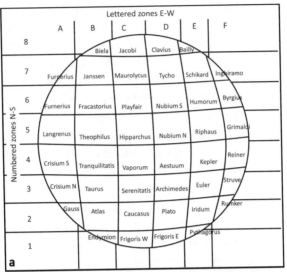

(a) For the *Photographic Lunar Atlas*, the Moon was divided into forty-four regions and two or three images for each region were published in the atlas. (b) An example sheet from the *Photographic Lunar Atlas* is region D-3, near the eastern rim of Mare Imbrium. The crater Archimedes is in the lower-left corner.

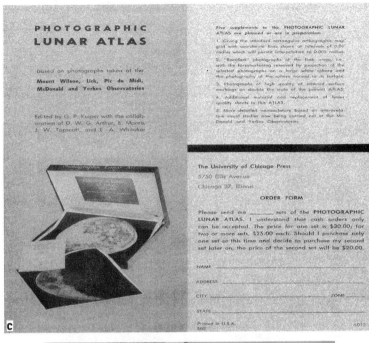

(c) Flyer advertising the *Photographic Lunar Atlas*. UASC, box 38, file 22. (d) Kuiper with sheets from the *Photographic Lunar Atlas*. Courtesy of Dale Cruikshank.

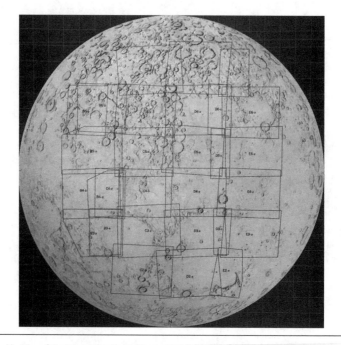

Orthographic Atlas of the Moon index map. Courtesy of University of Arizona Press.

The formal full title is *Photographic Lunar Atlas. Based on photographs taken at the Mount Wilson, Lick, Pic du Midi, McDonald and Yerkes Observatories,* and the authors' credits read "Edited by G. P. Kuiper with the collaboration of D. W. G. Arthur, E. Moore, J. W. Tapscott and E. A. Whitaker." The printers were D. F. Keller Company, Chicago. An introduction explained the scale, size, and usage of the images, the organization of the photographs, and plan for five supplements: the first would add a grid of geographical coordinates; the second would produce "rectified" images, namely, images produced to remove the curvature of the Moon; the third would consist of more detailed images of selected areas; the fourth would replace lesser-quality sheets in the original volume; and the fifth would provide a detailed nomenclature based on work then underway at Yerkes and McDonald Observatories.

After the introduction was a lengthy description of the history of the project, and Kuiper took great care to acknowledge those who had helped in the process by providing access to the observatory collections and technical support. In all, 1,200 prints were made, from which 230 were chosen for the atlas. The atlas was in three parts: an introduction of 11 sheets showing subdivision of the lunar surface into 44 fields, the 184 sheets of the atlas itself, and 35 supplementary sheets. Kuiper recommended that

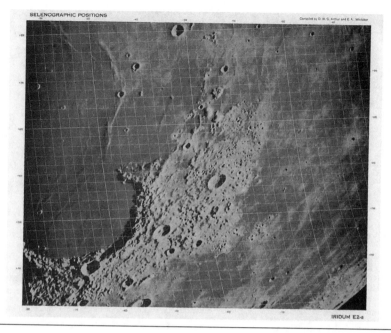

Sample sheet from the *Orthographic Atlas of the Moon* showing the coordinates superimposed on a region of the Moon containing Sinus Iridum and the Jura Mountains. This is sheet Iridum E2-a. Courtesy of University of Arizona Press.

observers buy two copies (at $30 per copy, $250 per copy in 2018 dollars) and keep one for reference and one for marking up and using at the telescope.

The prints of the lunar surface Kuiper had collected were not used in the state they came. Instead, prints were shaded during the photographic printing to remove strong gradients in grayness and then retouched to remove defects. The effect of scratches on the negatives was minimized by rubbing the negatives with Vaseline. Kuiper carefully notes that there were on average fifty defects per image and that it took forty-five minutes for Tapscott and photographic specialist Frank E. Manning to retouch each print. Kuiper checked every print. Meanwhile, Whitaker and Arthur checked the existing IAU nomenclature and found several errors because "Mr. Whitaker was able to draw here on his unpublished studies of the history of lunar nomenclature." These errors were listed in the atlas.

The reviews of the published atlas were ecstatic. Jules Brouet considered the work "quite remarkable" and told readers of *Ciel et Terre* where they could access a copy in France.[14] Frank Edmondson, who was instrumental in creating AURA and was its president from 1962 to 1965, wrote in *Science*, said, "This atlas is a magnificent

(a) Frank Edmondson. "Deceased—Frank Kelley Edmondson," Philosophy of Science Portal, December 13, 2008, http://philosophyofscienceportal.blogspot.com/2008/12/deceased-frank-kelley-edmondson .html. Courtesy of the Edmondson family. (b) Donald Menzel. Courtesy of Harvard-Smithsonian Center for Astrophysics.

achievement, prepared with loving care."[15] He later concludes, "they deserve the gratitude of astronomers and others who are interested in the lunar surface."[16]

Orthographic Atlas of the Moon

When Whitaker and Arthur arrived at Yerkes, the talk was of the first supplement, which was to place a coordinate system on the photographs. Whitaker applied the grid system to the central regions of the Moon while Arthur and a new assistant from Williams Bay, Ruth Horvath, worked on the more difficult outer edge. The Air Force's Aeronautical Chart and Information Center, located in Flagstaff, assisted in computing the grids. The atlas, titled *Orthographic Atlas of the Moon*, was published in 1961 under Kuiper's editorship, with Whitaker and Arthur credited as compilers. This atlas consisted of the best photos for each of the forty-four fields in the earlier atlas overprinted with the rectangular grid at a spacing of 0.01 lunar radii. Again, there were civilian and Air Force versions, which appeared in 1960 and 1961, respectively.[17] The Air Force version included colored lines of latitude and longitude at intervals of two degrees. This time the publisher was the University of Arizona Press, and the book was supported by NASA and the Air Force Cambridge Research Laboratories.

Edmondson reviewed the book for *Science*, describing it as a "very useful tool to be used in conjunction with the magnificent Lunar Atlas," which he had reviewed earlier.[18]

Rectified Lunar Atlas

The second supplement of the *Photographic Lunar Atlas* was the *Rectified Lunar Atlas* by Whitaker, Kuiper, Hartmann, and Spradley. The work was begun at Yerkes by Arthur, but when the group moved to Tucson it was taken over by William Hartmann, a graduate student, and L. Harold Spradley from the ACIC. The "rectification" refers to removing the curvature on regions near the edge of the Moon by projecting photographs onto a matte-white globe and then photographing the edges of the globe from the side. Seeing the images projected this way revealed the large number of craters in need of naming and the concentric nature of the structure of large impact basins, most noteworthy, Mare Orientale. Arthur and Whitaker gave sixty-five craters new names that were approved at the 1964 IAU General Assembly.

As with the first supplement, there were two versions, an Air Force version distributed by the ACIC and a commercial version sold by the University of Arizona Press.[19] Again, the acknowledged funding sources were NASA and the Air Force.

The scale of the *Rectified Lunar Atlas* was slightly smaller than that of the earlier atlas, 1:3.5 million instead of 1:1.37 million, and the face of the Moon was divided into only thirty regions. The photographs were also overprinted with a coordinate system. A sepia image of each region with the coordinate system, adopted from the *Orthographic Atlas of the Moon*, and names of features (30 sheets) was followed by several images of the same region under different lighting conditions (118 sheets). Tables listed corrections to previously approved IAU names and proposed new names. Further tables listed the plates used and details for the plates, such as date taken, observatory, and location on the Moon.

The introduction to the *Rectified Lunar Atlas* described in some detail how the images were made. Details of the projector, the three-foot matte-white globe, and the camera were given, with schematic diagrams. Whitaker selected plates from the original atlas, and the photographs of the projected images were made by Spradley using a camera with a focal length of 10.75 inches (27 centimeters) placed 5.5 feet (1.7 meters) from the globe. The dynamic range of the final prints was increased by printing twice on each image; one print favored dark areas and one favored light areas. Modern cameras perform this high-dynamic-range operation automatically.

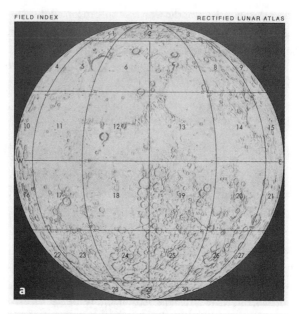

(a) Index for the *Rectified Lunar Atlas*. For the purposes of this atlas, the face of the Moon was divided into thirty regions. Courtesy of the University of Arizona Press. (b) Region 21 of the *Rectified Lunar Atlas* showing a highland region with crater Petravius in the lower-left central region. Courtesy of the University of Arizona Press.

Edmondson again reviewed the atlas for *Science*.[20] He opened his review by writing, "This atlas maintains the high quality set by Kuiper and his collaborators in its predecessors—the *Photographic Lunar Atlas* and the *Orthographic Atlas of the Moon*." He then describes the atlas in detail, including the number of sheets, their contents, and how the images were made, having previously described the photographic technique for increasing dynamic range. Edmondson was obviously very impressed with the results of correcting for foreshortening, mentioning particularly that a region that appeared to have only one crater (the crater Struve) is now seen to contain three craters (the new craters being named Russell and Eddington by the Kuiper group). He ends, "In summary, this atlas is a piece of magic. The master magicians who have produced it deserve the gratitude of astronomers, astrogeologists, and (eventually) those who will someday stand on the terrain shown in these remarkable photographic charts." Could the author of any book want more from their reviewers?

The Lunar Orbiter Program and the Consolidated Lunar Atlas

While the *Photographic Lunar Atlas* and its supplements appeared in press, the United States' attempt to land humans on the Moon continued. In the summer of 1966 the nation started the Lunar Orbiter program to observe the Moon's surface with orbiting spacecraft. They initially looked for suitable landing sites near the equator and on the near side of the Moon, but after their initial success the objectives became more science-based and gave wide high-latitude coverage.[21] The first of five missions launched on August 10, 1966, with subsequent missions every three or four months until August 1, 1967. Each spacecraft functioned for about a month and then was crashed into the surface. Nearly seven hundred photographs were made by the program, which are now available online.[22]

As the flow of images from the Lunar Orbiter program overwhelmed the lunar-science community,[23] the 61-inch telescope in Tucson was taking some of the best photographs ever taken of the Moon from the ground. Thousands of pictures were taken, from which 227 were selected, and 225 copies of each print were made by Kuiper's favorite local photograph processors, Ray Manley Commercial Photography in Tucson. These sets—titled the *Consolidated Lunar Atlas*, with the authors listed as Kuiper, Whitaker, Robert G. Strom (who joined the LPL faculty in 1963), and John W. Fountain and Stephen M. Larson (both having recently joined the LPL as a research assistants)—were privately distributed.[24] The *Consolidated Lunar Atlas* combined the promised fourth and fifth supplements (replace lesser-quality images and update nomenclature), thus completing the series announced in the original atlas. Special emphasis was placed on images for which Earth-based photography was still preferred: low-oblique photography (which brings out low domes, lava beds,

The Lunar Orbiter spacecraft. NASA photograph.

and other structures of considerable horizontal dimension but of low vertical relief) and full-moon photography (which obtains photometric information not otherwise possible).

Concurrent with the atlases, Arthur and Whitaker produced a series of publications in the *Communications of the Lunar and Planetary Laboratory* between 1963 and 1966 under the collective name "The System of Lunar Craters."[25] This catalogue described craters larger than 3.5 kilometers (about 1,700 total), listing diameter, latitude and longitude, rim freshness, whether it was located on highlands or maria, and if it had central peaks or terraces.

NASA's response to the *Consolidated Lunar Atlas* was not what Kuiper would have wanted.[26] On April 1, 1968, Kuiper wrote to Brunk, "I am sorry that you feel that the NASA contribution to the Atlas has been understated." After listing the citations to NASA in the atlas, Kuiper points out that "no reference was made to the fact that the University of Arizona supported more than half the salaries of the staff that produced the atlas."

Also of concern to Kuiper was that Brunk had received comments on the atlas from the Lunar Orbiter Project Office in Langley, Virginia. They wanted a better and more complete representation of the Lunar Orbiter photographs and greater reference to their publications.[27] Brunk concurred with the comments and informed Langley that they were being forwarded to Kuiper. Kuiper replied on April 2, 1968, "We did slip up on not plotting two of the 700 odd Orbiter fields," and ended, "Anyway, as authors, we used our best judgment as to what was significant and where to draw the line about including details."[28]

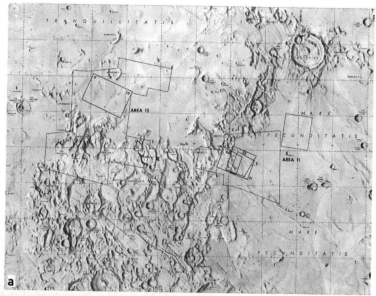

(a) A section from the Aeronautical Chart and Information Center map, which used lunar photographs as its basis with the addition of fine detail noted by visual observers. The fields of view of Lunar Orbiter images were also indicated. Courtesy of the University of Arizona Press. (b) The Consolidated Atlas contained new images of the Moon taken with the 61-inch telescope. Courtesy of the University of Arizona Press.

While Kuiper and his colleagues could point to several features and regions seen by the ground-based observations that the spacecraft missed, the truth is that the robotic spacecraft had overtaken the ground-based telescope in its ability to see the lunar surface, achieving resolutions way beyond anything possible from Earth. The wealth of data pouring from the Lunar Orbiter photographs, which included not only much better pictures of the near side but also the whole far side (except for the southern polar region), needed cartography and new names. Kuiper was to be heavily involved in this process long after the atlases had been completed.

A Naming Binge

Kuiper struggled for some time to prevent what he considered a naming binge for features on the Moon (and Mars). To this day, the IAU urges astronomers not to name more features than necessary. Kuiper believed that names were being created in reckless fashion. On March 7, 1973, he wrote to Brunk, "I am leaving for Houston for 2 days of meetings on lunar nomenclature. I do not look forward to this, but at least I succeeded in stopping the Menzel extravagance with 2000 new names, just as I was able to stop completely the Soviet extravagance of (inappropriate) new names on the lunar far side in Prague in 1967."[29] On May 14 Kuiper mentioned that he hoped that lunar nomenclature was "under control" and Martian nomenclature was hopefully also "getting some final coordination," but "frankly, I am so tired of the two subjects."[30] It was as if a monster Kuiper had created had turned on him.

But what Kuiper had achieved was significant. Before his work on the lunar atlases, pictures of the Moon were scattered in the vaults of the observatories, mostly unpublished and widely varying in quality. Several authors, mostly amateur, had tried to survey the images and systematize their content, but the task was huge and previous efforts seldom amounted to anything more than a few detailed studies of some regions and topics. Now a base existed on which to discuss landing sites for the Apollo mission and, perhaps more importantly, a scientific group was established with the expertise to support the many steps in landing on the Moon. The Ranger, Orbiter, Surveyor, and Apollo missions all required some participation by scientists. Kuiper and his colleagues were not the only group to be involved, and their involvement was stronger earlier rather than later, but they did make an important contribution to understanding the Moon and preparing NASA to land humans on its surface.

The race to put humans on the Moon, with all that it implied for Kuiper, was one of the more positive outcomes of the Cold War. There was, of course, another side of the Cold War, the nervous, palm-sweating sense that the Soviets were threatening any number of underhand operations against the West and that a nuclear World War III could break out any minute. This, too, was reflected in Kuiper's over-packed and diverse career.

Cold War Monitoring of Soviet Astronomy

Running up to the summer of 1959, Kuiper sent a proposal to the CIA. For many years he had been monitoring Soviet science for the U.S. government. He sent analyses of Soviet papers and books, reports of visits by U.S. scientists to Europe, and lists of participants, papers presented, and institutions visited at meetings, all with a liberal sprinkling of his personal knowledge of Soviet scientists involved. In the proposal, dated April 10, 1959, Kuiper offered to perform these functions but, contrary to requests made to him, he would not deceive Soviet astronomers.[31] The performance period would be November 1959 to October 1960, with Kuiper making a 5 percent time commitment. To execute the project, Kuiper hired a Croatian astronomer, Leo Randić, who would commit 75 percent of his time. Not only did Randić comb through the open literature, but the two men also spent considerable time talking to Soviet astronomers while entertaining them at Kuiper's home.[32] The work seems harmless enough, but in the tense days of the Cold War it left Kuiper so concerned that he later prevented his daughter Lucy from visiting Moscow because she might be harassed by Soviet authorities.

Kuiper's contacts with the CIA were through their field offices in Milwaukee before 1960 and Phoenix after his move to Arizona.[33] The agents he dealt with, G. R. van Doren in Milwaukee and Raymond O. Mitchell and Q. Maurice Hunsaker in Phoenix,[34] made frequent visits to the Kuipers and became good friends of the family. The CIA agents followed the fortunes of the family as the children grew, and they observed Kuiper's progress with landscaping the woods around the house and maintaining his garden. The relationships were cordial, with references in their correspondence to a "delightful interlude for tea" and observations that "I am not usually so graciously received." Kuiper's daughter has similar fond memories of the gatherings[35].

All these activities were known to the universities Kuiper served at the time. His security clearance at Chicago, up to the category of "secret," was handled by the university's contracts office, and the memos were copied to Strömgren as chair of the department.[36] The agents sometimes talked to others in the department if Kuiper approved and thought it useful. After his move to Arizona Kuiper asked that his government security clearance not be renewed, since he had never seen any classified documents.[37]

Military Interest in the
Moon and Soviet Astronomy

In the late 1950s Kuiper also was involved in a proposal to explode a nuclear bomb on the Moon. Leonard Reiffel was a physicist at the Illinois Institute of Technology who had worked on nuclear problems with Enrico Fermi at the University of Chicago and with Germans brought to the United States by Project Paperclip.[38] Throughout the 1960s he held many important appointments with NASA and the Apollo program. In 1958 he was asked by the Air Force to assemble a twelve-member panel to investigate the possibility of detonating a nuclear bomb on the Moon. The Soviets, who were leading the space race at the time, had a similar program. The project, known as Project A119, would be a highly visible demonstration of the United States' technical abilities because, if detonated on the dark side near the terminator (the boundary between illuminated and dark sides of the Moon), the effects would probably be visible from Earth.

The panel Reiffel assembled included Carl Sagan, then at Chicago and recently graduated with his doctorate, who was to perform calculations on the size of the mushroom cloud and the detonation effects. Also consulted was Kuiper because of his extensive knowledge of the lunar surface. In fact, while Sagan was listed as a panel member in the final report, "A Study of Lunar Research Flights, volume 1 (AFSWC-TR-59–39),"[39] Kuiper was the first person to be acknowledged: "We wish to thank Dr. G.P. Kuiper, Director of Yerkes Observatory and consultant to the Physics Division of ARF [Armour Research Foundation, a previous name for the Illinois Institute of Technology], for a number of very enlightening discussions of this work centering especially around interpretation of various theories of the moon's structure and origins and descriptions of the probable nature of the lunar surface." The report did acknowledge in a footnote that Harold Urey held differing views on the nature of the lunar surface. Two chapters were removed from this report and placed in an unpublished second volume, so the bulk of the text concerned the scientific value of the "experiment." The scientific topics were optical properties, seismic measurements, lunar surface and magnetic fields, plasma and magnetic field effects, and organic matter on the moon. We can surmise that the missing chapters concerned the secret military aspects of launching and detonating the bomb. Their plans to explode a nuclear bomb on the surface of the Moon provides insight into the military's feelings about Earth's nearest neighbor during the height of the Cold War. As ominous as this may seem, the same feelings gave rise to generous Air Force funding of lunar-atlas work.

CHAPTER 13

Momentous Changes

WHEN EDMONDSON WAS ASKED why Kuiper left the University of Chicago, he answered, "Well, part of the problem was that the directorship of Yerkes was a University of Chicago popularity contest. Every third year or something, the staff of the observatory had to vote on the observatory director. As far as I could make out, their tendency always was to vote against the guy who was in the job."[1] However, other writers have pointed to several reasons behind the decision, personal and professional, with some even claiming that this move somehow captured momentous changes in planetary astronomy at that time.[2] In the aftermath of the war, with Sputnik and the run-up to the Apollo program, astronomy was changing. Rocketry would come to determine our exploration of the planets, and this meant big budgets, strong government oversight, a new paradigm for solar-system science, and separation from traditional astronomy. While strong individual personalities were certainly involved, in important respects Kuiper's move from Chicago symbolizes the emergence of modern planetary science. It was a painful and complex process.

Kuiper's Second Directorship at McDonald Observatory

Kuiper assumed the directorship of the Yerkes and McDonald Observatories and chair of the Department of Astronomy of the University of Chicago on September 1, 1957, under conditions very different from his first stint in this role. This time there was no Struve reluctant to hand over power and constantly second-guessing Kuiper's

1	H. Straehler	11	B. Middlehurst	21	R. Schorn	31	E. Moore
2	B. Perkins	12	L. Schott	22	W. Hubbard	32	C. Smith
3	M. Wells	13	G. Kuiper	23	W. W. Morgan	33	R. Wickham
4	Hiltner	14	E. Whitaker	24	A. Behr	34	L. Fray
5	C. Siebert	15	M. Limber,	25	N. Limber	35	Hubbard
6	F. Woods	16	J. Chamberlain	26	R. Kraft	36	W. A. Hiltner
7	M. Schultz	17	F. Roach	27	P. Kuiper		
8	E. Sandberg	18	D. Bates,	28	D. Arthur		
9	A. Binder	19	P. Pesch	29	D. Cruickshank		
10	M. Mundschau	20	W. Beardsley	30	W. Matthews		

Yerkes Observatory staff in the summer of 1959. Courtesy of Special Collections Research Center, University of Chicago Library. Key by the author. Identifications as I have been able to determine them; some are uncertain. Whitaker (1985) identifies members of Kuiper's group, and others, in this photograph on page 23. The presence of a W. Hubbard in this image is interesting. Is this a coincidence or is this William B. Hubbard who joined LPL in 1972?

decisions. This time Kuiper had real power, and his primary goal was to increase the involvement of the University of Texas in the running of McDonald Observatory.[3] When Kuiper started his second directorship, he suggested that Chicago and Texas form a joint department to run McDonald Observatory and grant graduate research degrees in astronomy.

The research at McDonald and Yerkes was proceeding well. Kuiper was well underway with his survey of asteroids with the Cook telescope. The *Orthographic Atlas of the Moon* was also well underway, with Whitaker and Arthur joining the group in 1958. Also in 1958 the opposition of Mars had led to considerable excitement as Kuiper postulated the possibility of life on Mars. George Van Biesbroeck was working on comets, double stars, earthlight (for the International Geophysical Year, July 1, 1957, to December 31, 1958, during which the international community focused research efforts on space science) and he visited van den Bos in South Africa to make observations of double stars from the southern hemisphere. In 1958 Robert Kraft joined the group after getting his PhD at UC Berkeley and spending a year at Indiana University. He was setting out on an important career studying Cepheid variables and novae. Most notably, McDonald Observatory was achieving a reputation as the leading center for galaxy research with William Morgan working on galaxy classification and Geoffrey and Margaret Burbidge working on the mass distribution and rotation of galaxies.[4]

The contract for Chicago to run McDonald Observatory was to expire in 1962. The fear in Chicago was that Austin would be reluctant to continue to pay the observatory's bills because it was Chicago astronomers who really enjoyed the benefits. There was reticence about continuing to build up the facility, and there was also a notion that much of their observational work was moving away from McDonald as bigger telescopes came into operation elsewhere.

A Crisis at the University of Chicago

Just before he assumed the directorship and chair, Kuiper wrote to his friend in the Netherlands, Jan Oort, describing his plans. The mood was decidedly upbeat. The proposal to NSF to support McDonald Observatory, submitted through the University of Texas, was expected to be funded. The Office of Naval Research (ONR) was expected to fund new dorms, offices, labs, and improvements to the telescopes.[5] The ONR was also expected to support an infrared facility ($50,000, the equivalent of $440,000 in 2018), and the galactic studies of Morgan, Hiltner, Kraft, and Helmut Abt, the first PhD graduate from Caltech who was on the Chicago faculty from 1953 until 1959, were to be supported by a $50,000 grant from the Air Force. Chandra was to stay at Yerkes Observatory, rather than move to campus, and Kuiper expected the university to appoint an engineer for infrared work, for which he asked Oort for candidate suggestions. A celestial mechanics group would soon join Yerkes as well. It was all very exciting.

For Kuiper there was more good news. His proposal to create the center or institute of planetary and lunar sciences at the Universities of Chicago and Texas was sent to the chancellor in July 1958 and received a favorable response. In ten pages the proposal explained why centers or institutes were needed for focused research efforts. Departments tend to hire just single astronomers, he wrote, as determined by their teaching needs. Institutes are assemblies of multidisciplinary scientists hired to solve problems. The nature and origin of the Moon, which requires geologists and astronomers, is a good example, especially since surface experiments were likely to become possible in the next few years. Kuiper then summarized his current research on the Moon, planets, asteroids, and then the facilities at McDonald and Yerkes Observatories and how the new institute would function. Four or five "competent men" with research associates and graduate students with resources committed 2:1:1 toward the Moon, planets, and asteroids. Personnel would consist of Whitaker (expected to join the group soon), Arthur, Tapscott, French astronomer Audouin Dollfus, one or two people for the asteroids work, and temporary research associates from geology, meteorology, and upper-air physics. The university had already agreed to joint geology-astronomy degrees. The proposed budget was $40,000–50,000 per year, which was about half of Kuiper's current funding in two grants from ARDC, which was roughly $100,000 per year ($1 million in 2018 dollars). Kuiper ended his proposal by writing that this "would make an adequate and substantial effort possible in a field that has been neglected by professional astronomers for decades."[6] The chancellor replied that he liked the proposal and would encourage its approval by the various offices of the university.

Like Struve, Kuiper found himself running a university department populated by smart young men. Geoffrey Burbidge earned his first degree in physics at the University of Bristol and his PhD at University College London in 1951 working with Harrie Massey, a mathematician and head of the physics and astronomy department.[7] While in graduate school, in 1948, he met and married Margaret Peacher. Margaret obtained her BSc in 1939 and PhD in 1943 also from University College London. She went to the United States in 1951 to work at Yerkes Observatory and returned to England in 1953, where she worked with her husband and Fred Hoyle in Cambridge and William Fowler at Caltech to produce the seminal work on nucleosynthesis in 1957.[8] In 1957 Geoffrey and Margaret joined the faculty of the University of Chicago.

Kevin Prendergast was another young member of faculty who was hired in 1958. He received his PhD in astrometry at Columbia University in 1954 and two years later held a postdoctoral position at Yerkes working with Chandrasekhar on stellar magnetism.[9] After publishing on the elastic tumbling of asteroids and other topics, he began collaborating with the Burbidges on the rotation of galaxies.

Also hired in 1958 was D. Nelson Limber. Limber received both the AB and MSc degrees in physics from Ohio State University in 1950 and a PhD in astronomy from the University of Chicago in 1953.[10] Thereafter he held a postdoctoral fellowship at Princeton in 1953 and had joined the faculty of the University of Rochester in 1956 before moving to Chicago.

But with intelligence came strong wills, and those individuals demanded a say in running the department. Inevitably resentments arose, the resentments led to acrimony, and acrimony led to outright displays of anger and personal animosity. In a replay of how Struve left Chicago in 1947, Kuiper would leave the University of Chicago.

On November 10, 1958, Morgan, elated with his own progress on cataloging galaxies, wrote to Kuiper from Palomar to say how impressed he was with the Burbidges.[11] He described them as "as a powerful factor in modern astronomy" and asked more of Mrs. Siehert's time for them—presumably she was one of the computers. Kuiper agreed, but he used the opportunity to describe his agenda as chair and director. He was going to propose a joint department of astronomy with the University of Texas starting September 1, 1960, with himself as chair, Abt as assistant chair in Chicago, and Frank Edmonds as assistant chair in Austin. The budget for McDonald Observatory should be increased from $200,000 to $250,000 per year, taking just over $61,000 from the endowment, most for telescope improvements. Three Chicago faculty members would be given adjunct status in the University of Texas, and Chicago and Texas would equally own the "southern station" in Chile. The Yerkes faculty had agreed to this plan, and Kuiper would soon discuss it with Dean Whaley in Austin during an upcoming visit.[12]

With a general satisfaction at having the Burbidges in the department, and their promotions going through the university system, all was well. Then on May 11, 1959, the first signs of dissatisfaction emerged.[13] The problem concerned traveling between Williams Bay and the campus in Chicago, how trips were paid for, and what could be done about the excessive paperwork. The tone of the memo, cosigned by Geoffrey Burbidge and Joseph Chamberlain, an astronomer who joined the Chicago faculty in 1952, was aggressive: "Probably a few heads need to be knocked together." But more serious for Kuiper was a dig that would not have gone unnoticed: "Some individuals [obviously referring to Kuiper] in the past have managed to make frequent trips to the university without their own financial loss either by their personal good fortune or ingenuity." The response from Kuiper was not sympathetic, and he asked for more details. He also challenged some of the costs quoted.[14] One can imagine the groaning among the "luncheon group." Most university departments have a luncheon group,

(a) Geoffrey Burbidge in 1966. Courtesy of Special Collections, University Library, University of California Santa Cruz, Lick Observatory Records. (b) William Morgan. Courtesy of Special Collections Research Center, University of Chicago Library, accessed May 31, 2017, http://astro.uchicago.edu/vtour/history/oldresearch.html. (c) William A. Hiltner. Courtesy of Special Collections Research Center, University of Chicago Library, accessed May 31, 2017, https://www.lib.umich.edu/faculty-history/astronomy. (d) Joseph Chamberlain. From Donald Hunton's obituary. Courtesy of the American Astronomical Society.

(e) Lawrence Kimpton, Chancellor of the University of Chicago, 1951–1960. Courtesy of Special Collections Research Center, University of Chicago Library, accessed June 1, 2017, https://www.lib.uchicago.edu/projects/centcat/centcats/pres/presch06_01.html. (f) Margaret Burbidge. Special Collections and Archives, UC San Diego, La Jolla, 92093–0175, accessed March 26, 2018, https://en.wikipedia.org/wiki/Margaret_Burbidge. (g) Kevin Prendergast. Courtesy of Catherine Prendergast, accessed March 26, 2018, http://enculturation.net/pooka-of-literacy. (h) Jan Oort. Courtesy of Dutch National Archives.

a group of faculty members who take lunch together every day and usually end up discussing departmental affairs. Often these discussions ignite grievances that escalate with time. They are often directed at the chair of the department, the dean, or the "upper administration." The astronomers at Chicago had such a group, and Geoffrey Burbidge was a prominent member. Kuiper referred to the group as the Trojan horse of the department. The incident over travel expenses was a warning of bigger problems coming over the horizon.

In 1959 the Burbidges and Prendergast spent the summer in Southern California, where they could have ready access to the telescopes at Mount Wilson and Mount Palomar. On June 6, 1959, Kuiper used their absence as an opportunity to discuss them at a faculty meeting—particularly Prendergast, who he felt was under performing.[15]

A few weeks later another routine faculty meeting was called to discuss Kuiper's plans. Of immediate concern was hiring astronomers for the joint department that was to be located at the University of Texas. Kuiper wanted to hire Harold Johnson and William A. Sinton. Johnson was an assistant professor at Yerkes between 1950 and 1952 and was thus well known to the faculty there. In 1952 he left for a position at Lowell Observatory. Sinton was an infrared astronomer at Harvard College Observatory with a strong record of planetary astronomy, especially concerning organic compounds on Mars.[16] The Johnson case was somewhat urgent because he had visited Austin and been disappointed with the facilities. However, his decision to leave his current position at Lowell Observatory had made conditions there uncomfortable, and Johnson had started to look around. Kuiper felt that an early offer to Johnson would coax him to Austin despite his reservations. The matter was settled by the faculty. Accepting Kuiper's arguments and feeling comfortable about the candidates, the faculty agreed to make offers to both Johnson and Sinton. In the usual fashion, minutes of the meeting were circulated to all members of the department, including the absentees.

Geoffrey Burbidge went ballistic. Deeply offended that faculty members were being hired for the joint department in the absence of at least three members, he wrote, "We wish to protest in the strongest possible terms" and "what you have tried to do is morally unethical."[17] He asked about the procedure, how the rest of the faculty knew how the absentees would have voted, at what level the new hires would be appointed, whether they would reside and teach in Austin, what they had been promised, and what Texas had been told. "We feel extremely badly about the methods employed in this case, and plan to get in touch with the dean to find out whether retractions can be made if necessary." The message, dated August 14, 1959, was signed by all three and copied to the rest of the faculty. Kuiper forwarded the letter to Warren

Johnson, vice president of the university, with the message, "You will recall our discussions about a year ago about Mr. Burbidge. That my misgivings had some basis will be shown by the enclosed correspondence. Please destroy after you have read it."[18]

Kuiper immediately called a faculty meeting to discuss the letter. Chamberlain, Chandrasekhar, Hiltner, Kraft, Limber, and Morgan were present.[19] After the discussion, each was asked to write down his views, which Kuiper later forwarded to the dean. Kuiper defended himself, arguing that all members in residence approved (which included five of six tenured faculty), absentees' opinions were not required, and, in any event, the absentees had a record of not responding to memos. Furthermore, Harold Johnson had previously been promoted and tenured in the department, Texas wanted recommendations quickly, Johnson was looking at other offers, and Whaley was already scheduled to meet with Johnson and Sinton.[20] Chandrasekhar wrote that he thought the matter was a little rushed but that this was necessary and strongly supported the decisions. There were no university rules requiring faculty participation, but nothing would work unless "members of the department show mutual respect for each other, and are willing to show the others the same consideration they expect for themselves." Morgan supported the decision, accepting that it was urgent, but regretted that Burbidge had not been contacted. Limber supported the appointments but thought everyone should be consulted; however, he recognized that on this occasion the procedure followed was appropriate. If blame existed, it belonged equally to all present. Hiltner supported the complaints of the absentee faculty, arguing that the department should be a democracy. Kraft thought that the Texas appointments should be handled by the entire faculty but that all the questions raised by the complainants could have been answered. In other words, besides Hiltner, the faculty was supportive of Kuiper.[21]

The next day, August 18, 1959, Kuiper replied to the Burbidges and Prendergast, copying the department.[22] The letter was without the customary courtesies, essentially repeating the arguments he shared with the faculty the previous day. Needless to say, a few days later the antagonists rejected these arguments and continued to complain that the chair had abandoned customary procedures.[23]

By the time of the October 30, 1959, faculty meeting, the summer absentees were back in the department and still fuming. Kuiper wrote extensive notes in preparation for the meeting.[24] He planned to discuss the Chile observatory, the McDonald Observatory building and the improvement program, relations with the University of Texas and the Austin appointments for the joint department, and various small problems at Yerkes involving students, staff, the library, the building, and laboratory facilities. We can assume Kuiper opened the meeting as he planned in his notes: "It has come to my attention that some members of the department feel that they are insufficiently informed on certain observatory problems. I thought therefore we should have a review of our chief pending problems and take time to have a full discussion." He then reviewed his topics in detail.

The meeting was a disaster for Kuiper. In a long letter to the dean dated November 12, 1959, he describes the course of events.[25] Expecting to be greeted with applause because it seemed NSF was going to support an observatory in Chile, the meeting was instead "dominated by the persistent and hostile questioning of two faculty members; the atmosphere was no longer academic but violently political." Almost certainly the two faculty members were Geoffrey Burbidge and Al Hiltner. Kuiper wrote that after four hours he felt disheartened that all the accomplishments of the last two years was so little appreciated. "The October 30th meeting was a shocking experience which I shall find it difficult to forget."

Hurt and surprised, Kuiper sent a series of long letters to the administrators at the Universities of Chicago and Texas, rehashing the details, justifying his actions, and trying to understand what had happened.[26] He never did understand. Provost Roland Wendell Harrison at Chicago said simply that he agreed with Kuiper's analysis and had no answer, except that he was worried.[27] Chicago's chancellor, Lawrence Kimpton, publicly asked, "Why does the astronomy chair change so often?" Kuiper was quick to send a detailed analysis.[28] Kuiper explained what he saw were organizational defects, problems with the composition of the department, and problems with faculty recruitment. The chair and the director have fundamentally different roles, he argued. The chair has little power and should be short-term. The director should have much power, be appointed long-term, and have considerable university backing. He then went through the senior faculty members individually—Chandra, Morgan, Hiltner, and Chamberlain—candidly identifying strengths and weaknesses. Then, emotions taking control of his pen, he added a long and acidic attack on Burbidge, which included references to his body odor and the way he treated his wife. It was indeed the low point of Kuiper's amazing career.

With himself as chair and Morgan as acting director, Kuiper let the incident go. Despite the hurt feelings, Kuiper continued to push hard for his agenda until the day he left Chicago in September 1960. He called a faculty meeting for November 11, 1959, at which they set up a committee to write a departmental policy document and report on a meeting they had with NSF on the Chile Observatory two days earlier.[29] During the meeting Morgan, Chamberlain, Hiltner, and Burbidge were chosen to be the policy committee. "Thus, some measure of peace and agreement returned on Armistice Day," wrote Kuiper to the dean the next day.[30]

Two weeks before the disastrous October 30 faculty meeting, the dean had circulated a memo to all departments pointing out that the appointments of several chairs and directorships were expiring, and he needed recommendations for new appointments before December 2.[31] This included the directorship of Yerkes and McDonald Observatories. On January 4 the chancellor called Kuiper to his office to tell him that the faculty members of the astronomy department had voted to offer the chair and directorship to Morgan, over Morgan's objections.[32] Although the department was generally supportive of Kuiper actions, the vitriol of the October 30 meeting had persuaded the department that a new director was needed. Burbidge and Hiltner had won. Almost immediately, Kuiper made the decision to leave Chicago.

The Effect of Temperament?

The story of what happened to precipitate Kuiper leaving the University of Chicago was discerned from contemporary documents. However, these facts clearly only tell part of the story, and probably only a small part. A considerable amount has been written about why Kuiper left the University of Chicago, and, as shown above, views ranged from the simple popularity-contest perspective of Edmondson to the complex and historically significant opinions of professional historians like Ronald Doel.

Several other relevant facts can be identified with some confidence. Kuiper had achieved the status of one of the leading astronomers of his day. His income from research grants was greater than the rest of the department put together ($56,000 compared with $40,000, roughly $480,000 compared with $350,000 in 2018 dollars). In the competitive environment of the department, his success must have been difficult for some to bear.

He did not make it easy. In pursing his agenda, praiseworthy as it was, he applied all his considerable strength, energy, and determination. However, he failed to ensure that his colleagues were comfortable with his efforts. They had no sense of ownership for the decisions, particularly those who were not Kuiper's close confidants. Faculty buy-in was critical when facilities were involved that affected their careers, such as access to the telescopes at Yerkes and, more importantly, McDonald. However, Kuiper presented his ideas to the faculty as a fait accompli and his attitude to his colleagues within the department was insensitive and reflected by his constant discrimination between senior and junior faculty. Whether this started before his troubles with Burbidge, then an assistant professor, or after, is unclear, but it certainly goes against his own career experience, in which he was consulted at length by Struve on the appointment of new faculty even before he moved to Chicago. By contrast, his letters to the upper administration in Chicago and Austin are long, painstaking, and superbly written, reflecting his considerable efforts in thought and action. At times they were overkill, and one senses he was using these letters to sort, clarify, and document his own thoughts. They were almost a journal. Sharing such thoughts in real time must have endeared Kuiper to his correspondents who, one can imagine, came to feel part of the adventure. He even shared such letters with persons outside the university, such as Jan Oort in Leiden.

His lack of sensitivity to others haunted him throughout his career, but it was a problem he could not see, far less correct. Kuiper was so engrossed in his work and progress that he became highly insensitive to others. It was the trait he showed during his postdoc at Lick that upset the highly sensitive Luyten. It was the trait that caused Urey, with a recent Nobel Prize and confident in his mastery of the nascent field of solar-system science, to burn huge amounts of time and energy attacking Kuiper. Now it was Burbidge who could not cope, and it caused him to turn the luncheon group into a Trojan horse. Kuiper's friends and supporters, of which there were many, understood this characteristic of Kuiper's and admired his accomplishments. His enemies could not get beyond it and nurtured hatreds.

One manifestation of this temperament was his choice of research associates. Kuiper's team reflected the needs of his large, nontraditional program and was not composed entirely of researchers with the normal pedigree of graduate school, publications, and PhD degrees. Kuiper was overturning the customary way in which university research teams were constituted, which required team members to have a doctorate. Kuiper hired Lenham, Whitaker, and Arthur for their individual talents, specific to his needs, regardless of their academic standing. Another such person was Alika Herring.

It has been said, by Robert Doel for example, with justification, that there were other, bigger forces acting to determine the course of events at the University of Chicago in the summer and fall of 1959. These forces had to play out if planetary science was to come into being, and it fell to Kuiper to stand at the interface. Geoffrey Burbidge once called the process the "splintering of astronomy." Kuiper started out as a mainstream astronomer interested in the topics recognized at the time as being at the center of modern astronomy research: binary stars, white dwarfs, and stellar evolution. This work put him among the top astronomers and led to Struve hiring him to one of the best universities at the time. Kuiper's discovery of the atmosphere on Titan and the power of infrared spectroscopy to help us understand planetary atmospheres set him on a major program of the study of planetary atmospheres, which branched out to the Moon and asteroids. He became a veritable institute by himself. It comes as no surprise that he proposed the planetary and lunar institute for the University of Chicago with graduate degrees awarded jointly by the astronomy and geology departments. It is often said that he is the father of planetary science. What he did was to show that success required being interdisciplinary and multidisciplinary, no longer solely a branch of astronomy. But no one said the process of creating the new discipline was going to be painless. To the mainstream astronomers—not just at Chicago, but worldwide—there was a cuckoo in the nest. The new field, whether represented by a new institute or not, was going to compete for telescope time, funds, and personnel. With its new partnership with the space race, planetary science was going to glean not just funds but attention, and that attention would be from powerful people.

Against this background of ambition, personality, and intellectual pursuit, there is always the critical influence of funding. Profound changes were underway and driving the course of events. Before World War II, the funding of scientific research was the province of charities and foundations, usually channeled through one of the handful of elite universities. Charles T. Yerkes paid for the telescope and John D. Rockefeller paid for the observatory at Williams Bay, and William Johnson McDonald funded the observatory at Fort Davis. James Lick donated the observatory on Mount Hamilton to the University of California.

During the wars the government established agencies to support research. In 1915 the National Advisory Committee on Aeronautics (NACA) was established,[33] and during World War II the National Defense Research Committee (NDRC) and its successor the Office of Scientific Research and Development (OSRD) were supporting war-related research.[34] The success of these led to the establishment after the war of the National Science Foundation (NSF) in 1950 (as well as the National Institutes for Health and the Nuclear Regulatory Commission). Concurrently, the military had research programs that would sometimes support civilian research, examples being the Office of Naval Research (ONR) and the Air Force Office of Scientific Research (AFOSR). The launch of Sputnik in October 1957 galvanized this process. In 1958 NACA was disbanded and replaced by the National Aeronautics and Space Administration (NASA); also in that year the military programs were consolidated into the Advanced Research Projects Agency (ARPA), later renamed the Defense Advanced Research Projects Agency (DARPA). With his background in scientific research for the military during the war, and the pressures of the Cold War afterward, Kuiper was quick to exploit the military agencies as well as NSF in support of his various programs. Most spectacular were the NASA programs to land a man on the Moon and NSF's projects to create observatories that were accessible to the entire astronomical community.

Thus Kuiper represents this major evolution in scientific research, especially planetary research, which is the shift from reliance on private charities and foundations to sponsorship by the U.S. government. For example, at various times his lunar-atlas project was funded by NSF seed money and major grants by the Air Force and later NASA. In 1959 Kuiper obtained NSF funds to support operations at McDonald Observatory, and he acquired an NSF grant to build an observatory in Chile. He brought in NASA funds to build a 60-inch telescope in Tucson, and he was instrumental in landing NASA funds for an observatory on Mauna Kea. The availability of large grants from various agencies of the government shook astronomy to the foundations in the 1950s and '60s and largely (but not exclusively) favored planetary science. To traditional astronomers, accustomed to small grants and absolute control, the transition to large grants with considerable government oversight presented problems, and historians have argued that this was another factor forcing its way through the crisis at Chicago in the fall of 1959.[35] Kuiper loved the new order.

But Where to Go?

With these political projects and the lunar-mapping and asteroid work underway, Kuiper was faced with the prospect of leaving Chicago. Now fifty-five, he showed no signs of wanting to slow down. But where to go? He thought of the college down the road from McDonald Observatory, Sul Ross State College, which had been so supportive of the observatory in the early days. The president of Sul Ross had actually offered the astronomers office space on his campus. The college is situated in Alpine, Texas, the Brewster County seat with a population of about six thousand at the time. Sul Ross was a teacher's college when it opened in 1920, but by 1960 offered its one thousand students a wide spectrum of courses at both undergraduate and graduate levels.[36]

In early December, while Kuiper investigated the possibility of an appointment at Sul Ross State College, he telephoned his old colleague Aden Meinel to sound him out on a move to Arizona, close to the Kitt Peak National Observatory. Meinel, who began his career at Caltech and was hired at the University of Chicago in 1949, served as the associate director of the Yerkes and McDonald Observatories between 1953 and 1956. In 1958 he moved to Arizona to become the director of the Kitt Peak.[37]

Initially thinking Kuiper wanted to move to Kitt Peak, Meinel consulted the AURA board but met with resistance. They had heard about the problems at Chicago. But Kuiper was thinking about the University of Arizona. The two met at the Quadrangle Club at the University of Chicago on December 15, and Meinel agreed to take the idea to the administration of the University of Arizona.[38] The next day he spoke to David Patrick (coordinator of research), who took the idea to University President Richard Harvill. Harvill probably saw federal funding for big projects, like nearby Kitt Peak Observatory, as a means of strengthening American southwest universities, especially because their departments were not as entrenched in tradition as they were in older universities. As a result, Kuiper was invited to visit the campus on January 14–19, 1960, to meet with the relevant people, presumably Harvill, Patrick, Edwin Carpenter (chair of the astronomy department and director of Steward Observatory), and Richard Kassander (director of the Institute of Atmospheric Physics, or IAP).

Seeing that a move was in the works, on February 15, 1960, Provost Harrison at Chicago wrote to Kuiper to wish him luck and stress that Kuiper would be welcomed

(a) University of Arizona President Richard Harvill and his wife. Courtesy of University of Arizona Libraries, Special Collections. Photograph from the program for the recognition dinner for his retirement which is in UASC, box 8, folder 34. (b) Newspaper picture of the group unpacking on their arrival in Tucson from Whitaker (1985). Courtesy of LPL.

back to Chicago if the Arizona position did not work out.[39] A few days later the *Lake Geneva Regional News* ran an article describing Kuiper's resignation from the University of Chicago.[40] The same day, the *Tucson Daily Citizen* ran an article announcing Kuiper's decision to move to the University of Arizona.[41]

A puzzled Chandra took Kuiper to task for not discussing matters with him before resigning. The dean had plans for promoting Kuiper to Distinguished Service Professor and setting up a committee of the three members of the National Academy of Sciences in the department to make recommendations to the chairman concerning the department and its observatories. But the Rubicon had been crossed. In a letter to Oort, Kuiper confided that he no longer trusted Chandra because he was one of the main reasons that the Burbidges had been hired.[42] The crisis in the department had driven a rift through one of the longest and most productive friendships in astronomical research. Although Kuiper was instrumental in securing Chandra's position at Chicago, and the two were the closest of friends and colleagues for their twenty-five years on the faculty together, their correspondence virtually ended when Kuiper moved to Arizona, and there is no mention of Kuiper in Chandrasekhar's very long oral history.[43]

Harvill, Patrick, Carpenter, and Kassander were critical to the next—and very important—phase in Kuiper's career. Harvill served as university president from 1951 to 1971, a period in which the university underwent its furious postwar growth.[44]

Harvill was born in Centerville, Tennessee, in 1905 and obtained his BS in 1926 from Mississippi State College, his MA from Duke University in 1927, and his PhD from Northwestern University in 1932. He was hired by the University of Arizona as an economics professor in 1934, married George Lee Garner two years later, and rapidly rose through the professorial ranks. In 1946 he was appointed dean of the graduate college and the following year dean of the Liberal Arts College, and then president of the university in 1951.[45] Patrick was coordinator of research (and graduate dean) at Arizona between 1947 and 1959. Carpenter, known for his research on white dwarfs, supernovae, and galaxies, was chair of the department and director of the observatory.[46] He had a BA and MA from Harvard and a PhD from UC Berkeley. After being hired as an instructor at Arizona, he became head of astronomy in 1936 and director of the observatory in 1938, holding that post until he died in 1963. Kassander, director of the IAP, was a well-known meteorologist with a BA from Amherst College (1941), an MS from the University of Oklahoma (1943), and a PhD from Iowa State University (1950).[47]

Meinel had disagreements with the AURA board himself and was ready to leave Kitt Peak, so he brought up with Kuiper and Carpenter the subject of him also moving to the university. The initiative was successful, and Meinel moved to the University of Arizona as professor of astronomy the next year. After Carpenter's death in 1963, Meinel became director of Steward Observatory.

Before an offer was made to Kuiper, Harvill received a call from the *Chicago Tribune* asking why a distinguished scientist at the University of Chicago wanted to join the faculty of a comparatively unknown university in the Southwest. His well-known reply was that the reporter should, like Kuiper, come to Tucson and see the university.[48]

Kuiper and Kassander got on well. Feeling encouraged by his reception at Arizona and convinced of the likelihood of an offer, Kuiper viewed two houses and paid earnest money on one of them, 721 N. Sawtelle Avenue, a few blocks from the campus. Then Kassander brought up the subject of a name for Kuiper's group, suggesting "Planetary Studies Group of the Institute of Atmospheric Physics." Kuiper counter proposed "Planetary Physics Laboratory of the IAP" and then later suggested "Lunar and Planetary Laboratory of the IAP." Kuiper had been on his way to creating a planetary institute at the University of Chicago, but after the October 30 crisis meeting, events began to overtake him. Now he had an institute, or at least a laboratory, falling into his hands.[49]

It is significant that it fell to Kassander to make all the arrangements for accommodating the new group at Arizona. Harvill, distressed and perplexed at Carpenter's lack of enthusiasm, approached Kassander, who welcomed the idea of an association with Kuiper. Carpenter's reservations were primarily concerned with available space while

Kassander had recently acquired ample new facilities for the IAP. In any event, at this point Kuiper probably felt more comfortable associating with atmospheric scientists than the astronomers, especially since Kassander's institute had recently made the decision to add planetary atmospheres to its research activities. Thus Kuiper was given the west end of the fifth floor of new Physics and Atmospheric Sciences Building. Kassander wanted the space on the fifth floor for his institute but did not expect to get it, so these developments were doubly satisfying to him.[50]

The space was square, with approximately 2,200 square feet, so there was room for a large corner office for Kuiper and large offices down the corridor for the group that was to move from Chicago to Tucson with him: Whitaker and Arthur. Barbara Middlehurst, editorial assistant, and Miss Geoferion, graduate student, would have window offices on the opposite side of the building while graduate students Elliot Moore, Carl Huzzen, and Toby Owen would have small windowless offices nearby. Large rooms were set aside for spectroscopy and geophysics laboratories and record keeping. Kassander also offered Kuiper Quonset hut T6, which offered 2,000 square feet, room for a darkroom, a laboratory, and the students, Dale Cruikshank, Alan Binder, and William Hartmann. Later the hut was modified to include a corridor running the length of the building, which enabled the installation of a projector and white sphere for the rectified lunar-atlas work.

An Offer from Struve

Whether or not he knew of the unrest at Yerkes, Struve wrote to Kuiper on January 28, 1960, to invite him to spend a year at the new National Radio Astronomy Observatory in Green Bank, West Virginia, where Struve was now director.[51] He went on to describe the possibilities for collaboration, with an emphasis on the topics he thought would interest Kuiper—Drake's studies of Jupiter, for instance. Struve mentioned he would also value Kuiper's thoughts on radio instruments that could be carried by balloons and spacecraft. He ended his letter, "I should be glad if you would let me know your ideas on the work that you might be interested in and you might perhaps also give me an indication of the necessary salary."

The reply describes Kuiper's situation as of February 2, 1960.[52] "Your invitation is very much appreciated and I would be delighted to accept if I had not already made firm commitments in another direction." Morgan was taking over his administrative duties and Kuiper was leaving Chicago on September 1, 1960, to move to Arizona. He wrote that his time would be split between the LPL and the astronomy department (only later was the full role of the IAP to become apparent). President Harvill had made a fine offer of space, and Kuiper had already determined how it would be distributed between his various interests. He then ran down the status of his editorial work: volume 3 of the *Solar System* series and volume 1 of the Stellar Systems series were with the printers, and volume 4 of the *Solar System* series was not far behind. The lunar atlas would be finished in early March.

"I most heartily congratulate you upon your appointment in Arizona," replied Struve, adding as an afterthought, "Yerkes will never be the same again without you! I regret in many respects the gradual dispersal of our 1937 group!"[53]

He must have learned about the true story of why Kuiper was leaving Chicago much later, because in June 30, 1960, he wrote, "I am moved to write to you and tell you how sorry I feel that the situation at Yerkes has not developed in accordance with your wishes. . . . I want you to know that your fine astronomical work has secured for you a permanent place in the history of astronomy."[54]

Three years later Kuiper's long-term supporter, Otto Struve, died. On July 3, 1963, Kuiper received a request from T. G. Cowling, editor of *The Observatory*, for information on Struve that could be published in an obituary. On August 7, Kuiper wrote a long letter that began, "Writing about Struve is no easy task. He was a very unusual man, a man with enormous energies, at times highly emotional, a man who was very conscious of being the last member of the Struve family which placed a great responsibility on him."[55]

The Road to Arizona

While Kassander arranged for the move at the University of Arizona, Kuiper called his group together to tell them the news.[56] The group took it in their stride, being fully occupied with their various projects—the lunar atlas, the application of sensitive image orthicon TV system, and photometric properties of the Moon. In any event, the move was six months away. On August 17, 1960, Ewen and Beryl Whitaker and their family left Williams Bay, driving their Rambler American station wagon loaded with cardboard boxes, packed in such a way as to keep the children separated. "The move cost three week's salary," recalled Whitaker.[57] By arriving first, he could supervise the unpacking of the laboratory's precious materials, especially the thousands of glass plates. Members of the group followed, staying at the Kuiper house on Sawtelle until they could find local accommodation. At the end of the month the Kuipers arrived in Tucson, their furniture, personal belongings, and laboratory supplies arriving after some delay and anguish as their moving van broke down in Tucumcari, New Mexico. In a letter to Kuiper on March 16, Bart Bok welcomed the move as "fine and sensible" because it would free Kuiper of administrative duties and because Bok planned to retire to Tucson. He could then visit the Kuipers, the Carpenters, and the Meinels all in the same spot.

Kuiper's career at Chicago suddenly ended. In some ways, it was a replay of the process by which Struve had left Chicago. Both Struve and Kuiper were driven out by younger faculty who wanted more control over their department and the observatories. Certainly, personalities were involved—they always are. Strong, opinionated, younger faculty members were hired because they make good scientists. However, in Kuiper's case other factors of great importance were at play. Kuiper was morphing from astronomer to planetary scientist, and this was not possible at Chicago. Kuiper did not choose to leave Chicago, but his ambitions made it necessary.

CHAPTER 14

The Lunar and Planetary Laboratory at the University of Arizona

A LOT HAS BEEN WRITTEN about how the Lunar and Planetary Laboratory at the University of Arizona came into being. Mostly LPL's history has been written by its participants. Ewen Whitaker has written a fine, detailed account from the perspective of a loyal and enthusiastic colleague of Kuiper's, although objective enough to note Kuiper's authoritative management style.[1] At the request of Mike Drake, then director of LPL, Melissa Sevigny has compiled oral histories from virtually everyone associated with the laboratory and published a book *Under Desert Skies*, with two out of five chapters dealing with the founding of LPL and the Kuiper years. To me, Kuiper's years at Tucson demonstrate what happened when the dam broke and the constraints placed on him by being a member of an astronomy department were removed. At Tucson Kuiper soon had his own institution and reported directly to the president, just as he wanted in Chicago. The multiple forces that were at work to create planetary science could play out in their full measure, and one of the world's centers for planetary science could come into being. It is a story of how a federal agency, a supportive university, and an individual of extraordinary vision and capability successfully worked together to create the field of "planetary science" while the nation struggled to put a man on the Moon.

A New University, New People, and the Creation of an Institute

When Kuiper arrived at the University of Arizona in the summer of 1960, he had an appointment in three academic units.[2] He was a professor of astronomy, an astronomer

(a) William Hartmann, Dale Cruikshank, Alika Herring, Charles Wood, and James Marshall outside the new wing of physics building that housed LPL in the mid-sixties. Jim Marshall, a student, died in a car crash on his way home to Texas for a break from studies. Courtesy of Dale Cruikshank. (b) Elizabeth Roemer, the first woman member of faculty in LPL at the University of Arizona, was an expert in comets. Courtesy of LPL, accessed June 1, 2017, http://www.flickr.com/photos/smithsonian /4405668977/. (c) Steve Larson with Hazel Sears, November 2014. Author's photograph. (d) Harold Johnson. Courtesy of LPL.

in the Steward Observatory, and a research professor in the IAP, with one-quarter of his $12,600 salary coming from the astronomy department, one-quarter from Steward Observatory, and half from the institute. He took a considerable pay cut to make the move; his salary had just been raised to $16,000 at Chicago. His formal duties, as laid out in the job offer, were to teach stellar spectroscopy and planetary astronomy in the astronomy department, perform observational planetary research for the observatory, and conduct experimental geophysical and planetary research for the institute. The nature of Kuiper's appointment speaks to the unique research field he was helping create.

Initially, the group had eleven members: Whitaker, Arthur, Middlehurst, Horvath, Carl Huzzen, Owen, Ann Geoffrion, and Elliott and Gail Moore, and two graduate

(a) Alika Herring. Courtesy of LPL. (b) Spencer Titley. Courtesy of Spencer Titley, accessed May 31, 2017, http://www.geo.arizona.edu/Titley. (c) Frank Low. Courtesy of Edith Low, accessed Jun 29, 2017, http://www.phys-astro.sonoma.edu/BruceMedalists/Low/index.html. (d) Brunk and Kuiper. Courtesy of Dale Cruikshank.

students.[3] Most of them worked on the lunar-atlas project, although Middlehurst was by then working primarily for the book projects.

Soon Tom Gehrels and Harold Johnson joined Kuiper in Tucson. Gehrels wanted to fly a high-altitude balloon with a telescope, as his colleague Dollfus had done, but with the loss of funding for a balloon pilot, manned balloon telescopes stopped. Gehrels was forced to build an unmanned telescope, which he called a polariscope and which successfully flew. In 1963 Gehrels hired an undergraduate student, Stephen

Larson, who, while still a freshman, had been drawn to planetary science by observing Comet Mrkos in 1957. Originally hired for the polarimetry program, he is still working at the observatory detecting asteroids.[4]

Johnson was hired in 1959 by the University of Texas at Austin after his time at the University of Chicago and Lowell Observatory. Known for his work on stellar photometry, and as the developer of the UBVRI system for describing stars, he also arrived at LPL in 1961.[5] The UBVRI system involves reporting the intensity of the star's light at specified ultraviolet, blue, violet, red, and infrared wavelengths. The system effectively describes not only a star's temperature, but also the temperature of gas or dust (or both) surrounding it, which is key to determining the chemical composition of the star itself. Johnson joked at the time that he was in the "stellar division" of LPL. In 1969, shortly after an energetic reaction to a memo from Kuiper asking observatory users to clean up the kitchen after use, Johnson moved to the Optical Sciences Center, and then Steward Observatory.[6] He then worked at the National Astronomical Observatory in Baja California, first part time and then full time, using the 61-inch telescope provided by LPL for infrared observations.

A year after Kuiper arrived in Tucson, three more graduate students joined him. Alan Binder and Dale Cruikshank, who had worked with Kuiper as summer (and sometimes Christmas) interns, were excited to be working shoulder to shoulder with the world-famous astronomer.[7] William Hartmann also started in 1961, fresh from a BS in astronomy at Pennsylvania State University, having been inspired by articles by Wernher von Braun and the paintings of Chesley Bonestell. Charles Wood, an undergraduate, had read about Kuiper in the local newspaper. All four became noted planetary scientists, always ready to recount their time with Kuiper.[8] After leaving LPL, Binder spent time on lunar research at the University of Kiel, and the Max Planck Institute for Nuclear Physics in Heidelberg. Upon returning to the United States, he started his own institute and became principle investigator of the Lunar Prospector mission, whose main task was to search for water on the Moon.[9] Cruikshank left Tucson to become an astronomer on Mauna Kea until he moved to NASA Ames Research Center in California, where he continued his interests in infrared spectroscopy, especially as a member of several mission science teams. Cruikshank wrote the official National Academy biography of Kuiper. Hartmann spent his career at private research institutes in Tucson, first Illinois Institute of Technology Research Institute and then the Planetary Science Institute, making important contributions to, among other topics, crater counting as a means of dating planetary surfaces. He also became an accomplished space artist.[10] Chuck Wood went on to earn a PhD in planetary sciences at Brown University and then had a successful career at NASA's Johnson Space Center, the University of Arizona, Biosphere II and the Planetary Science Institute.

Also in 1961, Kuiper hired Alika Herring, an optician with artistic bent.[11] Herring was able to grind and polish his own mirrors and make good telescopes, and he was a good observer with keen eyesight. He liked to draw lunar features. Despite working so hard on photographic atlases of the Moon, Kuiper saw much value in Herring's drawings.

With no program in planetary sciences, the graduate students were compelled to apply for and take the programs in astronomy or geology; the fact that some joined the astronomy program while others joined the geology program again speaks of the multidisciplinary nature of LPL. Kuiper continued to be known for his lack of involvement with students, but with their background in physics and astronomy and their growing interests in planetary science, they found an unofficial mentor in Spencer Titley.[12] Titley was an economic geologist with a career-long interest in the copper deposits that have been so important to the Arizona economy. He obtained his first degree in mining geology at the Colorado School of Mines in 1951 and then a PhD in geology and chemistry at the University of Arizona in 1958.

When Kuiper's students came to his office, Titley was cooperating with Gene Shoemaker at the U.S. Geological Survey in Flagstaff.[13] He was working on mapping Meteor Crater (an impact crater east of Flagstaff) and an infant program in lunar mapping, and he was involved in astronaut training. "There was a coincidence of these three fellows coming in and my involvement with the Survey on this new, exciting thing with lunar mapping, lunar geology," says Titley in an oral history recorded by Melissa Sevigny. Sevigny also recorded the oral histories of Hartmann, Binder, and Cruikshank in which they recognize the important role Titley played in their doctoral education.[14] Remarkably, Titley and Kuiper never exchanged words about the education of the first planetary-science students.

Eighteen months after Kuiper arrived in Tucson LPL had twenty-two people, five in faculty appointments, five research associates, and twelve graduate students. In addition to the students associated with Kuiper, other faculty associated with LPL had students too: Gehrels was the major professor for Ben Zellner and David Coffeen, Frank Low advised Susan Geisel and Doug Kleinmann. These students were in the Department of Astronomy and Steward Observatory, but they had offices and labs in LPL. Regrettably, there was very little interaction between Kuiper's students and those of the other faculty, even though they were in the same building. Also at LPL were four secretaries and one draft person, and there were three people helping with data reduction and one helping with the balloon flights, presumably undergraduate students. Kuiper listed Stuart Hoenig, an associate professor in the engineering college, as a part-time member of LPL.

The accounts of students who worked with Kuiper suggest that he was a demanding mentor, and that some could not handle the relationship.[15] Gehrels recalls how he went to other faculty members for guidance as a student and avoided Kuiper, even after Gehrels was appointed professor.[16] Other students valued Kuiper's encouragement. Cruikshank remembers encouragement to participate in a few field trips to, for example, perform a gravimetric survey of lava flows in New Mexico in search of collapse depressions that may be similar to low-profile collapses in the lunar maria, or to obtain visual, photographic, and spectral information for volcanoes in the mainland U.S. and Hawaii. Hartmann, Cruikshank, Robert Strom, and Kuiper witnessed several Hawaiian eruptions. Kuiper's appreciation of student initiative meant that he encouraged the independent use of facilities, paying for the construction of student spectrometers and allowing access to the 61-inch telescope for independent student papers.[17]

In the summer of 1961 LPL was granted independent status from the IAP. On July 25 Harvill wrote to Kassander, "I am sure there are great opportunities at the University of Arizona which we shall be very glad to develop."[18]

The purpose of the lab and the accomplishments of the first eighteen months were proudly laid out in the first article in Kuiper's new *Communications of the Lunar and Planetary Laboratory*.[19] The purpose of the laboratory, he wrote, was to conduct teaching and research on the Moon and planets, and the University of Arizona was favorable for this because of the number of researchers in related fields, clear skies, humidity, telescopes and computers, nearby national and industrial laboratories, well-developed graduate school, and local geology that could help in understanding lunar geology. The group was to be concerned with research, publication of data, and graduate instruction. It is significant to note that in this article Kuiper put graduate-student instruction high in his priorities, presumably thinking of his position in the University of Arizona.

The research programs of LPL were to concern lunar studies, not just the atlases but polarization studies, geomorphology, and lunar nomenclature. There were also programs of research into planetary atmospheres, primarily using infrared spectroscopy (still using the lead sulfide detector); satellite systems dynamics, shape, masses, and surface properties; asteroids, light curves, an asteroid survey down to twentieth magnitude, their number and space distribution, fragmentation, and production of meteorites; zodiacal light and its polarization properties; stellar spectroscopy, photometry, and polarization; development of infrared instruments and telescopes; laboratory work on infrared spectra; and balloon-borne observations in the ultraviolet and infrared. It was quite a list. Finally, Kuiper mentions that he and Gehrels had been appointed "experimenters" on lunar and planetary space missions.

In 1963 Steward Observatory moved its 36-inch telescope to Kitt Peak, and Carpenter died that same year. Meinel took over as director of Steward Observatory and collaborated with LPL in spectroscopy. Three years later he resigned that post to create

The Lunar and Planetary Laboratory—I

GERARD P. KUIPER, *Lunar and Planetary Laboratory, University of Arizona*

IN THE FALL of 1960, the Lunar and Planetary Laboratory was organized at the University of Arizona to create, in a favorable academic setting, a research and teaching center for studies of the moon and planets.

The word *favorable* meant many things to us. First, it referred to the presence of research meteorologists, geologists, geochemists, physicists, and electrical engineers. Second, we needed clear skies, low humidity, and access to modern telescopes and electronic computers. Third, proximity to national laboratories was important. Fourth, we wanted a developing graduate school. Last, for comparative studies of Moon and Earth, we wanted interesting geological terrain in our neighborhood.

Our group was formed to conduct research, publish scientific records, and instruct graduate students. The accelerating national space programs made our plans particularly timely. Now, after three years of operation, it is worth describing the Laboratory and its facilities, telling about some of its results.

In problems and methods, the study of the moon and planets has some close parallels to geophysics. Geophysicists must concern themselves with the application of physics, chemistry, dynamics, hydrology, aerodynamics, photochemistry, and other basic sciences to interpret observable geophysical phenomena. Similarly, a wide range of physical sciences must be exploited in the study of planets and their atmospheres. But because of the remoteness of these bodies, the methods used must be indirect, based on phenomena observed from a distance.

In this respect, the techniques are analogous with and often identical with those of stellar astronomy. The use of telescopes — ground-based, airborne, and in spacecraft — is very similar. The problems themselves are, however, often very different. On the whole, planetary astronomy is concerned with cold matter, gases or solids in molecular or crystalline form. Thus, while our problems parallel those of geophysics, our techniques are those of stellar astronomy.

The interdisciplinary aspect of planetary astronomy requires special organizational effort. Astronomical telescopes are required, equipped with photometers, polarimeters, spectrographs, and spectrometers ranging over the entire accessible spectral region. There must be equipment for direct photography and for direct visual observation. In addition, matching laboratory programs must be developed in photometry, polarimetry, and spectroscopy to yield quantitative data and interpretations of observed phenomena. This is possible because planetary conditions can be simulated in the laboratory.

Planetary astronomy has traditionally been limited by the terrestrial atmosphere. The lack of high resolution in planetary and lunar photography has discouraged many observers and has led to comparative stagnation in past decades. Recently, with a better understanding of the causes of astronomical seeing, and of the advantages of going to increasingly high altitudes, substantial progress has been achieved, but much more can be done. The enormous cost of the national space program has led to a fresh com-

Harold Johnson works on a digitized infrared three-color photometer inside the dome of the 28-inch telescope. Photograph by Jan Miller.

parison of ground-based, balloon-based and rocket-based research, with the instrumentation of each leading into that of the next. This is a new and exhilarating experience in astronomy. Already it is apparent that determined effort can improve earth-based observations by at least an order of magnitude.

Left: G. Van Biesbroeck measures a photographic plate with the Mann engine at the Lunar and Planetary Laboratory. Photograph by Dennis Milon. Right: E. A. Whitaker (left) and D. W. G. Arthur are two British experts on lunar mapping, now on the LPL staff. Photograph by C. A. Federer, Jr.

In early 1964 Kuiper described the laboratory in a two-part article in *Sky and Telescope*. (Courtesy of *Sky and Telescope* ©1964).

the College of Optical Sciences. Bok then became Steward's new director and chair of astronomy.

The status of affairs by 1964 was described by Kuiper in a two-part article in the January and February issues of *Sky and Telescope*, the first dealing with people and equipment, the second with research.[20] He could already claim important research results: measurement of the isotopic ratios and the discovery of several "hot" bands of carbon dioxide on Venus[21]; experimental verification of the low atmospheric pressure on Mars[22]; detection of new absorption bands in the spectra of Alpha Orionis, Chi Cygni, and other cool stars[23]; and the first observations of the infrared lines of hydrogen in the spectra of early-type stars.[24] The cool stellar spectral bands were soon identified as carbon monoxide and, in the case of Omicron Ceti, water vapor.[25]

Part 1 of the *Sky and Telescope* article shows Johnson at the 28-inch telescope, Van Biesbroeck at the Mann engine (for precise position measurements), Whitaker and Arthur in front of images and maps of the Moon, Herring standing alongside the wooden dome of the 21-inch telescope at the Catalina Station observing facility on Mount Bigelow, a group of students with Herring outside the new LPL wing of the physics building, Robert Wayland (a Scotsman recruited by Kuiper from St. Andrews Observatory who arrived in March 1963) with the large mirror grinding machine, and finally the roll-away shelter for the 28-inch on Site II, a slightly higher site in the Catalinas. The pictures say it all.

Part 2 of the *Sky and Telescope* article describes the lunar work, with figures of Hartmann and his projection setup; Kuiper and Whitaker examining a printed copy of the rectified atlas; Strom with his crater-modeling equipment; spectra of Chi Cygni, Vega, and Betelgeuse obtained by Johnson and his group; and Cruikshank and Owen with the long gas tube for infrared spectroscopy of gas-mixtures. From the beginning, Kuiper had emphasized laboratory studies of gases in long-path absorption cells as an essential part of planetary spectroscopic work, an approach clearly resulting from his association with Gerhard Herzberg at Yerkes in the late 1940s and early 1950s. The expanded LPL facilities included a forty-meter multiple-path cell, one of the largest in the world. The final paragraph stresses the strength of a research institute attached to a university and acknowledges the funding agencies: NASA, NSF, ONR, and the Air Force. Kuiper's new institute had hit the ground running, and he was obviously very proud of it.

The Atmosphere of Venus

Kuiper made several attempts to detect water in the atmosphere of Venus, initially without success. However, in 1962 the group replaced the prisms in their infrared spectrometer with a diffraction grating.[26] This device separates the wavelengths of light more effectively than prisms and there is no loss of light from absorption in the prism. With the new spectrometer the group could determine isotope ratios for carbon and oxygen, which matched those in Earth's atmosphere, and they showed that the carbon dioxide in Venus's atmosphere was in an energized state, suggesting high temperature.[27] We now know that the surface temperature of Venus is nearly 500°C. For many years Kuiper could not decide whether there were water bands in the spectrum of Venus, and he published papers arguing each way.[28] The final verdict was negative.

Kuiper believed that he had identified the cause of the yellow haze on Venus as being due to incompletely hydrated iron chloride crystals, and he produced such crystals in the laboratory. They were crystallized from the vapor as hexagonal platelets. Thus Venus has a halide meteorology, compared with a water meteorology on Earth and an ammonia meteorology on Jupiter and Saturn.[29] This conclusion has not stood the test of time, and we now believe that the clouds on Venus are made of droplets of sulfuric acid.

The Atmosphere of Mars

In 1964 Kuiper and his student Tobias Owen were able to obtain a spectroscopic determination of the surface pressure of Mars.[30] This required laboratory experiments in which long tubes were filled with gases representing the atmosphere of Mars and radiation passed through the tubes to the spectrometer. These measurements were made for various pressures until the Mars spectrum was reproduced. The atmosphere pressure on Mars was determined to be 0.006 atmospheres, the value currently accepted.

The Rings of Saturn

In his chapter in *The Atmospheres of the Earth and Planets* in 1948, Kuiper found surprising infrared absorption bands for the spectrum of the rings of Saturn. He had expected the rings to be composed of rocky materials with bland spectra, but instead the spectra suggested that either the rings were composed of water ice or rocks coated with water-ice frost. In 1966, using new techniques for spectroscopy, L. Mertz and I. Coleman, of Block Associates in Cambridge, Massachusetts, found an absorption band at 1.66 micrometers that could not be due to water but which they suggested was due to formaldehyde.[31] These authors were using a new technique they had developed,

Fourier-transform spectroscopy, in which the entire spectrum is recorded simultane-ously and then a sophisticated mathematical analysis derives the spectrum. Cruik-shank, then a postdoctoral researcher with Kuiper, thought that a more likely expla-nation for the 1.66-micrometer band was that it was from frozen ammonia. The group performed the appropriate measurements of frozen ammonia in the laboratory and then sent a short paper to *Sky and Telescope* that appeared in the January 1970 issue announcing their discovery that the rings of Saturn were made of frozen ammonia.[32]

A month later Kuiper and his colleagues published a letter to the editor of *Sky and Telescope* that began, "A correction is needed to our preliminary announcement concerning the composition of Saturn's rings." They then went on to present several spectra of water ice at low temperatures and concluded that the spectrum of water ice at -190°C "appears to match the Saturn ring spectrum all the way." There is a signifi-cant footnote: "Before the LPL laboratory program was completed, Dr. Cruikshank was informed by C.R. Chapman, C.B. Pilcher, L.A. Lebofsky and H.H. Kieffer that the 'identification of NH_3 frost is not unique' and that 'H_2O frost spectra . . . pres-ent better agreement with the ring spectrum.'" The Chapman group published their refutation of the LPL frozen methane hypothesis and their affirmation of water ice a month later in the journal *Science*.[33] The confusion in the interpretation of the LPL laboratory spectra arose because the spectrum of frozen water is strongly dependent on the temperature of the ice. The spectrum of water ice at a temperature just below freezing is significantly different from the same sample at much lower temperature (which doesn't match Saturn's rings), while the ammonia ice was a better fit. The LPL measurements of water ice were made with relatively warm ice.

The Moons of Jupiter

During his time with Kuiper, Daniel Harris worked on the photometry and colorime-try of planets and satellites, publishing a major review paper on the topic in volume 3 of the *Solar System* series. In the 1950s Kuiper and Harris observed the Galilean satel-lites (Io, Europa, Ganymede, and Callisto), and in 1957 Kuiper presented the results at a meeting in a paper published in *Astronomical Journal*.[34] He reported "striking differences between the four Galilean satellites of Jupiter." He went on to suggest that Europa and Ganymede are covered with snow, in the case of Ganymede with a liberal addition of silicates. He said nothing about Io and Callisto, but it had been known since the beginning of the twentieth century that Io was different to the others, having an unusual orange color. Kuiper's next discussion of the Galilean satellites was sixteen years later, after his move to Tucson, when he summarized the early results he obtained with Harris and added new observations that he, Uwe Fink, and Harold Larson (two

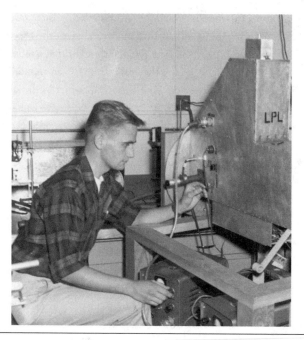

Student Dale Cruikshank adjusting an infrared spectrometer in 1963. Courtesy of Dale Cruikshank.

astronomers recently added to the Arizona faculty) made in 1970 using the 61-inch in the Catalina Mountains.[35] The major conclusion of these studies was that Io was unique among the Galilean satellites for having a thin surface deposit of sulfur and poles of a different color and composition, perhaps consisting of sulfur compounds.

A Strong Connection with NASA

At least one item was already on Kuiper's crowded calendar when he moved to Tucson: the first meeting of the Planetary and Interplanetary Sciences Subcommittee, which was to be held in Washington, D.C., in November 1960. During the busy time preparing for the move, on May 31, 1960, Kuiper received an invitation from Hugh L. Dryden, deputy administrator of NASA, to serve on this committee. It was to be chaired by Homer E. Newell, assistant director of Space Sciences. The relationship between Kuiper and NASA was to be especially important during the next decade or so.

Newell is one of the central figures in the management history of the U.S. space program.[36] Trained as a mathematician, he was hired by the Naval Research Laboratory in 1945 and by 1947 was head of the section organizing upper-atmosphere research with the V2, Aerobee, and Viking launch vehicles. The United States had announced its intention to place a satellite in orbit around Earth in 1954, and as the International Geophysical Year (1957) approached, Newell was responsible for the Navy's Vanguard project to accomplish this. During this time he chaired the advisory committee that resulted in the creation of NASA, the agency that would absorb all government space activities. Newell joined NASA in the year of its creation, 1958, and rapidly rose through the ranks. First he was assistant director for space sciences, then in 1960 was promoted to deputy director of space flight programs. A year later he was director of space sciences, and two years later he was associate administrator for space science and applications. He reached the rank of associate administrator of NASA in 1967, where he remained until retiring in 1974.

When NASA was formed in 1958, funds became available from the government civilian agency to support space and aeronautics research. In April 1960, President John Kennedy made his famous speech at Rice University, declaring that the United States should commit itself to landing a man on the Moon by the end of the decade. How does a nation power up for such a massive undertaking, especially when its track record for rocketry was so checkered? At the time, NASA was not considered an important government agency; in fact, Kennedy had to invite nine people before he found a successor to the first NASA administrator, T. Keith Glennan.[37] Fortunately for the nation and its space ambitions, when the finger came to rest, it was pointing at James Webb, arguably one of the most able administrators NASA has had. He came to office when the agency most needed inspired leadership. Webb made the decision to spend one-third of the agency's budget on space science and that the universities would be heavily involved. This way he would build up the academic force for space exploration. He also saw a major role for the robotic exploration of the solar system, rather than having a single-minded focus on getting men to the Moon. The focus on robotics was genius, and when the Congress and public lost interest in Moon landings after Apollo 17 in 1972, the agency, and the country, entered its golden age of planetary exploration.[38]

Another NASA administrator that was to be important for Kuiper was Bill Brunk.[39] Brunk obtained BA and MA degrees in astronomy from the Case Institute of Technology in 1952 and 1954, respectively, after which he obtained a position as a research scientist at NACA's Lewis Flight Propulsion Laboratory in 1954, which was renamed Lewis Research Center when NASA formed in 1958 (it is now named NASA John H. Glenn Research Center at Lewis Field). He returned to the Case Institute in 1963 to earn his PhD in astronomy and then accepted a staff-scientist position for the proposed Grand Tour program, in which NASA would send two spacecraft to Jupiter, Saturn, and Pluto and two to Jupiter, Uranus, and Neptune. Ultimately the

Grand Tour program was cancelled but replaced with the Voyager program to Titan and the outer planets in 1964. In 1965 he became the chief of planetary astronomy at NASA headquarters.

In this climate NASA offered several grants to universities to construct buildings that would house NASA-supporting projects, and Kuiper was well placed and motivated to respond. Proposals were not to come from individuals but from the universities, which were required to establish an infrastructure to make good use of the buildings.

A Dedicated Building

In mid-1962 a south wing was added to the Physics, Mathematics, and Meteorology Building at the University of Arizona, and the entire LPL moved into the top floor. The need for more space remained, and the mathematics department was persuaded to vacate 2,500 square feet immediately below LPL's spot, and the physics department gave up some space in the basement. On May 15, 1963, Kuiper wrote a proposal on behalf of the University of Arizona requesting funds for the construction of a new building to house LPL and parts of other departments with space-related interests.[40] The new Space Sciences Building was to be a five-story structure with 51,600 square feet at a cost of about $1.2 million ($12 million in today's dollars). The award was funded, and William Wilde and Dennis Brizee were hired as the architects. W. F. Conelly Construction Co. began construction 1965 and finished in fall 1966.[41]

The first floor was mostly an optical shop. A mezzanine allowed the optical shop to be two stories but also allowed space for photography facilities and a globe-projection tunnel. The second floor was electrical and mechanical shops and infrared laboratories. The third was a physics laboratory, library, conference room, and the director's office. The fourth housed the computer, astronomy, and optical laboratories. The fifth housed a lecture room and space for psychology, geochronology, and graduate-student offices. Sprinkled throughout the building were offices for the one hundred members of the laboratory. In April 1974, just eight years after the building opened, the Space Sciences Building was renamed the Gerard P. Kuiper Space Sciences Building.[42]

The building was occupied by October 1966, and on January 26–27, 1967, a dedication ceremony and a symposium were held.[43] On the first day, Jack Williams, the new governor of Arizona; George Chambers, the president of the board of trustees; Newell from NASA; and Les Hogan, chairman of the broad of Motorola (a representative of industry) gave greetings. Frederick Seitz, president of the NAS, gave an opening address, and David Patrick conducted the dedication ceremony. In the afternoon, after a tour of the building in the morning, there was a series of talks on space-science

(a) Architect's drawing for the new Space Sciences Building. Courtesy of University of Arizona Libraries, Special Collections. (b) Harvill, Newell, Kuiper and Brunk during the groundbreaking ceremony for the new Space Sciences Building. Courtesy of University of Arizona Libraries, Special Collections. (c) The new Space Sciences Building at the time of its opening. The elevator shaft and stairwell are a floor higher than the building in the hope that an additional floor might be added. Courtesy of University of Arizona Libraries, Special Collections.

(a) The formal proceedings and the technical seminars were held in the Students' Union. Courtesy of University of Arizona Libraries, Special Collections. (b) The visitors move from the Students' Union to the new building for a building tour. Courtesy of University of Arizona Libraries, Special Collections.

programs in Arizona, with speakers summarizing the activities of Kitt Peak, the Lowell Observatory, the U.G. Geological Survey, and the Smithsonian Astrophysical Observatory, which was building an observatory fifty-five miles south of Tucson on Mount Hopkins. The various units of the university—LPL, Steward Observatory, the Optical Science Laboratory, and the Space Science Committee, which handled NASA funds within the university—also had talks.

On the second day was a technical symposium, with sessions on the gross features of the lunar surface, surface structure by telescope studies, results from lunar missions, and a comparison of lunar and terrestrial features.

As the day came to an end, the astronauts preparing to fly the first Apollo spacecraft to the Moon were performing tests in the command module. Suddenly there were screams from the crew, and the ground crew quickly realized the interior of the spacecraft was a ball of fire. What came to be known as the Apollo 1 mission came to a tragic end with the loss of all three astronauts. It was sobering news to hear at the end of two days of celebration.

The building and the institutional grant from NASA associated with it were managed by the University Space Sciences Research Committee established by President Harvill on December 17, 1965.[44] The committee would report to David Patrick. The minutes of the committee's meeting of October 5, 1967, illustrates how it functioned.[45] In October 1967 the committee consisted of Albert Weaver (chairman of the physics department), Gerard Kuiper, Aden Meinel, Mattson (presumably Roy H. Mattson of one of the engineering departments), and Melvin Simmons (business manager for LPL and secretary of the committee). On that day the committee considered six proposals. One was declined because it consisted simply of travel for a student. Two were funded—one for laser noise measurements and one to purchase an interferometer for Kuiper, Johnson, and Low to use for infrared astronomy. The minutes do not say whether Kuiper recused himself during that discussion. One proposal to the committee was awarded seed money to enable some colleagues to prepare a proposal to NASA. Two more were deferred pending more information.

The same committee also determined how space was used, and this was the next item on the agenda. Kuiper submitted constant requests for more space as LPL grew throughout the 1960s, often resorting to clandestine tours to inspect how other departments used the building. One report, submitted to the committee on September 19, 1967, referred to one large room that had been assigned to the Chemical Engineering Department but only "has one student there (who looks like a soldier maintaining a distant outpost, sometimes asleep)."[46] The university believed the building had been acquired to cultivate interdepartmental programs and was reluctant to see departments ousted one by one to make room for Kuiper's LPL, but it was a process that was hard to resist.

More Hiring

From 1960 until about 1967, the number of members of LPL increased from ten by about fifteen per year until it leveled off at around one hundred.[47] Strom joined LPL in 1963 from the UC Berkeley, where he had been a research geologist in the Space Science Laboratory.[48] His BS (1955) from the University of the Redlands and MS (1957) from Stanford University were in geology, and after graduating he obtained a position in the oil industry in Pakistan. There he saw a copy of Patrick Moore's *Guide to the Moon*, which attracted him to space science.[49] He partnered with Whitaker to work on problems related to the lunar surface and found himself involved, with others in the laboratory, in astronaut training.

Also in 1963 the eighty-three-year-old Van Biesbroeck, long since retired, followed Kuiper to Arizona to continue observational astronomy.[50] He remained active until his death in 1974.

An insight into Cold War conditions at the time is an anecdote concerning Wieslaw Wisniewski, a Polish astronomer who also joined LPL in 1963.[51] After obtaining his degrees and working for several years at various universities and observatories in Poland, he approached Kuiper about a position in Tucson. Under Cold War conditions communications about leaving Poland were difficult, so Kuiper agreed to send a postcard to Wisniewski, black and white if he did not have a job, color if he did. Wisniewski worked at the LPL between 1963 and 1967 and then again from 1972 until his death in 1994. Mostly he was interested in infrared galaxies and the photometry of asteroids. His photograph of the fragmentary nature of the nucleus of comet Shoemaker-Levy 9, before it impacted Jupiter in 1994, became famous.

While Kuiper pioneered infrared astronomy through his use of the lead sulfide detector, which had a range of about 1–2.5 micrometers, Low took it to the next stage using his solid-state, gallium-doped germanium infrared detector, which extended the range to 0.7–1,200 micrometers.[52] Low joined LPL in 1965. Born in Mobile, Alabama, he obtained his first degree at Rice University and a PhD from Yale University, both in physics. After graduation, he was employed by Texas Instruments, where he developed the new detector. Now infrared astronomy could include not just the bands caused by molecules but the whole range of thermal emissions, and infrared astronomy in its broadest sense could truly be said to have been born. After testing the detector at the Green Bank Radio Telescope, Low realized that the full strength of the detector could not be realized at ground level because of the absorption of infrared radiation by water in the atmosphere. Thus in 1965 he moved to the University of Arizona to fly the instrument on the Skywarrior aircraft being used by Kuiper, with some additional experiments on a Lear Jet.[53] The instrument also found its way onto space-borne observatories, such as the Spitzer Space Telescope.[54] Arnold Davidson of the National Radio Astronomy Observatory accompanied Low, and together they used the telescopes at Mount Bigelow in 1963 and McDonald Observatory in 1964 to make the first infrared observations of Mars, Jupiter, Saturn, Titan, and twenty-four stars, and they expanded the stellar color description to create the UBVRIJKLMN system.[55]

They set up a laboratory to build sensitive detectors and added infrared detectors to the 60-inch telescope. The data were gathered on punch-paper tape and then transferred to IBM punch cards. Kuiper bragged that the system was "so highly automated that a pencil is never used."[56]

The first woman professor was hired by LPL a few years later, in 1966. Elizabeth Roemer completed a BA and PhD at UC Berkeley and spent several years at various observatories.[57] In forwarding a proposal from her to NASA in July 19, 1966, Kuiper remarks, "We regard ourselves very fortunate that Dr. Roemer has joined our staff here. She is probably the most active astronomer in the United States working on comets and is extremely well informed on both the physical phenomena of comets and the relationship between comets and the methods of orbit determination."[58]

George Rieke was a Harvard physicist who joined LPL in 1970 and focused on galaxies. He worked with Johnson and once remarked that in those days you could build something, observe anything, and find something interesting.[59] Rieke and Low published a book chapter on infrared telescopes and detectors that was essential reading for a generation of infrared astronomers.[60]

In 1972 William Hubbard, who was an assistant professor at the University of Texas at Austin, was hired as associate professor at Tucson. He obtained his first degree at Rice University, his PhD at Berkeley, and was interested in the interiors of the major planets.[61] He has had a distinguished career, being director of LPL between 1978 and 1981, and is now an emeritus member of faculty.

In just over a decade, the LPL attained independent status within the university and had its own building, thriving research projects, graduate students, and a distinguished staff.

CHAPTER 15

Telescopes and the Editorial Projects

THE AGILITY AND VERSATILITY of Kuiper's mind was probably never better demonstrated than in his years at the University of Arizona. Contrary to the trend at the time, which was to campaign for ever bigger telescopes that he actively shared, he built a nest of modest telescopes around Tucson that were well suited to planetary science. Following the example of the grand nineteenth-century astronomical observatories, he created a publication for primary research produced by the LPL. He also maintained a major effort collecting review papers for publication in books, a tradition that survives at the University of Arizona to this day. Ultimately, he grew LPL until it became a teaching department offering degrees. Moreover, he was an engaged citizen, giving public lectures, writing articles on matters of public interest, and even writing to the city about poor construction methods being used for a local hotel. However, the stress of all these efforts set him on a road to serious health problems, leading to his resignation as director of LPL in 1973.

The Telescopes of Tucson

The history of the Tucson telescopes was described in several articles by Kuiper and by Whitaker in his history of the laboratory.[1] In 1961 Kuiper began planning for a 60-inch telescope, but recognizing the learning curve this represented, he obtained funds from the Naval Ordnance Test Station in California to build a 21-inch.[2] The design was by Johnson, with an emphasis on low fabrication and maintenance costs, and followed that of a similar instrument at Lowell Observatory. The telescope,

(a) Dome housing the 21-inch telescope at the Catalina Observatory. Courtesy of LPL. (b) The 61-inch telescope at the Catalina Observatory. Courtesy of LPL. (c) Robert Wayland and Ed Plamondon in LPL Optical Shop. Courtesy of LPL. (d) Kuiper's private observatory. Courtesy of LPL. G. P. Kuiper, "No. 172. The Lunar and Planetary Laboratory and Its Telescopes," *COMM* 9 (1972): 199–247.

the first built by the LPL, was placed on the Catalina site. When replaced by the 0.7-meter Schmidt telescope in 1972, the 21-inch was moved to Tumamoc Hill for pre-spacecraft synoptic (general, or survey-mode) imaging of the planets. It sits there today, unused. Johnson came to the LPL in 1962 after a career developing photoelectric photometry and becoming a specialist in low-cost telescopes.[3] After working in the Radiation Laboratory at MIT during the war, he undertook a graduate degree at Berkeley developing photoelectric methods for stellar photometry. This was followed by short periods at Lowell and the University of Wisconsin–Madison, and eventually a position at Yerkes for two years (1950–52), where he met William Morgan. Johnson introduced the UBV system for stellar photometry in collaboration with Morgan and Harris, a system that was rapidly and widely adopted.[4] Not liking teaching or the Wisconsin skies, Johnson sought to return to Lowell Observatory, where he moved in August 1952. When Kuiper was given the go-ahead to hire astronomers for the new astronomy department at the University of Texas, he recruited Johnson, who moved to McDonald Observatory in 1959. Four years later, Kuiper persuaded Johnson to join him in Tucson.

The task of finding a site for these telescopes fell to Kuiper. Whatever space Kitt Peak and Steward Observatory had was reserved for their growth. The highest two peaks in the Santa Catalina Mountains were already occupied, Mount Lemmon by an Air Defense Command radar station and Mount Bigelow by microwave towers belonging to Mountain Bell. Kuiper chartered a small aircraft to fly over the Catalinas and other mountains to the south and southwest of Tucson looking at air turbulence, ground cover, haze, and smoke. He looked at weather records, considered closeness to campus, accessibility (especially during winter snows), height of the inversion layer, and height of water vapor in the atmosphere. He settled on the Catalinas and asked Herring to perform site tests. Herring spent four months on the task.

After a small fire caused Mountain Bell to abandon the Mount Bigelow site, Kuiper obtained a lease from the Forestry Service and started preparing the site for the 21-inch, which was to be housed in a twenty-foot dome. Kuiper referred to this site as Site I.

In 1961 Kuiper submitted a proposal to NASA to fund the construction of a 61-inch reflector. The proposal was successful and site selection began. A flimsy twenty-foot-high tower supported a ten-by-twelve-foot wooden shed with slide-off roof into which Herring's 12.5-inch telescope was placed to test the seeing. In the fall of 1963, while the mirror was being prepared on campus, smoke bombs and hydrogen balloons were used to evaluate wind conditions. The tower was dismantled in mid-1964 and the dome for the 61-inch was constructed in 1965. The telescope was installed in the first week of October. In 1966–67 a ten-room dormitory was constructed downhill of the dome. The ninety-thousand-dollar building has six rooms, three with their own bathrooms, living room, and kitchen. On one side are single rooms labeled U, B, and V, and on the right side are K, L, and M (K was Kuiper's room), a nodding tribute to Johnson's UBVRIJKLMN system for star characterization. Kuiper thought it would be a good place to come up and relax, but he never had the time. The dormitory is accessible from the dome through a tunnel that was essential during the winter when eight-foot snows are common. At times the dorm and the boundary gate would be buried.

The telescope and its mount were designed by Johnson and Sam Case, an engineer with no previous experience in telescope design, using commercial items as much as possible. Construction was by Western Gear in Lynwood, California. The mirror was manufactured in-house by Wayland, who arrived at LPL three months before the Pyrex blank did. At that time the optical shop was located at 1530 E. Broadway, about three miles from campus and far from perfect. Wayland and his assistants would have to work unusual hours to avoid mid-day heat that would distort the glass. Construction of the observatory was started in September 1964 and completed the following May, and the telescope was installed. On October 7, 1965, the telescope was completed. Kuiper, Wayland, and Whitaker drove to the dome with whatever eyepieces they could find to give the telescope "first light," looking at the brightest star, Vega.

Other nearby sites were leased and telescopes installed. A 28-inch (71-centimeter) telescope for Johnson's group was installed in a corrugated-steel roll-off shelter, and a second-hand trailer was purchased for observers. In 1969 a 40-inch (102-centimeter) telescope was installed on Soldier Peak, about three hundred feet higher than the earlier sites. In 1970 the peak of Mount Lemmon became available when Air Defense Command abandoned a radar station. In competition with several proposals, Kuiper's proposal to make the site an infrared observatory was accepted, and the 40-inch (102-centimeter) and two 5-foot (1.52-meter) reflectors moved to the site. Johnson's 28-inch (32-centimeter) observatory was moved to the Mount Lemmon, and its previous site was returned to the Forestry Service.

Kuiper owned a private observatory that was built alongside the pool in the back yard of his home on Sawtelle Avenue. The structure is still there forty-four years after his death. The dome housed a 12.5-inch (32-centimeter) reflector built by Cave Optical Company that was "quite satisfactory." Kuiper explained the purpose of the observatory in a letter to the president of the Ash Manufacturing Company, which manufactures observatory domes.[5] "The purpose of my observatory is 2-fold: a serious aspect of being able to watch the cloud developments, particularly on Jupiter, and alert LPL staff to events that might otherwise have escaped them; and a more personal one of enjoying the views of moon, planets, and stars or nebulae, with guests."

It is interesting to note that no one else in the United States was building small telescopes on multiple sites. This direction was contrary to current trends in astronomy, which were toward large telescopes built at remote sites such as Hawaii and Chile. However, as Kuiper demonstrated, many projects are best served by small telescopes, such as the search for asteroids.

The Editorial Projects

An ever-present entry in Kuiper's progress reports and plans were something he called "the editorial projects." Since his success with *The Atmospheres of the Earth and Planets* in 1949 there was seldom a time when Kuiper was not editing a book or LPL's own publication, the *Communications of the Lunar and Planetary Laboratory*.[6]

For the first year or two after Kuiper moved to Arizona, he remained in contact with Chandrasekhar, but their correspondence rapidly tapered off. Chandra was upset that Kuiper had left Chicago without discussing it with him, but Kuiper probably also wanted separation from the department that had rejected him, even though this included his longtime friend. In one of their last exchanges Chandra wrote that he thought it was a mistake to publish the *Communications* and instead research should be published in the established peer-reviewed journals. On October 1, 1963, he wrote to Kuiper, "I must express my disappointment at your decision to publish much of your work and that of your associates in your private communications. I think this is a mistake. I strongly believe that scientific work, particularly that of importance, should be recorded in standard scientific periodicals which are accessible in all relevant libraries."[7]

Kuiper was insensitive to Chandra's words. Eight days later he replied, "I had a feeling you might disapprove of my decision of setting up our own series of publications."[8] He went on, explaining that "large plate collections could not be published in current periodicals," and neither could "extensive catalogs such as Communications No. 11 and 30" or the large collection of infrared spectra. There were also "subsidiary reasons": The *Communications* serve as progress reports to send to NASA and other funding agencies. They are library-exchange materials, giving free access to numerous publications. Publication is rapid. Control over the quality of reproductions is greater. And probably most important for someone rapidly building up a massive empire at Arizona, "We are merely following the traditional example of observatories in the past which found it necessary to publish their bulky manuscripts and catalogs in their own series." But as is often the case with arguments of this type, the strongest reason probably went unspoken: self-publishing avoids the tedium and frustration of having to deal with external reviewers.

Chandrasekhar was not the only person to object to Kuiper's self-publication procedures. NASA, his major funding agency and the body to which he had to make his progress reports, also felt uncomfortable with the *Communications*. On June 30, 1966, Kuiper wrote a longer and more carefully crafted justification to Bill Brunk, chief of planetary astronomy at NASA.[9] The reasons had changed somewhat, now consisting primarily of cost. The peer-reviewed journals had instigated page charges whereas the *Communications* were not only cheaper, but also the cost was partly carried by the University of Arizona Press, who were selling copies. The arguments about exchange, reporting, the traditions of the great observatories, and the added value of having all their work in one place, were repeated and embellished. This time the previously unspoken reason was spoken — the problems of peer review. Incompetent referees had recently prevented Owen from publishing in *Astrophysical Journal* and Arthur from publishing in *Icarus*. Thus in one breath, one of the major tenets of scientific publishing, peer review, was swept aside.

Brunk was extremely important for Kuiper's efforts. Not only was he the source of NASA funding for most of his research, which for 1966 was half a million dollars (nearly $4 million in 2018 dollars), but Brunk also was someone he shared thoughts and news with. Kuiper's letters to Brunk were almost a diary. Struve and Chandrasekhar had served the same purpose in earlier years. The Brunk folders of the special collections at the University of Arizona show that in 1967, for instance, Kuiper sent thirty letters to Brunk and received three.[10] He shared images and spectra, discussions of their interpretation, reports of every Convair flight and what was achieved, every new instrument put on the plane, air conditions, altitudes, flight durations, and problems. Kuiper explained to Brunk that his own main research endeavors during the 1960s were obtaining high-resolution infrared spectra of the Sun and planets from NASA's aircraft and learning about the lunar surface from the Ranger and Surveyor space missions.[11]

Most of Kuiper's papers during this period concerned his progress with high-resolution infrared spectroscopy of the stars, planets, and the Sun. He took spectra of Venus (in 1962, 1967, and 1969 to determine the composition of the atmosphere and the particularly vexing question of whether water was present), stars (A0–B8 stars in 1963, looking for the completeness of hydrogen lines), the Sun (he produced a series of papers on the spectrum of the Sun from 1967 to 1970), Mars (in 1963 to determine the minor constituents of the atmosphere), and Jupiter (in 1972, in particular the Great Red Spot and the white ovals). But the letters flowing between Kuiper and Brunk went further. They reported events such as the new academic programs at Arizona, public talks Kuiper had been invited to give (with detailed itineraries), and frustrations.

Whatever the downsides of self-publication, nowhere is the depth and breadth of the activities in LPL, and the vigor of the group Kuiper assembled, better illustrated than in the pages of the *Communications*. In eleven years (1962–73) Kuiper published 195 articles in the *Communications*, authoring or coauthoring 45 himself, and ten or more were authored or coauthored by Arthur, Cruikshank, Johnson, Gehrels, Hartmann, Van Biesbroeck, and Larry Bijl, a young Dutchman who was with Kuiper for a year or two working on the solar spectra. In all, eighty-one persons published in the *Communications*. Many of the papers could have been published in the major peer-reviewed journals, some were reprints of articles that had appeared elsewhere, and many were short records that most research groups would have been happy to see only in student theses.

Kuiper became convinced that the particulates in the atmosphere of Venus were halides.[12] On June 28, 1969, he wrote to Brunk, "We have made what I regard as a breakthrough. It is clear now that the extremely low water-vapor content of the Venus atmosphere, discovered in our 1967 CV-990 operations, has caused Venus meteorology to be a halide meteorology, instead of a water meteorology. . . . The existence of a halide meteorology is predictable on the basis of a composition of terrestrial volcanic gases if water were to be subtracted." He then pointed out that ferrous chloride has the right optical properties to also explain the polarization data.

He also published articles describing the structure and management of the laboratory and his hunt for high-altitude sites for ground-based infrared astronomy.[13] Many of the sites visits he conducted doubled as opportunities to study volcanic features he could compare with features on the Moon.

Three-quarters of the articles in the *Communications* did not have Kuiper's authorship. Arthur published twenty-four papers, mostly on selenography—the nature of the lunar surface and its cartography. Cruikshank mostly obtained laboratory spectra of gas mixtures to compare with planetary spectra. Johnson specialized in instrument development. Gehrels focused on polarization measurements. Hartmann on craters and crater statistics. Van Biesbroeck studied comets mostly to determine their orbits. Bijl played a major part in determining the solar spectrum. Herring published a series of papers detailing his sketches of the Moon's limb areas. Whitaker published on the Ranger VII and Surveyor I data.

One paper Kuiper rejected from the *Communications* was a paper by Barbara Middlehurst and her colleagues (Jaylee Burley of the Goddard Space Flight Center, Patrick Moore of Armagh Planetarium, and Barbara Welther of the Smithsonian Astrophysical Observatory). The topic was transient lunar phenomena, a topic of considerable interest throughout the 1960s. Reports of short-lived glows or color changes from the Moon go back to the twelfth century and have been reported by astronomers as distinguished as William Herschel. Middlehurst published several papers, including some in *Nature* and *Science*, the most prestigious scientific journals.[14] The paper she gave Kuiper was a catalog of reported observations, but his reception of the paper could not have been worse. Kuiper thought the majority of the entries had simple explanations or were clearly unreliable.[15] Kuiper's final advice on the matter is classic Kuiper: "As a practical matter, I would suggest that your list be very drastically reduced in length after consultation with one or two competent observers, like Whitaker, and that a vigorous program of 'event' watching and analysis be started, not by amateurs but by professionals." Undaunted, Middlehurst and her coauthors sent the paper to NASA, who published it as a technical report, and as such it has become well known.[16]

Barbara Middlehurst was born in Penarth, Glamorgan, Wales, in 1915 and attended Girton College, Cambridge, where she studied mathematics.[17] A year later she returned to teach. During World War II she served as an ambulance driver and became interested in astronomy. In 1951 she began work at the St. Andrews's University Obser-

(a) Barbara Middlehurst. Courtesy of LPL. (b) Ralph Turner in his studio producing a model of the central peak of lunar crater Alphonsus for the NASA Ranger IX report. Courtesy of LPL. G. P. Kuiper, "No. 172. The Lunar and Planetary Laboratory and its Telescopes," *COMM* 9 (1972): 199–247.

vatory, and in 1959 moved to Yerkes Observatory and then to LPL, where she stayed until Kuiper died in 1973. After a short stint as astronomy editor with Encyclopaedia Britannica in Chicago she moved to the Lunar and Planetary Institute in Houston, where she worked during 1973–74. She died in Houston in 1995. During her time with Kuiper she did succeed in authoring or coauthoring several papers in the *Communications*. One dealt with the photometry of galaxies and nebulae,[18] and two dealt with the compositions of stars.[19]

Reviewing Middlehurst's paper on transient lunar phenomena must have reminded Kuiper of a previous encounter with the suggestion that volcanism had been observed in Moon's crater Alphonsus.[20] In November 1958 Western newspapers reported the Soviet astrophysicist Nicolai Alexandrovich Kozyrev had obtained a spectrum of active lunar volcanism. The historian Ronald Doel has written at length about this incident in his article "Evaluating Soviet Lunar Science in Cold War America."[21] Doel argues that the incident demonstrates the many socioscientific pressures of the time and Kuiper's mixed roles as scientist, administrator of a major observatory, and government advisor during the Cold War. Kuiper was convinced that lunar craters were produced by impact, yet he had observed what he interpreted as solidified lavas that had once flown in the great basins. Kozyrev's claim was therefore extremely important to him. However, after considerable effort to evaluate the claim, which involved consultations with numerous Soviet astronomers, the matter faded without satisfactorily resolution.

Two unusual papers in the *Communications* were written by Ralph Turner, an artist hired by Kuiper to help visualize the lunar landscape by making models. He described his process in some detail. He used ground-based and Orbiter images of features on the Moon under a variety of lighting conditions. Next to his desk he had a large vertical tray of clay. He would mold the features of the lunar surface into the clay to reproduce the features and shadows. He then repeated this with a second photograph under different lighting conditions. His task was finished when he made a model consistent with all the photographs and all lighting conditions. His articles described models of the eastern portion of Schröter's Valley and the northeast rim of the crater Tycho.[22] One of Turner's models hangs in the Kuiper Space Sciences Building to this day. Turner was also responsible for making planetary globes using the best photographs available. These were described by Kuiper in his two-part 1972 article on LPL in *Sky and Telescope*.[23]

Stars and Stellar Systems Books

Kuiper had long been interested in a publication to replace, or compete with, *Handbuch der Astrophysik*, which was originally published 1928–33.[24] This time, he would identify topics and then commission others to serve as editors to help find authors and organize peer review and revision. There would be ten volumes of roughly 750 pages each, and with about one thousand copies sold, the cost of each would have to be $150. A new edition of the *Handbuch der Astrophysik* was published in the 1950s, but Kuiper continued undaunted. Kuiper wrote to Struve on this topic on November 29, 1954.[25] "I have given considerable thought about the future of the Stellar Systems books," he began. He went on to ask for suggestions for editors for the series, hoping Struve would edit the first volume. "It would assist me greatly in my further dealings with NSF if provisional acceptance of the editors were assured." The process of identifying editors was slow and complicated. For instance, Kuiper wrote Struve on January 5, 1956, complaining that Jesse Greenstein had still not decided on whether to edit volume 6 and Albert Whitford, then director of the Lick Observatory, had not replied to him about editing volume 4. Kuiper mentioned that he planned a trip to Lowell in connection with one of the chapters in volume 3.[26] He was so dedicated to the book editing that he "asked Miss Groeneveld to make the asteroid observations during the first session at McDonald" and has "abandoned any plans for a trip to California this winter."

Kuiper approached NSF to pay to produce the series, and NSF turned to the astronomical community to determine whether the foundation should be supporting book production.[27] The decision was positive, and the NSF funded the series.

The title of the series was *Stars and Stellar Systems: A Compendium of Astronomy and Astrophysics*, with Kuiper and Middlehurst as executive editors. Seven volumes were published during his Tucson years, and one posthumously.[28] The first volume, *Telescopes*, was edited by Kuiper and Middlehurst and appeared in 1960. Volume 6, *Stellar Atmospheres*, edited by Greenstein, appeared in the same year. Two years later volume 2, *Astronomical Techniques*, appeared under Hiltner's editorship. Volume 3, *Basic Astronomical Data*, edited by K. A. Strand, appeared the following year. Then in 1965 volume 5, *Galactic Structure*, edited by Adriaan Blaauw and Maarten Schmidt, at Mount Palomar, and volume 8, *Stellar Structure*, edited by Lawrence Aller and Dean Benjamin McLaughlin of the University of Michigan, were published. Volume 7, *Nebulae and Interstellar Matter*, edited by Middlehurst and Aller, appeared in 1968. The last volume, *Galaxies and the Universe*, edited by Allan Sandage, Mary Sandage, and Jerome Kristian, appeared in 1975. While listed in the preface to the other books, volume 4, *Clusters and Binaries*, appears never to have been published.[29]

The reviews of the series were generally favorable. For instance, John Irwin of Indiana University described in detail the contents of volumes 2 and 3 and thought them a "must read" for both professional astronomers and graduate students.[30] He ended one of the reviews with a prophetic observation that was to be an important theme in twentieth-century planetary science: "We enter the expensive 'Space Age' caught in a strait jacket. Suddenly astronomy has become much too important for its instrumentation to continue to be supported only by the occasional generosity of a millionaire," referring to the history of observatories being funded by millionaires and noting that space missions will require high levels of government funding.

An Engaged Citizen

Kuiper was always an engaged citizen. During the war he had paid as much attention to the state of astronomy and the European observatories as he did his official tasks and his personal search for infrared detectors. He wrote an article on German astronomy during the war and reported on the state of French and Dutch astronomers.[31] It's no wonder that when described by a newspaper as one of the Unites States' "Top-Secret spies" he protested, saying he was simply a "historian or a sociologist."[32] His Tucson years saw an increase in his interest in the world around him. His concerns ranged from research and education to politics, air pollution, building safety, and noise pollution.

Within days of the Moon landing, in response to an article that had appeared in the *Arizona Daily Star*, Kuiper wrote to the editor to point out that the landing was a political enterprise and not a scientific one.[33]

In September 1971 Kuiper was asked to address the students at the U.S. Army Intelligence Center and School at Fort Huachuca in Sierra Vista, about ninety minutes southeast of Tucson.[34] University Provost Al Weaver had identified a series of lecturers that suited the course required by the army, and Kuiper was asked to talk about "science intelligence." Kuiper described his wartime experiences, German preparations for warfare, aerodynamics, aviation, rocketry, submarine warfare, nuclear physics, and his own specialties of infrared detectors and radar countermeasures. He also showed off a Nazi flag and whip he had "captured" from "a German concentration camp whose inmates produced the German V2's under the general direction of Wernher von Braun and the SS." The event was not a success. Kuiper was furious at the poor reception he was given by the students. Some fell asleep. Only three asked questions. This was despite them being senior officers with ranks major to colonel. They showed a "level of knowledge somewhere between high school and junior college,"[35] Kuiper recounted. When invited to join the speaker for lunch, only three "volunteered" out of a class of twenty-five to thirty. He wrote lengthy letters to Weaver. Kuiper argued that the university should reconsider supplying speakers for this purpose. Kuiper also wrote to his local U.S. representative, Mo Udall, who sympathized and promised to speak to the appropriate people in D.C. about the quality of the army officers.[36] After a decade or so of campaigning for better education in the United States, Kuiper felt that his encounter with the students at Fort Huachuca was a validation of his concerns. President Harvill of the University of Arizona shared Kuiper's concerns to the point of wanting to discuss the matter over lunch.[37]

On February 1, 1971, Kuiper wrote to Harvill that Arnold Evans, superintendent of LPL observatories, and Melvin Simmons had observed that the University Inn being built on the corner of Campbell and Speedway was unsafe.[38] "There was insufficient steel put in the building to reinforce the thin and rather flimsy-looking walls," wrote Kuiper. He called the city manager and a few days later heard from the city inspector's office that they agreed and had already required modifications to reinforce the structure.

On August 16, 1972, it was the noise from helicopters that troubled Kuiper. He fixed this with a letter to the mayor. He later wrote about this to Harvill: "There have been no close helicopter passes since then and no low-level tight circular passes at all. I suppose the average noise level due to helicopter has been cut by a factor of more than a thousand in my area."[39]

Two days later it was a lunch invitation that prompted a letter to Harvill, but he took advantage of the opportunity to offer congratulations on the president's new home, which he had "noted in past evening walks with Mrs. Kuiper as very attractive. . . . You will thus be able to enjoy the soft street lights in our district for which I fought so hard with the Mayor and Council some years ago."[40]

Kuiper's concerns were not just restricted to local matters. On March 15, 1971, the *Arizona Daily Star* ran an article on the state convention of Young Republicans that had been held the previous evening.[41] The article concluded with a paragraph about Kuiper: "The University of Arizona astronomer Gerard P. Kuiper spoke earlier on foreign policy implications of the U.S. space program and showed unpublished photographs from the Apollo 14 mission. Kuiper's appearance emphasized the convention's theme, 'Make the Moon Republican.'" Kuiper retorted in a way that revealed his attitude to the Vietnam War versus the space program. He wrote to Harvill,

> I must say that I was somewhat taken aback by the nightclub atmosphere of the meeting and had a feeling that few paid attention to what I said. I did point out that the Vietnam venture cost the country 10 times the budget of the space program and that from my personal experiences abroad, in both Western and Eastern Europe, I know that the U.S. space program is considered an enormous asset to our country, off-setting to some extent the unfavorable aspects of our Asian venture. I also stressed the revolutionary progress made in industrial production and quality control as a result of the U.S. program, as well as the benefits to the physical sciences practiced in all major U.S. universities.[42]

Kuiper was never more engaged as a citizen than when it came to air pollution around Tucson. The air quality over the observatories was being threatened by haze and smog. The cause was not obvious, but some blamed the copper-smelting activities in the area. As the Spanish explorers found, Arizona had a lot of copper deposits, and they are often associated with gold, silver, and molybdenum. There was little commercial gain in mining these elements until the Southern Pacific Railway came in 1876. Then Arizona became the leading copper-producing state in the country. But the citizens of Arizona, especially its astronomers, paid for this production in the form of air pollution.[43]

Others disagreed that the mining operations were to blame for the Tucson smog. By the 1960s the population of Tucson had reached a quarter of a million, and the downtown streets were filled with large gas-hungry automobiles pouring pollutants into the air. Others, including an anonymous meteorology professor quoted in the *Tucson Citizen*, continued to blame the mining operations. In 1954 the Arizona legislature created the Institute of Atmospheric Physics to consider such issues,[44] and in 1958 (two years before Kuiper's arrival) the university created the Department of Atmospheric Sciences.

Kuiper was a member of the Pima County Air Pollution Advisory Council from 1965 until 1973, for which he wrote annual reports and took photographs of the air pollution in the Tucson basin. The authorities in Tucson set up a committee to study the causes of air pollution and invited three researchers from the university to join the committee: Kuiper, Hartmann, and Roger Caldwell, a professor in the College of Agricultural and Life Sciences.[45] The group wrote several reports, and Hartmann even published a paper in the *Journal of the Arizona Academy of Science*.[46] Gathering data from the U.S. Weather Bureau and the Pima County, Arizona, Health Department, Hartmann plotted the number of days with visibility less than sixty miles for 1957 to 1970. The data scattered badly, but there was a steady increase in the number of low-visibility days over this period. The amount of sulfur in the atmosphere showed a similar trend. However, more dramatically, Hartmann found an abrupt drop in low-visibility days between July 1967 and March 1968. It took little effort to discover that between these dates there was a strike at the San Manuel Mine that shut down operations. Hartmann also showed that the Tucson traffic also played a part in polluting the Arizona air, since visibility improved between the daily rush hours and on weekends and holidays. The newspapers reported Hartmann's work, not always with the sober fashion of the technical discussions. On one occasion Kuiper wrote a letter to the editor of the *Tucson Daily Citizen* correcting their exaggerations. "We feel that the objective of the Pima County Air Pollution Advisory Council and our report to it, namely, to keep the Tucson Basin air reasonable clean, is not served by stirring up controversies."[47]

It Has Come to My Attention . . .

In addition to research and being a good citizen, Kuiper had the important role — not always to his taste — of managing LPL. Kuiper's autocratic management style was one factor, among many, for his problems at the University of Chicago. When he moved to Arizona his management demands did not go away, but rather slowly grew with LPL.[48] Whitaker has commented on Kuiper's autocratic management style at Tucson and reproduced some sample memoranda.[49] The University of Arizona collections include fifty internal memos from 1970 to 1973, essentially two such memoranda per month. They might be called reprimand memos, and they usually began, "It has come to my attention . . ." Such memos concerned illegal modification of equipment, taking jobs to the photo lab without going through his office, moving equipment without permission, tampering with the declination drive on the 61-inch telescope, leaving lights on in the dome, taking mail room supplies, unauthorized use of workshops at the weekend, unauthorized parking outside the building, and so on.[50]

However, some memos did not read like reprimands. A few years earlier Kuiper wrote a very touching memorandum concerning the annual Christmas party: "I wish I could write all a personal letter of thanks for the exquisite book that you presented to me at the Laboratory Christmas Party."[51] The gift was a book of photographs of Kauai that Kuiper thought highly appropriate. "You know I have a fondness for Hawaii and I love good photography," he wrote. But there was a twist. Kuiper had had a long and busy day and was too tired to attend the party, so somewhat apologetically he ended his memorandum of thanks, "You have touched me deeply with this wonderful gesture." Another undated memo invited LPL personnel to a Christmas party in the conference room at 3 p.m. "Please bring a 'funny' gift under $1 for a drawing. Your wife or husband is also welcome."[52]

Most of his internal memoranda from this time were signed with his initials, and he came to be known as "GPK" by his students and some associates. One magazine biography even has "GPK" in its title.[53] However, he is seldom if ever referred to in this way in his enormous correspondence or any other contemporary documents. In his rapidly evolving relationship with Harold Urey, for instance, his signature ranged from "Gerard P. Kuiper" to "Kuiper" and back to "Gerard P. Kuiper." Urey's mode of address for Kuiper was either "Kuiper" or "Gerard," depending on the state of their relationship.

A Graduate Program, a Department, and a Resignation

The realization that a graduate course would be needed for students in LPL happened in spring 1968. In that year a committee was established, with Kuiper as chair and Elizabeth Roemer as co-chair, to manage an "intermediate-type" graduate program comprising existing graduate courses prior to the formation of a formal program and department.[54] "A gradual approach was needed for administrative reasons," wrote Kuiper in describing the plan to Brunk on May 7, 1968. It was not until March 24, 1972, that Kuiper wrote to Harvill, retired and now emeritus, explaining how he thought it time to move ahead with the graduate degree program with an academic department to manage it.[55] The proposal, "A Graduate Academic Program in Planetary Sciences," had the support of Ted Ringwood, "one of the most brilliant geochemists of this period," and was expected to get support from Francis Birch, "the most outstanding geophysicist."

Also in the March 24 letter, Kuiper makes the case for a new academic department at the University of Arizona. "I feel strongly that a Department of Planetary Sciences is much needed in this country," he wrote, and "it would appear to be the best way to define a clear-cut future program for the Lunar and Planetary Laboratory."

On April 11 Harvill replied that he was "tremendously pleased" and that it "should be given high priority."[56] He added, "I will make a point to let it be known again how favorably disposed I am toward a graduate degree program, under your direction, in this field. . . . I am impressed with the quality of the students you can attract and by the quality of additional faculty for which you are planning." By July 14, 1972, the proposal for a graduate program in planetary sciences and a department of planetary sciences was formally approved.[57]

Around Christmas 1972 Kuiper made the decision to resign as director of the laboratory he had worked so hard to create and build. It must have been an extremely difficult decision, and we can only surmise why he did so since there appears to be little or no documentation concerning his resignation. The pace he set himself since he arrived at Tucson required a superhuman. He earned the admiration of the university president, who heaped praise on him and his accomplishments at every opportunity. LPL had grown at an extraordinary rate, and the number of activities Kuiper was responsible for similarly increased. Furthermore, his management style was not universally appreciated—many of those who worked with him found it difficult, and some left, which must have added to the stress. Harvill had long been concerned that Kuiper was endangering his health with his energetic pursuits of his interests. As early as 1964 he had written to Kuiper, "I do hope that you will be alert to your own welfare and not overtax yourself, even though the cause is exciting and of great importance."[58]

News of Kuiper's resignation as director of LPL reached NASA rapidly. On February 22, 1973, Brunk wrote to Kuiper, "I am aware that there will be a change in the administration of the Lunar and Planetary Laboratory in July. As the total amount of money available from my office for all programs at the Lunar and Planetary Laboratory will not be as much as in the past, I feel it appropriate at this time to request that you make a careful analysis of those programs for which support will be requested from my office to ensure that the research balance is appropriate for the entire laboratory."[59]

Kuiper had to identify a person who could succeed him as director of LPL when his resignation became effective on July 1, 1973. Furthermore, the Department of Planetary Science was to launch in the summer of 1973, so the new director would also be the head of the department. Over the next few months Kuiper identified Charles P. Sonett to be his successor.

Sonett had a distinguished war record when he obtained his first degree in physics from UC Berkeley in 1949, an MA in 1951, and a PhD in nuclear physics from UCLA in 1954.[60] For the next six years he worked with Space Technology Laboratories while also lecturing at UCLA. Between 1960 and 1962 he was appointed chief of NASA's Lunar and Planetary Sciences office. During this period he actually handled Kuiper's grants. In 1962 he was appointed head of the Space Sciences Division at the NASA Ames Research Center. At Ames he did not work directly with Kuiper, but he must have been familiar with the infrared astronomy flights on the Convair 990 that flew from their airstrip. It is significant that Sonett was not an astronomer, and neither were his successors as LPL director. A new discipline had surely arrived.

Unlike Kuiper's departure from the University of Chicago, the details over Kuiper's resignation as director of LPL and Sonett's struggle to take over the reins are not well documented. The memories of those who were around at the time suggest Kuiper identified Sonett as a strong person who could keep LPL an independent entity.[61] It

(a) Charles Sonett. Courtesy of University of Arizona Libraries, Special Collections, accessed June 2nd, 2017, https://uanews.arizona.edu/story/charles-p-Sonett-the-legacy-of-a-pioneering-space-scientist. (b) Rare image of Sonett and Kuiper together. Courtesy of Dale Cruikshank.

is not clear what exactly he feared now that the university had created a department of planetary sciences—maybe he worried that the department would replace rather than enhance LPL. In any case, problems began as soon as Sonett took over. Sonett was a strong individual who would look after LPL, but that same strength was turned against Kuiper by a new director who wanted to be his own man. At the heart of the matter was Sonnet's treatment of the observatories. Kuiper complained that Sonett had shown no interest in the observatories and argued that they should be made an independent unit in the university, with himself as director.[62] The new university president, John P. Schaefer, quickly agreed to make Kuiper director of LPL observatories but not to make them a separate unit.[63] They were, at least for the moment, to remain a component of LPL. This put Kuiper into the position of answering to Sonett for the observatories, a move that was not going to bring peace to LPL.

The oral histories of those around at the time suggest that the department split into two factions, one sympathetic to Kuiper and another to Sonett. Those sympathetic to Kuiper have suggested that the stress of his fight with Sonett caused the final heart attack that killed him. One contemporary document, a letter from Kuiper to Schaeffer, provides an indication of the levels of animosity by revealing that Kuiper felt it necessary to send Schaeffer a copy of his publication list: "Partly in self-defense against some recent inane writing from this Building, I asked the assistance of my secretary to compile a list of my publications. A copy of this list will be sent to you later today under separate cover. I had no idea of the number (275) that would be on the list, of their time distribution (enclosed graph). I certainly have enough observational material, ideas, and plans for another 100 or so publications, including 1 or 2 books."[64] Sonett, in his oral history, described the atmosphere at LPL when he arrived as "poisonous." His first task was to make stationery freely available[65]; his second was to remove the counter in the mail room.[66] At least one person suggested that Kuiper's lack of popularity caused him to be forced to step down as LPL director.[67]

It is often said in university centers that the true quality of the founding director is whether the center outlives them, so in this sense Kuiper could not have been more successful. Facing the extraordinarily difficult task of taking over the laboratory, Sonett had the additional task of running the new graduate program and the new department. Sonett was chair of the department and head of the laboratory for four years, after which he had a successful career analyzing spacecraft magnetic-field measurements to understand, for example, the Moon's interior. He was responsible for a Lunar Surface Magnetometer placed on the Moon by the Apollo program. This work revealed significant variations in the strength of the Moon's magnetic field on both local and regional scales, and it ultimately put an upper limit on the radius of a lunar core at 400 kilometers; the core radius is currently estimated to be about 330 to 350 kilometers.[68] LPL and Department of Planetary Sciences now occupy three buildings on the University of Arizona campus. The first was named after Kuiper. When the second was named after Sonett on April 30, 2004, then director Michael Drake said, "Bringing Sonett to UA was one of Gerard Kuiper's best moves."[69] Shortly after Drake's death on September 21, 2011, the third building occupied by LPL was named the Drake Building.

CHAPTER 16

A Man on the Moon

T WAS THE MOST EXTRAORDINARY TIME, one that is hard to imagine many decades later. In 1961, when President John Kennedy made the commitment to land a man on the Moon by 1970, rockets were exploding on the launch pad. The U.S. manned-space experience consisted of two suborbital lobs on rockets that claimed direct heritage from the German V2 missiles that dropped bombs on London.[1] Robotic spacecraft sent to the Moon were as likely to miss the Moon as they were to hit. Eight years later, highly sophisticated, complex, large, manned spacecraft would touch down on the surface of the Moon within feet of their intended landing site. It took a lot of small steps, a lot of dedication, and a lot of stress to make it happen. Astronauts died in the effort. To Gerard Kuiper and his small group of lunar specialists at Tucson, the task was to produce maps and interpretations of the lunar surface and help with decisions concerning landing sites. The program also took a toll on Kuiper, who in 1965 was entering his sixties, and his health started to suffer. Kuiper and his group participated in what has been called man's greatest adventure, an adventure that gave Kuiper his greatest moment and put him on a path of rapidly declining health.

Sputnik and the Coming of the Space Age

Ewen Whitaker left London on the evening of Saturday, October 5, 1957, just as the evening newspapers were splashing the banner headline "Sputnik Orbits the Earth." He later recounted, "I picked up a copy, little realizing that this event was heralding the dawn of the Space Age, an era that would land men on the Moon within a dozen years, give us thousands of close up views of Jupiter and Saturn and their satellites, and provide us with a better appreciation and understanding of our own planet Earth. I was met at Chicago's O'Hare Field (it was hardly more than a group of converted Air Force buildings then) by Kuiper, who was accompanied by French astronomer (Audouin Dollfus) and a Yerkes student (Clayton Smith). I handed him the newspaper; he was surprised and impressed by the news of Sputnik and amused that he should learn about it by what amounted to a personal courier."[2]

A month later came Sputnik 2, with its canine payload, and then the geophysical research satellite Sputnik 3 in May of the following year. Sam Pellicori described his first encounter with Kuiper one summer evening in 1960.[3] On a visit to Yerkes Observatory, he introduced himself as an astronomy student at Indiana University. He had been raised thirty miles from Yerkes in the city of Kenosha, Wisconsin. The staff showed him around, and they showed him Kuiper's office. At a certain moment everyone proceeded out to the south lawn to see the passage of one of the Sputniks. The group watched the blinking light slowly traverse the sky. Kuiper explained the blinking as probably due to the irregular shape of the rotating satellite. Suddenly, the tiny light went out! Everyone scratched their head in puzzlement. Then Pellicori suggested that it had entered Earth's shadow. Kuiper, quick on the uptake, said "yes, yes, that is it of course!" in his Dutch accent. Pellicori then records, "I recall gliding back into the observatory as everyone left the lawn, my feet never touched the ground."

The impact of Sputnik on the development of planetary research in the United States cannot be overstated and has been described in some detail by Doel.[4] Its primary effect was galvanizing the government to support space research through large, well-funded military programs and through the creation of NASA.

America's decision to land a man on the Moon affected Kuiper in two ways. It led to the construction of a new building at Tucson, eventually to become the Kuiper Space Sciences Building, discussed in a previous chapter. It also led to a series of robotic missions to the Moon. The science team led by Kuiper, who were amid publishing three atlases of the Moon, would be obvious candidates to participate in these programs.

The Ranger Program

The American robotic precursors for humans to land on the Moon consisted of three programs. The Ranger program was to be a series of spacecraft that would crash into the Moon and take close-up images as they did.[5] The Lunar Orbiter program was to be a series of spacecraft that would, as their name implied, orbit the Moon and

(a) The Ranger spacecraft. NASA photograph. (b) The television camera system of the Ranger spacecraft. NASA photograph.

take photographs of the surface.[6] Third, and the most sophisticated of the three programs, was to be the Surveyor program that would consist of robots that landed on the Moon.[7]

Kuiper became involved in the Ranger and Surveyor programs in 1961 when he was asked to serve on committees advising the Jet Propulsion Laboratory (JPL) in Pasadena, California, who were managing these programs.[8] This involved many meetings at JPL, which at the time consisted of tiers of trailers with a maze of interconnecting walkways.[9] In 1961, Rangers 1 and 2 failed on launch. Rangers 3 and 5 missed the moon. Ranger 4 hit the Moon but failed to return any data. In 1963 the project was reorganized, and Kuiper was asked to be the chief experimenter with a science team of Urey, Shoemaker, Whitaker, and Ray Heacock, a JPL engineer. The feud between Kuiper and Urey was at its height at this time, and there are accounts that meeting organizers, keen to keep their meetings peaceful, would seat them apart.[10]

Four redesigned spacecraft were prepared, each with six TV cameras. At Kuiper's suggestion, the TV system was tested in mock lunar landscapes at Goldstone Station in the Mojave Desert. Camera operations were carried out by Ralph Baker, who later joined the Optical Sciences Center in Tucson. The science team, especially Whitaker, played an important role in determining the impact sites for Ranger and the approach angles. Ranger 6 was another failure, but Ranger 7 was a spectacular success. It crashed just south of the Copernicus crater in a region now known, at Kuiper's suggestion, as Mare Cognitum.[11]

(a) Ranger 7 press conference. Kuiper with the microphone, Heacock on the left, Shoemaker and Whitaker in the background. NASA photograph. (b) Ranger 7 press conference from the NASA JPL movie "Lunar Bridgehead." NASA photograph. (c) Heacock, Shoemaker, and Kuiper at the Ranger 7 press conference at NASA headquarters. NASA photograph. (d) Urey, Kuiper, and Shoemaker discussing Ranger photographs. NASA photograph.

It is not difficult to imagine the elation caused by the success of Ranger 7 after so many failures and the prospect of human missions to the Moon just a few years away. While the engineers, planners, and publicity officers celebrated at fever pitch, the science team had the task of interpreting the pictures returned by Ranger. The pictures, said Kuiper, had a resolution one thousand times better than anything seen before. Kuiper and the other scientists had just hours after receiving these images before they would have to explain their results on television. At 9 p.m. on the evening of July 31, 1964, Kuiper stood up in front of the press and TV cameras for a coast-to-coast broadcast. "This probably marked Kuiper's finest hour," writes Whitaker.[12] With Shoemaker, Urey, and Whitaker seated behind him, Kuiper began in his unmistakably Dutch accent, "This is a great day for Science, this is a great day for the United States." The magic of the moment meant that Kuiper and Urey found it possible to be "positively amiable" as they climbed, with Whitaker, into Urey's huge Lincoln Continental for the ride back to JPL.[13]

Kuiper arranged for LPL to make loose-leaf albums from the Ranger 7 prints, which required a local company, Ray Manley Commercial Photography, to make fifty thousand prints. In the rapid-fire days leading up to the Apollo landings, things moved fast. Ranger 8 hit Mare Tranquilitatis in February 1965, and Ranger 9 crashed near the Alphonsus crater a month later. Both were completely successful.

The science team quickly published their conclusions, and the contribution of each team member and his interests can be seen in the list of conclusions published by the IAU.[14] Urey argued that Ranger pictures showed little or no volcanic activity and that the Moon was a captured object, similar to meteorites. Shoemaker did not comment on the origin of the Moon but emphasized the impacted nature of the surface and the difference in age of the maria and uplands on the basis of crater counts. Whitaker argued that the flow structures in maria were due to lava and therefore that volcanism was an important process on the Moon; he noted that the thickness of the fragmental debris layer was less than a meter. Kuiper emphasized the number of structures suggestive of volcanism, depressions in the maria, the crater rays (which he thought required volatiles), ridges, rilles, and lineaments. He wrote that the distribution of large boulders suggested a bearing strength of $1-5$ kg/cm^2 and thus the surface was suitable for a soft landing of a space vehicle.

Never had the LPL attracted such attention. With this success came attempts by the NASA centers and JPL to recruit LPL scientists. Whitaker was approached by JPL,[15] Kuiper was invited to take a position at NASA headquarters in Washington, D.C.,[16] and both Gehrels and Kuiper were invited to move to the Manned Spacecraft Center in Houston.[17] Kuiper took great pleasure in recounting the details to President Harvill.

Gold Dust

The early pictures from Ranger were important in resolving the issue of deep dust layers on the Moon, on which hung the success of the Apollo project. In 1955 Thomas Gold had reasoned that the surface of an airless and rather ancient body like the Moon would have a very deep layer of surface dust.[18] Thus Gold argued that any spacecraft landing on the Moon, like the Apollo Lunar Excursion Module with its two-man crew, would sink into the surface and become buried forever. Gold was a distinguished man.[19] Born in Vienna, his Jewish family fled to the U.K. before the war. After a period of internment, he entered Cambridge University and in 1948 and, with Fred Hoyle and Hermann Bondi, published the steady-state theory for the evolution of the universe. In 1952 he moved to the Royal Greenwich Observatory at Herstmonceux Castle in Sussex, where he proposed that ionized solar particles interacted with Earth's

Thomas Gold. Courtesy of Cornell University Digital Collections. G. Burbidge G. and M. Burbidge, "Thomas Gold 1920–2004," *BMNAS* 88 (2006): 1–14.

magnetic field to produce atmospheric effects. In 1956 he moved to Harvard University and then to Cornell University in 1959, where he set up the Center for Radiophysics and Space Research. In the early 1960s his views on the nature of the lunar surface were taken very seriously.

Whitaker and Kuiper mulled over the photographs of the lunar surface and eventually found a site at which large boulders were lying on the surface. Making reasonable assumptions about size, density, and thereby mass of the objects, Kuiper and Whitaker announced that the surface could support the weight of a spacecraft. In a telegram to Homer Newell dated April 4, 1965, Kuiper wrote, "Completed preliminary analysis of bearing strength of Alphonsus floor from dozen rock masses each about one meter diameter thrown 50–100 meters by known crater 50 meters in diameter. Stop. Rock penetration estimated from observable shadows. Stop. If rock bulk density assumed twice water then average bearing strength upper two feet equals one to two tons per square foot but if density less then strength proportionally less. Stop. Believe one ton per square foot reasonable but actual value could be three times less."[20] Their conclusion was to be confirmed the following year when the Surveyor touched down and turned its cameras to its own feet. It was clear that the foot pads on the spacecraft had penetrated the surface by only a few centimeters. "Gold dust," as it came to be called (somewhat derisively), was not a problem.

John Lear and Saturday Review

Even before the successful mission, Kuiper's involvement in the space program brought him much publicity. In April 1962 he appeared, with Cruikshank, Hartmann, and L. Harold Spradley, in the ABC TV show *Meet the Professor*, which was broadcast on Sundays. "The program with Dr. Kuiper has been acclaimed as one of the best and most successful in the series," wrote the show's producers.[21] After the Ranger 7 mission, the stack of letters from the public to Kuiper was three-inches thick. President Harvill wrote to Kuiper, "Congratulations on your terrific work."[22]

With success inevitably came the naysayers, and Kuiper became extremely upset when he received the *Saturday Review* for September 9, 1964.[23] In the issue was an article by the science editor, John Lear, who contributed articles to the *Saturday Review* regularly between 1957 and 1972 on topics at the interface of science and policy. His writing was lively and entertaining, but his criticisms were daggers to the heart. He began his article, "What the Moon Ranger Couldn't See," with following words:

> They made a robot the shape of a dragonfly and named it Ranger VII, thereby expressing their hope that it would range the face of the moon. For eyes, they gave it camera lenses, and taught it how to photograph by blinking the shutters. Near the place where its nose should have been, they put a radio antenna like a saucer, and taught the robot how to send pictures back through the saucer to earth. When this was done, they folded the dragonfly's wings, set the mechanical insect on top of a rocket, and fired the rocket into orbit around the earth. Finally they shot the robot out of orbit, told it to unfold its wings, and pointed it into a curving path across 243,665 miles of sky. The path ended in a dry lakebed that hadn't been thought important enough to be named.

Lear wrote about the lunar surface, the hillock into which Ranger VII crashed, craters, and rays, one of which passed through the hillock. "Of course I do not report the foregoing information on my own authority," wrote Lear, but on the basis of a ten-and-a-half-hour interview one Saturday with Kuiper. After complimenting the professor's "exquisitely trained eye," he describes his finding that heavily cratered regions seen by Ranger 7 are the bright areas seen in the 200-inch telescope. Now Lear begins on his criticisms. Is this not just a trick of the light? Can we assume that the "Garden patch" seen by Ranger is typical of the entire lunar globe? "There has been too much instant science in the American program for exploring the Moon." Kuiper was essentially a mouthpiece for JPL and NASA who wanted evidence that it was safe to land men on the Moon. It is impossible to believe, as Gold has pointed out, that surfaces as old as the Moon's, constantly impacted by meteorites, do not have a thick dusty layer into which spacecraft would sink. "The net sum evidence here recited is that Congress and the American public have been misled about the true significance of the Ranger VII mission. The pictures sent back home by the ingeniously constructed robot dragonfly contain no more assurance of the safety of a manned landing on the moon than existed before Ranger VII took off." Kuiper's role, Lear asserted, was to hold together the unique team of engineers at JPL.

Kuiper immediately dashed off letters to the editor of *Saturday Review*, to NASA headquarters, and to JPL, demanding right of reply in the first available upcoming issue. "Mr. Lear grossly abused my hospitality." His article was "an insidious mixture of reasonably correct quotations, his own gross exaggerations, unfair allegations and personal peeves."[24] In his usual reassuring and supportive manner, Harvill replied, "I regret that this work by Lear was so unsatisfactory. I am not surprised, though, because through the years I have heard many comments from persons of competence in science that were not very favorable to the writings on science in the *Saturday Review* by Lear." Enough said. Once again Harvill had demonstrated how important he was to Kuiper's sense of well-being and therefore his success.[25] Matter closed.

The Lunar Orbiter and Surveyor Programs

The Lunar Orbiter program consisted of five robotic missions between August 1966 and August 1967 to place cameras in orbit about the Moon to identify Apollo landing sites. Instead of television cameras the spacecraft carried film cameras for better image quality, and the film was developed on the spacecraft, printed, and then scanned for radioing back to Earth. The cameras had two lenses, a high-resolution lens and a wide-angle lens. The spacecraft was built by Boeing and the camera system, based on that used in military reconnaissance aircraft, was built by Eastman Kodak.

Kuiper was asked to be the chief experimenter for the Surveyor program, which he eagerly accepted.[26] Immediately he shared the news with Harvill.[27] Since funds were to be provided, NASA needed a proposal and a budget, and Kuiper provided a thirteen-page proposal describing the state of lunar cartography, the work to be done, the personnel required, and the budget. Ten members of staff were to be involved, and $178,140 ($1.4 million in 2018 dollars) was required.

With JPL managing and Hughes Aircraft building the spacecraft, Surveyor was to land on the Moon in preparation for the Apollo landings, addressing, among many other issues, whether the surface could support the weight of a spacecraft. A television camera would provide images from the surface. Between May 1966 and January 1968, seven Surveyor missions were flown, and five were completely successful. Four had landing sites spread along the equator—two at Oceanus Procellarum, one at Mare Tranquilitatis, and one on Sinus Medii. The fifth successful Surveyor landed in the southern hemisphere near Tycho Crater.

Important questions were resolved by the Surveyor program. NASA had learned enough to successfully land humans safely on the Moon. The Moon's surface could support the weight of a large spacecraft. The Surveyor program also resolved the argument between Urey and Kuiper on the nature of the Moon. The Surveyor spacecraft carried an instrument for analyzing the surface, and the result was unequivocal. The surface had a basalt composition characteristic of volcanism and distinct from the chondritic composition proposed by Urey. Kuiper was right.

Apollo 11 and a Landing Site for Apollo 12

A little over a year after the last Surveyor, the Apollo program took men to the Moon, first in a dress rehearsal by Apollo 10 on May 18–26, 1969, and then for the landing of Apollo 11 on July 20. Kuiper spent the time of the Apollo 11 landing at Flagstaff, where part of the CBS television coverage of the historic occasion was to be broadcast. George Herman, the well-known TV journalist, asked Kuiper to attend, but he was reluctant because he would have to share the stage with Urey. However, since it would only take a weekend and he was assured he would have some individual screen time, he agreed to participate.[28] While the world went euphoric over the most important event in the history of humanity, Kuiper could spare a weekend, provided he did not have to share a stage with Urey.

The landing of Apollo 12 came soon after Apollo 11. Whitaker's interpretations of the surface of the Moon from spacecraft images were critical in 1970, when it was decided to test the engineering precision of landing spacecraft on the Moon by having Apollo 12 land close enough to Surveyor 3 for the astronauts to walk over to the robotic spacecraft and remove a piece. For this the flight engineers needed to know the precise location of the Surveyor. Whitaker scanned the images of the surface returned by Surveyor 3 and then tried to find a matching landscape among the photographs taken by the Orbiters. After several long days, the effort was successful.

Apollo 12 provided some of the most memorable images of the Apollo program. With the Apollo 12 spacecraft perched high in the background, two astronauts

(a) The launch of Apollo 11. NASA photograph. (b) The crew of Apollo 11 on the Moon. NASA photograph. (c) Moon rocks arriving at the Manned Spacecraft Center, Houston. NASA photograph.

approached Surveyor 3 and removed the periscope. Whitaker received a letter of appreciation from President Nixon and the astronauts of Apollo 12. The framed letters hung on the wall of his home office for the rest of his life.

Moon Rocks

A second nail was driven into the coffin of a Urey-type chondritic Moon by the rocks that were brought back by the Apollo missions. A few days after Apollo 11 landed, the moon rocks were received at the Manned Spaceflight Center in Houston. In his oral history, Don Bogard, then a young sample scientist hired to study the Apollo rocks, described the heady atmosphere when the rocks from the Moon were opened and underwent their first examination.[29]

> The age of the samples would be critical. When the first lunar box from Apollo 11 was opened—in the F201 vacuum chamber—it was pretty obvious the rocks were volcanic.

The Surveyor 3 and Apollo 12. Astronaut Pete Conrad retrieves a periscope from Surveyor 3 for returning to Earth. The Apollo 12 Lunar Excursion Module can be seen in the background. NASA photograph.

They sure did not look weathered, they looked young and fresh. Harold Urey and the proponents of the old primitive Moon were very disappointed. This whole question was very well known to the members of the press. They were much attuned to these two views and what these samples might mean for one side or the other. Shoemaker was busy talking up the attributes of these volcanic samples. But Taylor [Ross Taylor, a geochemist from the Australian National University] and Schaeffer [Oliver Schaeffer, an expert in meteorite dating] knew the age was critical. Taylor's group had measured potassium, and Schaeffer's group had measured argon-40, so we had had a potassium–argon age for days. Schaffer and Taylor had only reported these values to one significant figure. The advisory group, wanting to know the rock ages, kept saying that you must know the numbers better than that, and they kept saying "no, no, we don't." They played cat and mouse and wouldn't tell the advisory committee and the outside world what the age was.

Being discouraged about the initial science finding, Harold Urey had already gone back to La Jolla. One day I decided this had gone on too long, and I was scheduled to give the daily report. I gave the K and ^{40}Ar results to two significant figures. Those scientists working with the samples had been instructed that we were not supposed to do science, only make measurements. How does one take new data on the first lunar samples without doing science? Well, the day I gave the scientific briefing, I said we are not supposed to do this, but if you combine these two abundances here is an age you get for the rocks. Immediately, Jim Arnold [a chemist from Urey's institution, UC San Diego] left the briefing room, went to the telephone, and called Urey. He said, "Harold come back to Houston, the lunar rocks are old! You were right after all!"[30]

By any standards, the Apollo rocks are old, but as time passed it became even clearer that the rocks were volcanic and that none were as old as the chondrites. There was no room for doubt. Maybe that is why, after so much vitriol, Urey wrote to Sarah after Kuiper's death and told her that Kuiper was a good scientist.[31]

A Heavy Workload

Signs of the stress on Kuiper, and the effect on his health, can be traced back to a letter to in 1965 to Newell in which he wrote that he was going to skip the IAU symposium for health reasons.[32] "I have reluctantly concluded that I would be endangering my health if I did [go]. This meeting could be a strenuous one, and could be a repetition of the IAU symposium which ended in a disorderly debate because of the insistence of some strong-willed individuals."[33] In an addendum he added that Menzel had called to say that there would be "unfortunate consequences" if he did not attend; however, Kuiper was confident that the people who mattered were sufficiently impressed with his Ranger publications that it would not matter. "I have been extremely tired during the last several weeks," he wrote, and ended his letter with "Ewen Whitaker, with his personal charm, may be more effective than I would be under the circumstances."

As the decade came to an end and the astronauts walked the Moon, Kuiper found himself under considerable stress. The heavy workload was mainly to blame, but another big cause was the competition he was receiving from others wishing to name craters on the Moon and Mars; Kuiper claimed sole right to do this through his activities with the IAU and experience with the lunar atlases. This is reflected in a letter to Brunk dated March 7, 1973, in which he writes, "I am leaving for Houston for 2 days on meetings on lunar nomenclature. I do not look forward to this."[34] On May 14, 1973, he wrote to Brunk that he thought the nomenclature of lunar and Martian craters was now under control, and he had received a letter of appreciation from Newell that he enclosed, but Kuiper confessed a level of frustration. "Frankly, I am so tired of the two subjects that I have decided not to go to Sydney," where the next General Assembly of the IAU was meeting.[35]

Kuiper experienced two heart attacks during the summer of 1973. Harvell wrote to him, "It was with great regret that Mrs. Harvill and I learned of your illness. We are delighted that abstention from heavy duties and more rest have brought about a recovery to normal health." Harvill continued, "We urge that you continue to be generous in the extent of caution against endeavors that might slow your recovery."[36]

To add to the stress, a new man was running LPL, chairing the new department, and running its new graduate program. Kuiper thought Charles Sonett would be strong enough to ensure LPL's independence, which he did. However, that strength meant distancing Kuiper from the management of the operation, and Kuiper protested loudly. In a compromise, Kuiper could remain director of observatories, reporting to Sonett. Some say the difficulties with Sonett caused the final heart attack.

Finally, as the end of 1973 approached, the news coming from NASA was not all that he would have wanted. On November 29 Brunk wrote to Kuiper that all future proposals for funding his research would be reviewed by two or more qualified persons outside of the program office.[37] Ongoing programs would be reviewed at least every three years. Furthermore, in the face of travel restrictions that kept the managers in Washington, D.C., mandatory progress reports needed to be adequate and prompt. Brunk goes on to lay out in detail what would be required in proposals. Background, value of research, past research success and progress reports, future plans, clear separation of work performed under which grants, sources of additional funding, a detailed budget, and budget justification. Kuiper would have been given this glimpse of the space-research world after Apollo by the time he boarded the plane for his last flight. To Kuiper, who created his own journal to avoid peer review, the news that his proposals were to be peer reviewed rather than evaluated by NASA program directors must have sent shudders through him.

It is impossible to overstate what was achieved by the U.S. space program in the 1960s. Nothing like it has been accomplished since, as the fate of the Constellation program, the United States' short-lived attempt to return to the Moon, demonstrates. Neither funds nor incentive were there. The Apollo project required a national imperative. It also needed a scientific face. Every major national voyage of exploration had a scientific face: Lewis and Clarke, James Cook, Alexander von Humboldt, Peter Simon Pallas. All had important scientific directives, and all resulted in major scientific advances. For the Moon landing, that role was played by many scientists, but Gerard Kuiper's certainly ranks among the greatest.

CHAPTER 17

Legacy

KUIPER MADE TWO TRIPS to Mexico in the last weeks of his life. The first was in a small plane with Laurel Wilkening, who had recently joined LPL as an assistant professor, and Godfrey Sill, an LPL chemist.[1] In a small grey notebook he noted every detail of the trip, from the make of car he rented to the raisins and orange he ate at the airport on his return. The task of the trio was to visit Tequila Volcano, near Guadalajara, to evaluate it as a potential observatory site. He thought it worth investigating, but nothing came of it. His second trip was with Sarah in a commercial plane at the media's expense; his task was to record excerpts for a television show with Fred Whipple, Jurgen Stock, and some Mexican astronomers, presumably including Guillermo Haro of the Observatorio Astrofísico de Tonantzintla. Kuiper died of a heart attack in their hotel room after a day's filming. A memorial ceremony was held at the University of Arizona, during which his friends and colleagues mourned his passing. Many memorials have been dedicated to him; however, the greatest memorial of all is the creation of a research field that has occupied the popular media more than any other, the exploration of the solar system.

A Flight to Guadalajara

Freeway Airport was a small private airport just a few miles northwest of the University of Arizona, known mostly for flying Cessnas. On his last flight out of Tucson to evaluate observatory sites, Kuiper left from Freeway Airport at 8:50 a.m. on December 13, 1973.[2] Sill piloted the plane. Sill suggested that they ask Wilkening, who shared

Laurel Wilkening and Kuiper at the Freeway Airport prior to their Mexico flight in December 1973. Courtesy of Godfrey Sill.

his laboratory and knew a little Spanish, to join them as interpreter and to help with hotel and other arrangements.

The destination of the small plane was Guadalajara, Mexico. As the aircraft climbed, Kuiper noted the conditions in the air around Tucson. Initially, the air was clear except for light cirrus clouds in the northwest. There was a low haze, which he considered "not excessive." However, as they climbed to several thousand feet they could see a belt of smog thirty to fifty kilometers to the north, with its southern edge running east–west. Kitt Peak National Observatory on Mica Mountain was clear from four thousand feet. The Catalina Mountains were clear from five thousand feet. By 9:18 they were flying over Mount Hopkins, and the air was fine. However, east of Nogales, on the Mexican border, flying just below five thousand feet, there was a heavy smog layer covering thirty square miles. By 9:34 the plane was five miles south of Nogales heading southeast to Cananea at seven thousand feet. Kuiper took some photographs of the smog, recording in his notebook that he was using Kodachrome II film. Staring intensely out of the window of the plane, while feverishly writing notes, Kuiper must have been a poor traveling companion for the rest of his party; however, the flight was the beginning of a relationship between Sill and Wilkening that led to their marriage.

"That was the trip where Godfrey and I decided that we liked each other," Wilkening recalls in one of her oral histories.[3]

An Observatory on Tequila Volcano

The little plane with its cargo of Tucson planetary scientists continued south, eventually reaching Guadalajara. The next day they met with Guillermo Haro, who had flown from Mexico City to meet with them. Haro was a Mexican astronomer hired in 1943 by the newly founded Observatorio Astrofísico de Tonantzintla.[4] Initially he was involved in the study of extremely red and extremely blue stars, but over the years Haro discovered several planetary nebulae in the direction of the galactic center and nonstellar condensations in high-density clouds now known as Herbig-Haro objects, George Herbig of the University of Hawaii having independently observed them. Haro and he coworkers discovered flare stars in the Orion nebula region and later in stellar aggregates of different ages. Haro was very influential in the development of astronomy in Mexico, and in 1959 he was the first person elected to the Royal Astronomical Society from a developing country. The observatory in Cananea was given his name after he died in 1988.[5]

The four astronomers shared coffee in the airport dining room while Kuiper reported on the flight and the visibility and weather conditions around Guadalajara. The main objective of this trip was to make a preliminary evaluation of a volcano near Tequila as a possible observatory site. The next day the group hired a rental car (a Plymouth Valiant) and drove to Tequila, arriving at 4:07 p.m. Then they drove to the summit of Tequila Volcano, noting the vegetation and the gradients of the road: 10 percent on the lower levels and 25 percent nearer the summit. Kuiper thought the climate was mild, there were no destructive winds, and looking at the sky he could see the Andromeda nebula "very plain," Alpha Eridani was "bright and high," Venus, Jupiter, Mars, and Saturn were "very bright," and the Milky Way was "excellent." Light pollution from Guadalajara was comparable to that of Phoenix seen from the Catalina Observatory. There were towers on the volcano rim, and the enclosure around one was a possible site for an observatory. It was past midnight when the group fell into their beds in the Guadalajara Hotel. After another day visiting the volcano and making water measurements, Kuiper felt the site was "very favorable" and suggested to Haro that they place a trailer on the rim so a more detailed evaluation of the site could be made, an action Haro wanted to defer.

On the December 17 Haro organized a visit to the Autonomous University of Mexico, where Kuiper was introduced to the rector (Kuiper noted that it was a nice and useful visit), the dean (could speak no English), and director of the computer laboratory (who was a University of Arizona graduate). At 4:25 p.m. the group took off for the return flight to Tucson. Using Frommer's *Mexico on Five Dollars a Day* and staying at cheap, seedy motels for overnight stops, the plane followed the coast, passing over Culiancán and Cuidad Obregón.[6] At Hermosillo they stopped for lunch and cleared Mexican immigration and customs. Kuiper made numerous pages of notes and sketches about the seeing conditions along the route, still in search of possible sites for observatories among the Sierra Madres. Finally they reached the familiar mountains around Tucson and the little plane landed at 3:10 p.m. Kuiper recorded in his notebook that he ate an apple in the customs hall at the airport, having previously eaten an orange and some raisins while still on the plane. Such was his mindset for recording every detail of his flight. After twenty-five minutes on the ground, Kuiper called Sarah to pick him up while Sill took the university's car back to the university. A week later, Kuiper was in Mexico again, this time on a commercial flight at Mexican TV's expense. Sarah accompanied him.

A Show for Mexican Television

The pyramids of Xochicalco, about an hour out of Mexico City, were built around 650 CE by a Mayan group of traders.[7] By about 900 CE the site had become home to almost twenty thousand people but was destroyed by fire. After excavation and restoration, especially in the 1950s and 60s, Xochicalco was declared a World Heritage site and became a major tourist attraction. It is famous for the Temple of the Feathered Serpent, whose sides are adorned with beautiful reliefs, and for its observatory where Mayan astronomers developed an astronomical calendar.

Kuiper was in Xochicalco to take part in a three-hour documentary on astronomy being produced for Mexican television. Sarah was present "as a guest, not a participant," she was quick to point out.[8] The producers of the show had also invited two astronomers from Mexico, one of whom was presumably Haro, and Whipple, who was also accompanied by his wife, Babette. Stock, then director of Cerro Tololo Inter-American Observatory, was also to participate in the recordings. All had gone well, and it was time to relax over dinner with friends.

A Call from the Embassy

In November and December 1973, Kuiper's daughter Lucy was a world away, in Indonesia.[9] Lucy had three dreams that her father had died, and each required a telephone call to the United States for reassurance. She knew that her father had experienced two heart attacks in recent months and, despite this and the fact that he was now in his sixties, he continued to work long, stressful hours. In fact, Lucy and her young family had flown to Tucson to spend time with her parents in the summer of 1973. She recalls her father's love for his little granddaughter. The visit was pleasant, although, like most of America at that time, Kuiper was glued to the television watching the details of the Watergate scandal unfold. But, according to Lucy, "He was such a good grandfather!" Their daughter Rosabelle would get bored at meals and start to complain, so her grandfather would take her to sit on the floor with him and play, or he would put her in the stroller and push her around. At the end of the visit the departure was emotional. The flight back to Indonesia stopped at Hawaii, so Kuiper thought of arranging to meet them during the stopover. "Oh Dad, that's unreasonable!" responded Lucy. Then he broke down.

Lucy had inherited a love for Indonesia from her father. She was teaching Spanish, English, and History at the Jakarta International School, which was founded in 1951 as a school for the children of United Nations staff posted in Jakarta. In 1969 it became the Joint Embassy School, sponsored by the Australian, British, Yugoslavian, and U.S. Embassies.[10] Lucy got married in Indonesia, and Kuiper felt he could not come all that way in the face of his current workload. So Sarah went alone. Back home, alone in Arizona, Kuiper paced around the garden, feeling badly that he had not been able to go. About a year later, when he learned his daughter was pregnant, he started to feel his age. He had never felt he was getting old until he became a grandfather. But he fell in love with his granddaughter, and that made up for the sense of aging.

On Christmas Eve, 1973, Lucy was at home with Jean-Marc des Tombe, a Canadian she met in 1962, started dating in 1968, and married in 1971. He had joined her and her father on flights along the Baja Peninsula looking for potential observatory sites. At that time they had one child, Rosabelle. Lucy was sitting with her husband in their home in Jakarta when the British Embassy doctor, who lived in their compound, came to say that he had just had a call from Mrs. Kuiper to say Lucy's father had died. Apparently, last evening, after dinner with the Whipples, Kuiper had collapsed in their hotel room and had died in a matter of moments.

A Man Who Truly Was Alive

The next day, Paul Kuiper arrived in Mexico City to be with his mother and help with the arrangements. On December 25 the *Arizona Daily Star* ran an announcement of the death of "the energetic University of Arizona astronomer," with quotes from Bart Bok, who said it was "a terrific loss for American astronomy" and that Kuiper was "recognized as one of the great men in planetary astronomy."[11] One week after

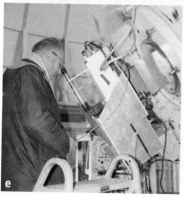

(a) Kuiper in his office at LPL. Courtesy of LPL. (b) Looking at the first lunar atlas with Whitaker and Herring. Courtesy of LPL. (c) Looking at volcanic rocks in Hawaii. Courtesy of Dale Cruikshank. (d) Delivering a speech at the dedication of the 107-inch telescope at McDonald Observatory, Bill Brunk in the background. Courtesy of Dale Cruikshank. (e) At the 61-inch telescope. Courtesy of LPL. (f) With astronaut Frank Boreman. Courtesy of Matthew des Tombe.

Kuiper succumbed to a heart attack, a memorial service was held at the University of Arizona. The program for the service featured a poem by the twentieth-century American poet, John Ciardi:

> *In the planetarium of an apple tree*
> *I shook some spiral nebulae*
> *And sent some systems reeling, just to see*
> *How that might be. Just between time and me,*
> *To see how it might be*
> *To shake a universe—or an apple tree—*
> *And see what fell, and think how it would be*
> *If any of what fell was me.*
> *None of it was—that day. But I could see*
> *It all would be. Some day. Whatever tree.*

Paula Fan, one of the premier performers in the university's music department and an accomplished pianist, played Beethoven's Ninth Symphony at the opening of the service, which was followed by some appropriate readings. Then there were remembrances from six friends and colleagues: Richard Harvill, chancellor of the university; Hermann Bleibtrue, dean of the college; Braulio Iriarte of the Mexican National Astronomical Observatory; Ewen Whitaker, his immediate colleague on the lunar work; Bart Bok, his friend from Leiden and now chairman of the astronomy department; and, finally, Reverend Russell Lincoln. Bok read some remembrances sent by Kuiper's lifetime friend and colleague, Subrahmanyan Chandrasekhar. After the remembrances there was a meditation and some closing words. Fan played music as the service ended. Immediately following the service was a reception in Goddard Hall, which was provided by friends in the Smith Club of Tucson, essentially the alumni association and center of social activities on campus.[12]

Dean Bleibtrau's remembrance was brief and moving, capturing Kuiper as he was perceived by his senior colleagues at Arizona:

> Just over two years ago—when Gerard Kuiper decided the time had come to initiate an academic counterpart to LPL—and when he decided that its administrative home should be in the College of Liberal Arts, that's when I first met him. Not long ago. But because he believed in educating his administrators, I learned to know and respect this incredibly dynamic individual whose productivity and scientific activities were truly outstanding, who kept the institution panting to keep up with his activities, whose energy and accomplishments were in evidence everywhere, in his own discipline, in that of

Bart Bok. Courtesy of AIP Emilio Segrè Visual Archives. J. A. Graham, C. M. Wade, and R. M. Price, "Bart J. Bok 1906–1983," *BMNAS* 49 (1994): 72–97.

others, in the affairs of the university and those of this community. He wanted everyone to know what he was feeling and doing, he wanted them to have his enthusiasm and spirit. He felt it the duty of scientists to educate and influence policy makers. He truly believed that rationality coupled with enterprise were the only keys to human freedom and human potential and now that he has gone we will continue—I know he would not want us to lose so much as one pace in carrying on—just as his pace never slackened. But we will quietly miss him. He was such a presence on campus among us it is hard to imagine that he is physically gone. I will miss his oddities, his wry humor, his freshness and intensity. A scientist and spokesman for science has been lost, a man who truly was alive and who therefore left us more of those things that do not die.

Bok's address was an eloquent summary of Kuiper's career and their relationship over many decades.[13] He recalled how they met on their first day as undergraduates and how Kuiper had passed three national tests in one year at high school, any one of which was considered a major accomplishment, and one that allowed him to enter university. He spoke of Kuiper's interest in the origin of the solar system at an early age, of his education at Leiden among some of the world's leaders in astronomy, of his time at Lick and Harvard, and the importance to his career of his marriage and appointment. Bok remembered the powerhouse at Chicago created by Hutchins and Struve, Kuiper's stellar and binary star work, his work on stellar masses and temperatures, and his growing interest in planetary atmospheres. He spoke of Kuiper's war work, his infrared spectroscopy, his discovery of the compositions of the atmospheres of Titan and Mars, the nature of Saturn's rings, the discovery of Miranda and Nereid, and his book-publishing efforts. Bok also listed some of the many honors Kuiper received while at Chicago: the Order of Orange-Nassau, the Janssen Medal, and the Henry Norris Russell Lectureship. Bok did not discuss Kuiper's accomplishments at Arizona because others would deal with that, but he spoke instead of how, from 1966, they were colleagues on the Arizona faculty, of how Kuiper spoke when receiving the Kepler Gold Medal and when talking to the AAS about celestial mechanics. "He looked rather askance at my retirement plans," wrote Bok, "and said that he could not possibly think of retirement now because he had three or four years of hard work ahead of him." Bok ended, "Today we mourn the passing of a great man and of a devoted husband. Our sympathies go to Sarah, to Paul and to Lucy. The name of Gerard Kuiper will go down in astronomical history."

Over the next few months, Ida Edwards, Kuiper's secretary from soon after he arrived in Tucson until his death, tidied up his papers—Kuiper always referred to them as his "archives"—and delivered them to the library that housed the University of Arizona's Special Collections.[14] She was aided in this by Ewen Whitaker, who published an appeal for any additional papers that other members of the community might had.[15] Ida delivered thirty-eight banker boxes of papers. By the time I came to write this book, the collection had grown to eighty-five boxes.

Monuments

At that moment Mariner 10 was on its way to Mercury, and soon would return images of the planet.[16] Among the first details seen as the spacecraft approached to give humans their first view of Mercury was a sixty-two-kilometer bright crater with rays of material radiating in all directions.[17] It is one of the youngest craters in the solar system and the freshest looking, and it stands out brightly against the darker surface of the planet. The camera-imaging team proposed, and the IAU accepted, the name "Kuiper" for that distinctive feature on the Sun's closest planet. It was one of

(a) Statue of Gerard Kuiper a few feet from his birthplace in Tuitjenhorn, Netherlands. The setting of the statue in a residential street. Author's photo. (b) Statue of Gerard Kuiper a few feet from his birthplace in Tuitjenhorn, Netherlands. The statue looks skyward, as befitting an astronomer. Author's photo.

three craters that were named after Kuiper in 1976, the others being on Mars (which is 82 kilometers in diameter) and the Moon (6.3 kilometers).[18]

This was just the start of the tributes. When the Convair aircraft Kuiper had used for infrared astronomy was replaced as NASA's aerial platform, the new facility was called the Kuiper Airborne Observatory. The belt of objects that accompanies Pluto outside the orbit of Neptune is the Kuiper Belt. There is an asteroid named 1776 Kuiper. The premier award given by the AAS for lifetime achievements in planetary science is the Gerard P. Kuiper Prize in Planetary Science. The building that houses LPL in Tucson is the Kuiper Planetary Sciences Building. His images have appeared several times on postage stamps.[19]

At a ceremony on June 19, 1999, the 61-inch telescope on Mount Bigelow in the Catalina Mountains was renamed the Gerard P. Kuiper 61-inch Telescope in honor of the founding director of LPL. A reception was held in the atrium of the Kuiper Space Sciences Building. Gerard's cremated remains were eventually interred in the concrete foundation of the observatory that houses the telescope. Sarah remarried twice, to Daniel Deleon Roth in 1976 and William Ross Lansberg in 1985, but when she died in 2000, her remains were interred with Gerard's.

On Friday, July 27, 2001, a statue by the Dutch artist Gosse Dam was unveiled in Tuitjenhorn, just a short distance from the site of the home Kuiper was born in. City officials and members of the family attended.[20] The event was organized by the Municipal Council of Harenkarspel and consisted of a reception in the town hall at 2:30, the statue unveiling by Lucy (now going by Sylvia) Kuiper at 2:45, informal meetings and an exhibition until dinner at 6:00, and then evening lectures on Kuiper's career by local historians and astronomers. Dam's statue stands in a quiet, tree-lined street in its own rectangular space surrounded by a boxwood hedge. The statue is in a simple, upright, and blocky style, so characteristic of Dam's work, and it is looking skyward.

The tributes made to Kuiper serve to list his scientific accomplishments: his work on binary stars and white dwarfs, his discovery of the atmosphere on Titan, the application of astronomical infrared spectroscopy, the identification of the atmospheres of Mars and Venus, the lunar atlases, the catalogs of asteroids, the discovery of Miranda and Nereid, his description of the origin of the solar system and the Kuiper Belt, and the existence of major observatories in Chile and Hawaii. It is a list of accomplishments that would satisfy a dozen other scientists. Most of all, his monument is the establishment of planetary science as a discipline worthy of the attention of professional scientists.

A Prodigious Energy

Kuiper's success was accomplished by excellent training, long hours, supportive universities and funding agencies, and a fortunate accident of timing as war technologies and the Cold War affected science. But most of all, according to his student, Dale Cruikshank, it was all achieved by a prodigious energy.[21]

Kuiper's humble beginnings played a major role in determining his life. It must have been a blow, if we are to judge from his later character, that he was not admitted to the gymnasium or the lyceum but instead went to the teacher's college with its lower expectations. The hostile reception by the wealthier students at Leiden continued to put pressure on the young Kuiper. It is not difficult to imagine that when he left undergraduate school he had grown into a man with a hard-nosed determination to prove his worth to himself and those around him, and with an ability to take such abuse without deflecting from the path he had chosen. Another important element of his early success was his association in graduate school with some of the world's best astronomers. By the time he emerged from Leiden he had a track record strong enough to be accepted into the top echelons of American astronomy.

Kuiper's manner of presenting his work, probably best described as an unrelenting focus on the work itself, was inevitably perceived by some as insensitivity to others in the field. At Lick Observatory this set Wilhelm Luyten on an acrimonious path that lasted a lifetime and frequently became public. Two decades later, Urey felt the same sense of outrage, but he dealt with it in a way that drove him to ever greater frustration while Kuiper apparently remained impervious. To withstand the attacks of Luyten, knowing he had the support of Lick observers and Otto Struve, was one thing. To absorb the attacks of Urey was another. Unlike Luyten, Urey was very popular. Urey also was a Nobel laureate who thought he had written the Bible on the origin of the planets and felt he had done likewise for the Moon. In fact, Urey regarded his own book, *The Planets*, as the most authoritative and thorough treatment of the subject, which was what so enraged him when Kuiper published his views on the origin of the solar system without the slightest mention of Urey's "masterwork." The feud between Urey and Kuiper highlights the entire point of planetary science as a discipline.

Urey was a first-rate chemist. He thought his chemical arguments irrefutable and judged that Kuiper did not have the background to understand or overturn them. But Kuiper never tried. He freely admitted that he did not have Urey's chemical knowledge. Instead, Kuiper used his professional training in astronomy, his personal training in geology, and his access to one of the biggest telescopes in the world to construct his own theories for the origin of the solar system and the nature of the Moon. Kuiper had a question, and he used whatever approaches he could to address his question. He did not take a chemist's approach, or an astronomer's, or a geologist's, he just asked questions and tried to find answers using his excellent training, strong will, and enormous energy. The thick skin he apparently acquired in the face of attacks from Luyten and Urey may have been essential to his ability to study subjects in a way not possible before. This included topics, like the planets, that were previously not thought to be worthy of a top astronomer's attention. Thus he founded the new discipline of planetary science. Kuiper's achievements in planetary science tower above Urey's because Urey refused to be anything but a chemist.

Kuiper's uncomfortable relationship with Geoffrey Burbidge is equally important in understanding Kuiper and the emergence of planetary science. Like Luyten and Urey, Burbidge disliked Kuiper's autocratic insensitivity. But this was not the main problem. Kuiper was breaking the mold, not just in building a team of smart artisans rather than traditional researchers with graduate-school and postdoctoral experiences neatly cataloged on a résumé, but he also failed to keep himself within the traditional topics of stellar and galactic astronomy. Burbidge complained about Kuiper "splintering" astronomy by getting involved with planetary topics and using the techniques of geology. Burbidge not only represents why Kuiper left Chicago and unleashed the decade of extraordinary productivity at the University of Arizona (and a legacy that lasts today), he also represents the problems of establishing a new discipline. The innate conservatism, the inertia, of highly trained, reputation-conscious researchers struggling to establish themselves as leaders in the traditional fields didn't want the ground moved under them. When a new field is created, others must give way. They give up scarce resources like personnel, funds, buildings, and access to shared facilities like observatories. Most hurtful of all, they give up prestige, especially when the new field is successful, and planetary science has been spectacularly successful since Kuiper made it respectable.

The Pillars of Strength

More numerous than the antagonists that influenced Kuiper were the pillars of strength he relied on. Otto Struve, Subrahmanyan Chandrasekhar, Richard Harvill, and Bill Brunk were persons who played an immense part in Kuiper's successes, and the role of each was very different. Struve, Harvill, and Brunk, and for a while also Chandra, played the role of seniors to Kuiper. They were the people to whom he reported as chairs, directors, presidents, or representatives of funding agencies. But this understates their importance, large though it would be. These were people in whom Kuiper confided every detail of his life—his likes and dislikes, his challenges, and his successes. They provided a platform for Kuiper to explore his thoughts and ideas, as much for his own sake as theirs. Charles Wood, Kuiper's graduate student, has noted that he would be summoned to Kuiper's office just to listen.[22] As active as he was, with so many projects underway at one time and international success of the highest order, what Kuiper needed most was someone to listen.

Bart Bok has stated that the most important year in Kuiper's life was 1936 and the most important events were his appointment to Chicago and his marriage. Surely this is true. Chicago brought him access to some of the best equipment of the day, the 82-inch telescope at McDonald Observatory. In the prewar period his research on binary stars and white dwarfs continued unabated, he discovered the atmosphere of Titan, and he discovered the moons Miranda and Nereid. He became more familiar with the surface of the Moon than any other astronomer through his use of the 82-inch. His appointment at Chicago also brought him into daily contact with some of the best astronomers of the day, especially Chandra. About Sarah it is impossible to know many details, but it is clear from snippets in correspondence and observations by his daughter that the couple took long evening walks, both at Williams Bay and in Tucson, during which some of his best brainstorming took place. The biggest pillar of all was Sarah.

There were of course many other leading scientists who played important parts in the emergence of planetary science. Whipple had immense influence at the time and is now famous for his insights into the nature of comets and his work on meteors. Baldwin convinced the community that the craters on the Moon were from impacts. Shoemaker, more than anyone else, established that meteorite craters do exist on Earth and popularized the study of near-Earth asteroids. The list is long and contains many

Europeans and Russians. However, none achieved the breadth and depth of research that Kuiper did. None had the productivity. None had the influence in determining the next century of research.

World War II

It is impossible to overstate the importance of World War II in sculpting Kuiper's career. Aside from bringing rockets and the space race to the Moon, the war contributed in two important ways to the next decade or two of Kuiper's life. First and foremost, it made him familiar with the great advances in infrared technology. Greenstein claims Kuiper brought infrared-imaging equipment back from Germany.[23] Certainly he tried to do this, but it seems the primary the role of the war was to impress on Kuiper the remarkable developments in infrared technology that had been made in Germany and bring him into contact with Robert Cashman, who was being funded by the war effort to develop near-infrared detectors that Kuiper could strap to telescopes.

The war also put Kuiper in contact with European and Soviet astronomers. Kuiper's role was to provide intelligence to the U.S. government on the progress of German science. He was given detailed briefing sheets on equipment to search for, but the allies also wanted knowledge on scientific progress in general, and he obtained this by interviewing scientists and visiting laboratories. He wrote papers on the state of European astronomy and observatories. After the war monitoring German scientific progress morphed into monitoring Soviet scientific progress. This involved summarizing Soviet scientific literature, the movements of Soviet scientists, and gathering information from his frequent travels and from the many visitors to the University of Chicago. It was a task Kuiper relished, and the pleasant nature of the visits to his home by Soviet scientists and CIA agents have been noted by all involved, including his daughter. One might argue that this willingness to cooperate with U.S. security forces may have even helped in his efforts to get Air Force and Navy funding.

The Father of Modern Planetary Science

The father of modern planetary science was a brilliant man with a simple vision and amazing drive. After the discovery of the atmosphere on Titan, he set a course to study planetary atmospheres and then everything in the solar system. Unlike many scientists, he remained hands-on until his death. He never entirely delegated the task of gathering data and making observations to his students and technicians. He took great joy in being in the laboratory, being in the observatory dome, or being on the aircraft alongside the spectrometers and telescopes. But the world swirling around him and within his mind made this a complicated, colorful, exciting, and action-packed process, albeit sometimes painful. He made great friends and some enemies, and even his friends sometimes found him difficult to live with; his students have placed on record their mixture of feelings. There were those who could not overlook what they regarded as his insensitive attitude, and when I started this book I was cautioned not to open old wounds. However, it cannot be denied that his list of accomplishments added up to a list that would be longer than a dozen careers added together. The community that has named so much after him places his accomplishments on the highest level.

It is also clear that he was a devoted and loving family man. He hated to be separated from his family for even short periods, and he always took them with him on his annual (sometimes biannual) trip to McDonald Observatory. They also attended the triennial IAU General Assemblies. His feelings about his family are well documented in the correspondence that passed between him and his colleagues. His family has recounted their knowledge of him to me with great love and respect.

As a student entering the field, I thought little about Gerard Kuiper and his flourishing group at the University of Arizona. I enjoyed his *Sky and Telescope* articles about the institute he was creating at Tucson, and I occasionally read an article of his, although he overlapped relatively little with the meteorite-research field I was struggling to enter. Toward the end of my career, I tasked myself with compiling a list of people important to our exploration of space for the colloquium series I was about to hold for the young minds in the University of Arkansas. There was Gerard Kuiper, standing out as an extraordinary scientist living at an extraordinary time and living life to the fullest. He was interested in everything and successful at everything he set his hand to, yet some of the personal setbacks and personal insults he had to bear were of the highest order. What better example to show our students and to contemplate for ourselves. In this short book, which has left out as much as it has included, I have tried to describe Kuiper's personal and professional life, his accomplishments, and how through his labors we have learned so much about the solar system and the modern scientific discipline of planetary science. It is a story worth telling and will remain so.

ABBREVIATIONS

AIP	Oral history archives, Niels Bohr Library and Archives, American Institute of Physics, College Park, MD USA.
AJ	Astronomical Journal
AN	Astronomische Nachrichten
ApJ	Astrophysics Journal
BAN	Bulletin of the Astronomical Institutes of the Netherlands
BMNAS	Biographical Memoirs of the National Academy of Sciences
COMM	Communications of the Lunar and Planetary Laboratory (available online at https://www.lpl.arizona.edu/sic/collection/journal)
MNRAS	Monthly Notices of the Royal Astronomical Society
MSOR	Oral histories collected by Melissa L. Sevigny as the basis for her book "Under Desert Skies: How Tucson Mapped the Way to the Moon and Planets." Copies can be found in UARSC (see below).
PA	Popular Astronomy
PASP	Publications of the Astronomical Society of the Pacific
PNAS	Proceedings of the National Academy of Sciences
PSS	Planetary and Space Science
PT	Physics Today
QJRAS	Quarterly Journal of the Royal Astronomical Society
RMP	Reviews of Modern Physics
RPP	Reports of Progress in Physics
UCSC	Special Collections of the library of the University of California, Santa Cruz, UA 36 (Lick 1), box 94.

UCSD   Special Collections of the library of the University of California, San Diego, MSS 44, box 51, folder 6–7.

UARSC   Special Collections of the library of the University of Arkansas, Fayetteville, Derek Sears papers.

UASC   Special Collections of the library of the University of Arizona, Tucson, Gerard P. Kuiper papers M-180.

UChSC   Special Collections of the library of the University of Chicago, Illinois, Subrahmanyan Chandrasekhar papers, box 19.

ZfA   Zeitschrift für Astrophysik

NOTES

PREFACE

1. Kuiper expressed these views in the foreword to *Planets and Satellites*, volume 3 of his four-volume work, *The Solar System* (Chicago: University of Chicago Press, 1961).

CHAPTER 1

1. T. Gehrels, *On the Glassy Sea* (Charleston, S.C.: BookSurge, 2007).
2. UASC, biofile, "Text of Remarks by B. J. Bok Delivered at the Memorial Service for G. P. Kuiper (December 31st 1973)."
3. UARSC, S. des Tombe, "Oral History Recorded at Half Moon Bay, California on February 20 and March 7, 2013 (with later additions)."
4. We refer here to apparent magnitude, the magnitude of the star as it appears in the sky. This should not be confused with absolute magnitude, which is the magnitude the star would have if it were 1 AU (i.e., the distance of the Sun from Earth) from the observer.
5. UASC, biofile, "Text of Remarks by B. J. Bok."
6. Gehrels translates this as "nerd" but my dictionary says as stated. (A similar slang term is *oik*, which the dictionary defines as a rude and unpleasant man from a low social class).
7. K. A. Strand, "Ejnar Hertzsprung, 1873–1967," *PASP* 80 (1968): 51–56.
8. This is mentioned in a later report Kuiper wrote for Aitkens, although I have found no information about this visit. UCSC, Kuiper to Aitken, progress report, January 21, 1934.
9. In addition to Kuiper (listed as a doctoral candidate in the Observatory at Leiden) there was the expedition leader, Van der Bilt, and astronomers Minnaert, Minnaert-Coelingh, Van den Bijllardt, Kreiken, Wallenquist, Knol, Keller, and van WijngaAden. There was

also a management and technical team of Bakker, Verstelle, Birfelder, Dorst, Waltz, Nieborg, and Cramer. Two students were taken on the expedition in addition to Kuiper, W. R. van Wijk (doctoral assistant in the Physics Laboratory in Utrecht) and C. J. Gorter (doctoral candidate in the Physics Lab at Leiden).

10. J. Van der Bilt and M. G. J. Minnaert, Veralag van de expedite, ultgezonden ter waarneming van de totale zonsverdulstering van 9 Mei 1929. *Verslagen van de gewone vergaderingen der Wis- en Natuurkundige Afdeeling* 38 (1929): 101–19.

11. D. P. Cruikshank, "Gerard Peter Kuiper," *BMNAS* 62 (1'993): 258–95.

12. "Solar Eclipse: Clouds Spoil Observations," *The West Australian*, May 16, 1929, p. 15.

13. UCSC, Kuiper to Aitken, December 24, 1929.

14. A good description of astronomy at the time is given by O. Struve O. and V. Zebergs, *Astronomy of the 20th Century* (New York: MacMillan 1962).

15. G. P. Kuiper, "Discourse Following Award of the Kepler Gold Medal at the AAAS Meeting, Franklin Institute, Philadelphia," *COMM* 9, no. 183 (1971): 404.

16. UCSC, Kuiper to Aitken, April 16, 1930.

17. By "small distances" Kuiper meant binary systems in which the stars have small angular separation.

18. "Astronomical Seeing," Wikipedia.org, last modified July 12, 2018, https://en.wikipedia.org/wiki/Astronomical_seeing.

19. UCSC, Kuiper to Aitken, October 20, 1931; Kuiper to Aitken, December 17, 1931.

20. G. P. Kuiper, "Visual Measures of Double Stars Made in the Years 1929.6–1931.3," *BAN* 7 (1933): 129–49.

21. D. E. Osterbrock, J. R. Gustafson, and W. J. S. Unruh, *Eye on the Sky: Lick Observatory's First Century* (Oakland: University of California Press, 1988). See also Anthony Misch and Remington Stone, "Building the Observatory," The Lick Observatory Historical Collections, accessed November 7, 2016, http://collections.ucolick.org/archives_on_line/bldg_the_obs.html.

22. UCSC, Kuiper to Aitken, December 24, 1929; Aitken to Kuiper, February 7, 1930; Kuiper to Aitken, April 16, 1930; Kuiper to Aitken, October 20, 1931; UCSC, Kuiper to Aitken, December 17, 1931; Aitken to Kuiper, February 27, 1932; Aitken to Kuiper, January 17, 1933; Aitken to Kuiper, March 7, 1933; Kuiper to Aitken, Radiogram, June 14, 1933; Kuiper to Aitken, July 20, 1933.

23. UCSC, Aitken to Leuschner, January 6, 1932.

24. UCSC, Aitken to Kuiper, February 27, 1932.

25. UCSC, Aitken to Kuiper, January 17, 1933.

26. UCSC, Aitken to Kuiper, March 7, 1933.

27 UCSC, Kuiper to Aitken, Radiogram, June 14, 1933.

28. UCSC, Kuiper to Aitken, July 20, 1933.

29. UARSC, S. des Tombe, "Oral History Recorded at Half Moon Bay."

CHAPTER 2

1. M. Hoskin, *Discoverers of the Universe: William and Caroline Herschel* (Princeton, N.J.: Princeton University Press, 2011).

2. S. L. Lippincote and M. D. Worth, "The Double Star Sirius," *Sky and Telescope* (January 1966): 4–6.

3. David DeVorkin, *Henry Norris Russell: Dean of American Astronomers* (Princeton, N.J.: Princeton University Press, 2000), 317–19.

4. In addition to Aitken, the astronomers were William Hammond Wright, Joseph Haines Moore, and Robert Julius Trumpler. There were five associate astronomers (H. M. Jeffers, F. J. Neubauer, G. E. Paddock, N.U. Mayall, A. B. Wyse, and an assistant, C. E. Smith), four Fellows (P. S. Riggs, P. Herget, S. Herrick, and Miss D. N. Davis), and six permanent staff (a photographer, foreman/carpenter, instrument maker, electrician, and janitor).

5. A. H. Batten, "Wright, William Hammond," *BEA* 2 (2007): 1243–45.

6. W. H. Wright, "Joseph Haines Moore, 1878–1949," *BMNAS* 29 (1956): 123.

7. H. F. Weaver, "Robert Julius Trumpler," *BMNAS* 78 (2000): 276–97.

8. D. E. Osterbrock, "Nicholas Ulrich Mayall 1906–1993," *BMNAS* 69 (1996): 187–212.

9. For details of Barnard's career, see William Sheehan, *The Immortal Fire Within: The Life and Times of Edward Emerson Barnard* (Cambridge: Cambridge University Press, 1995). It also contains a great deal of information about the early histories of Lick and Yerkes Observatories.

10. G. P. Kuiper, "The Visual Magnitude of the Companion of Sirius," *PASP* 46 (1934): 99–104.

11. UCSC, Kuiper to Torchiana, September 29, 1933.

12. UASC, Misch to Sears, email, September 11, 2013.

13. E. E. Barnard, "An Account of the Discovery of a Fifth Satellite to Jupiter," *AN* 131 (1892): 73–76.

14. W. W. Campbell, "Some Stars with Large Radial Velocities," *ApJ* 13 (1901): 98–99.

15. N. U. Mayall, "The Spectrum of the Spiral Nebula NGC 4151," *PASP* 46 (1934): 134–38.

16. UASC, Mayall to Cruikshank, September 22, 1982.

17. UCSC, Kuiper to Aitken, progress report, January 21, 1934.

18. G. P. Kuiper, "Measure of the Rapid Binary BD -8°4352," *PASP* 46 (1934): 360–61.

19. G. P. Kuiper, "A Triple Star of Large Parallax," *PASP* 46 (1934): 361.

20. G. P. Kuiper, "The Visual Magnitude of the Companion of Sirius," *PASP* 46 (1934): 99–104.

21. G. P. Kuiper, "Problems of Double-Star Astronomy I and II," *PASP* 47 (1935): 15–42, 121–50.

22. The session included George Louderback (UC Berkeley), The Age of the Earth from Sedimentation; Robley Evans (MIT), The Age of the Earth from Radioactive Disintegration

and Related Problems; B. Gutenberg (Caltech), The Age of the Earth from Changes in Temperature and Elastic Properties; G. P. Kuiper (Lick), The Age of the Galaxy from the Disintegration of Galactic Star Clusters and Binary Stars; Richard Tolman (Caltech), The Age of the Universe from the Red Shift in the Spectra of Extragalactic Objects; and Paul Epstein (Caltech), Attempts to Reconcile the Long and the Short Time Scales in Cosmology.

23. F. C. Leonard, "Report of the meeting of the Astronomical Society of the Pacific, Held in Connection with the Meeting of the Pacific Division of the American Association for the Advancement of Science, at the University of California at Los Angeles, June 26, 27, and 28, 1935," *PASP* 47 (1935): 194–97.

24. G. P. Kuiper, "Two New White Dwarfs of Large Parallax," *PASP* 46 (1934): 287–90; G. P. Kuiper, "A New White Dwarf of Large Parallax," *PASP* 47 (1935): 96–98; G. P. Kuiper, "The White Dwarf +70°8247, the Smallest Known Star," *PASP* 47 (1935): 307–13.

25. B. Lovell, "J. P. M. Prentice," *QJRAS* 23 (1982): 452–59.

26. W. Wright, "Comments on Nova Herculis 1934," *PASP* 47 (1935): 47–49.

27. B. Ricca, *Super Boys: The Amazing Adventures of Jerry Siegel and Joe Shuster—the Creators of Superman* (New York: St. Martin's, 2013).

28. UCSC, Ruby's Inn menu.

29. UCSC, Kuiper to Aitken, May 16, 1935.

30. G. P. Kuiper, "Note of the Light Variation of Nova Herculis 1934," *PASP* 47 (1935): 99–100; G. P. Kuiper, "The Light-Variation of Nova Herculis 1934: Second Note," *PASP* 47 (1935): 165–66; G. P. Kuiper, "The Duplicity and the Light Variation of Nova Herculis 1934," *PASP* 47 (1935): 228–30; G. P. Kuiper, "Observations of Nova Herculis 1934: Fourth Note," *PASP* 47 (1935): 267–71.

31. UCSC, Nova Herculis, press release sent to AP and the *Mercury Herald*, July 1935.

32. "DQ Herculis," Wikipedia.org, last modified September 7, 2917, https://en.wikipedia.org/wiki/DQ_Herculis.

33. G. P. Kuiper, "De Planeet Mars," pts. 1, 2, and 3, *Hemel en dampkring* 29 (1931): 153–61; 195–208; 221–36.

34. W. H. Wright and G. P. Kuiper, "Clouds on Mars," *PASP* 47 (1935): 92–93.

35. G. V. Schiaparelli, "Il Pianeta Marte," 1893, https://ia601305.us.archive.org/7/items/Marte1893/marte%20_1893.pdf.

36. There has been a whole book devoted to understanding this: William Sheehan, *Planets and Perception: Telescopic Views and Interpretations, 1609–1909* (Tucson: University of Arizona Press, 1988).

37. Kuiper, "De Planeet Mars," pts. 1, 2, and 3.

38. Kuiper, "A New White Dwarf of Large Parallax."

39. J. Tuominen, "Some Remarks on the Russell Diagram and on the White Dwarfs," *Meddelanden fran Lunds Astronomiska Observatorium Series I* 139 (1934): 1–4.

40. UASC, box 17, file 15, Luyten to Aitken, May 7, 1935 (as copied by Kuiper).

41. UASC, box 17, file 15, Aitken to Luyten, May 12, 1935 (as copied by Kuiper).

42. UASC, box 17, file 15, Luyten to Aitken, May 17, 1935 (as copied by Kuiper).

43. UASC, box 17, file 15, Aitken to Luyten, June 8, 1935 (as copied by Kuiper).

44. A. Upgren, "Willem Jacob Luyten 1899–1994," *BMNAS* 76 (1999): 15.

45. D. E. Osterbrock, *Yerkes Observatory, 1892–1950: The Birth, Near Death, and Resurrection of a Scientific Institution* (Chicago: University of Chicago Press, 1997); DeVorkin, *Henry Norris Russell.*

46. H. Shapley, "Henry Norris Russell 1877–1957," *BMNAS* 32 (1958): 352–78.

47. DeVorkin, *Henry Norris Russell.*

48. UCSC, Kuiper to Torchiana, March 29, 1935.

49. H. N. Russell, "Some Remarkable Double Stars," *Scientific American* 154 (1936): 314–15.

50. UASC, box 9, file 32, brochure, "A Visit to the Bosscha Observatory of the University of Indonesia."

51. Wyse was best known for his analysis of the spectra of novae, especially Nova Aquilae 1918, which, he demonstrated, expanded at a nearly constant rate for more than twenty years. Sadly, Wyse died at only in 1942 at only thirty-three in an accident involving two lighter-than-air Navy craft (i.e., dirigibles) over the Atlantic Ocean off the New Jersey coast. William Sheehan, personal communication to author, July 9, 2017.

52. "Award of Eddington Medal to R. Wildt," *QJRAS* 7 (1966): 120.

53. Shapley and Jean Harlow were both featured on the cover of *Time* magazine within a few months of each other. Some wit noticed this and said that the one showed "Harlow Shapley" and the other "Shapely Harlow." William Sheehan, personal communication to author, July 7, 2017.

CHAPTER 3

1. UCSC, Kuiper to Wright, October 30, 1935.

2. At the time, the largest was still the 100-inch Hooker telescope at Mt. Wilson (1917), followed by the 74-inch David Dunlop telescope (1935), the 72-inch Dominion Observatory telescope (1918), and the 69-inch Perkins Observatory telescope (1931), respectively.

3. UCSC, Kuiper to Wright, October 30, 1935.

4. UASC, biofile, "Text of Remarks by B. J. Bok."

5. UASC, box 1, file 11, Hindward to Kuiper, October 3, 1935.

6. D. West, *Harlow Shapley—Biography of an Astronomer: The Man Who Measured the Universe* (Portland, Ore.: C&D, 2015).

7. B. J. Bok, "Harlow Shapley 1885–1972," *BMNAS* 49 (1978): 239–91.

8. J. A. Graham, C. M. Wade, and R. M. Price, "Bart J. Bok 1906–1983," *BMNAS* 49 (1994): 72–97.

9. K. C. Wali, *Chandra: A Biography of S. Chandrasekhar* (Chicago: University of Chicago Press, 1991).

10. Chandra is also an example of a young scientist's findings failing to gain acceptance in the face of criticism (some say ridicule) from a senior scientist. See K. C. Wali, "Chandrasekhar vs. Eddington—An Unanticipated Confrontation," *Physics Today* 35 (1982): 33–40.

11. G. Srinivasan, ed., *From White Dwarfs to Black Holes: The Legacy of S. Chandrasekhar* (Chicago: University of Chicago Press, 1997); K. C. Wali, ed., *Chandrasekhar: The Man Behind the Legend—Chandra Remembered* (London: Imperial College Press, 1997); R. Ramnath, ed., *S. Chandrasekhar: Man of Science* (New York: HarperCollins, 2012). K. C. Wali, ed., *A Scientific Autobiography: S Chandrasekhar* (Singapore: World Scientific, 2011). D. M. Salwi, ed., *S. Chandrasekhar: The Scholar Scientist* (New Delhi, India: Rupa, 2004).

12. G. Field, "Fred Lawrence Whipple 1906–2004," *BMNAS* 89 (2007): 1–23.

13. Interview of Frank K. Edmonson by David deVorkin, April 21, 1977, and February 2, 1978, AIP, https://www.aip.org/history-programs/niels-bohr-library/oral-histories/4588-1 and https://www.aip.org/history-programs/niels-bohr-library/oral-histories/45881-2.

14. UCSC, Kuiper to Wright, October 30, 1935.

15. UASC, box 1, folder 11, lecture notes.

16. G. P. Kuiper, "Double Stars," *The Telescope* 3 (1936): 2–7.

17. G. P. Kuiper, "The White Dwarf A.C. +70°8247, the Smallest Star Known," *PASP* 47 (1935): 307–13; G. P. Kuiper, "The Visual Binary with the Shortest Known Period, B.D. -8°4352," *PASP* 48 (1936): 19–21.

18. G. P. Kuiper, "On the Hydrogen Content of Clusters, Binaries, and Cephei," *Harvard College Observatory Bulletin* 903 (1936): 1–11.

19. Kuiper, "The White Dwarf A.C. +70°8247."

20. W. S. Adams and F. H. Seares, "Annual Report of the Director of the Mount Wilson Observatory" (1938): 28.

21. Kuiper to Chandrasekhar, October 12, 1937, UChSC, box 19, folder 20.

22. G. P. Kuiper, "List of Known White Dwarfs," *PASP* 53 (1941): 248–52.

23. N. F. Comins and W. J. Kaufmann, *Discovering the Universe*, 9th ed. (Gordonsville, Va.: W. H. Freeman, 2011).

24. A. J. Cannon and E. C. Pickering, "The Henry Draper Catalogue: 21h, 22h, and 23h," *Annals of the Astronomical Observatory of Harvard College* (1918–24): 91–99.

25. H. N. Russell, "Relations Between the Spectra and Other Characteristics of the Stars," pts. 1, 2, and 3, *Nature* 93 (1914): 227–30; 252–58; 281–86.

26. C. Payne-Gaposchkin and K. Haramundanis, *Cecilia Payne-Gaposchkin: An Autobiography and Other Recollections* (Cambridge: Cambridge University Press, 1984).

27. UChSC box 19, folder 20, Kuiper to Chandrasekhar, March 15, 1936.

28. UChSC box 19, folder 20, Kuiper to Chandrasekhar, March 15, 1936.

29. UChSC box 19, folder 20, Kuiper to Chandrasekhar, March 15, 1936.

30. B. Strömgren, "On the Interpretation of the Hertzsprung-Russell Diagram," *ZfA* 7 (1933): 222–48.

31. Kuiper, "On the Hydrogen Content of Clusters, Binaries, and Cepheids."

32. R. J. Trumpler, "Preliminary Results on the Distances, Dimension and Space Distribution of Open Star Clusters," *Lick Observatory Bulletin* 420 (1930): 154–88.

33. G. P. Kuiper, "On the Hydrogen Content of Clusters," *ApJ* 86 (1937): 176–97.

34. Strömgren, "On the Interpretation of the Hertzsprung-Russell Diagram."

35. UChSC box 19, folder 20, Kuiper to Chandrasekhar, May 11, 1936.

36. UASC box 17, file 15, Kuiper to Struve, April 18, 1936.

37. UARSC, S. des Tombe, "Oral History recorded at Half Moon Bay."

38. UChSC box 19, folder 20, Kuiper to Chandrasekhar, May 11, 1936.

39. UASC box 17, file 15, Kuiper to Struve, June 18, 1936.

40. Wedding invitation, UChSC, box 19, folder 20.

41. UASC box 17, file 15, Kuiper to Struve, August 20, 1935; Struve to Kuiper, August 22, 1935; Struve to Kuiper, October 9, 1935; Kuiper to Struve, October 20, 1935.

42. William Sheehan sent me some comments on Edwin Frost:

> Frost was blind the last decade he was at Yerkes, until his retirement—in those days, there was no social security, of course, so one had to plug away until the last. He used to navigate his way to the office (Brantwood, very close to where Kuiper had his house in late years) through the woods by means of a guide wire—bits of which still could be traced when I was last at Yerkes a few years ago—and there was a sign on his door that said "please close door." Obviously the carelessness of colleagues had led to more than a few accidents in which he had bumped into it.
>
> He was a humane person, a well-trained Dartmouth spectroscopist, but lost a lot of the "first string" people, like Ritchey and Adams, to Hale's Mt. Wilson adventure. He coped as best he could and would have grad students and the like read to him from the astronomical journals, and noticed—a discovery that a blind astronomer *would* make—that the rate of cricket chirrups were linearly related to the temperature, that they chirruped more slowly in the cold weather than the warm weather.
>
> Speaking of weather, it's interesting how many of the figures in the early history of the observatory had weather names: Gale, Hale, William Rainey Harper, Snow (for the Snow telescope), Frost. (William Sheehan, personal communication to author, July 9, 2017)

43. K. Krisciunas, "Otto Struve 1897–1963," *BMNAS* 60 (1992): 349–87.

44. UASC box 17, file 15, Struve to Bok, September 30, 1935. Why Struve would write to Bok about hiring Kuiper is interesting. I assume it is because it was Bok, with Shapely, that first encouraged Struve to consider Kuiper (UASC box 15, file 15, Struve to Kuiper, August 12, 1935) and because Bok and Kuiper were students together in Leiden. Bok's encouragement might have helped Kuiper accept Struve's offer.

45. UASC box 17, file 15, Struve to Bok, September 30, 1935.

46. President Robert Maynard Hutchins moved to Chicago in 1929 to become the youngest university president in the United States at the age of thirty (and who is best remembered now, perhaps, for discontinuing intercollegiate football at the University of Chicago. At the time, the University of Chicago was part of the Big Ten, and the first Heisman Trophy winner was its own Jay Berswanger.) In the mid or late 1930s Hutchins was sued by Charles Rudolph Walgreen, the founder of Walgreen's, who accused some university faculty of promoting communism (citing that pernicious influence on Walgreen's niece). Hutchins nobly defended academic freedom. Another interesting byway: Walgreen's granddaughter, Ruth Walgreen Stephan, was interested in poetry, moved to Tucson, and in 1960 (or just about the same time Kuiper left Yerkes to found the LPL) founded and endowed the Poetry Center at the University of Arizona. William Sheehan, personal communication to author, July 7, 2017.

47. UASC box 17, file 15, Struve to Kuiper, October 9, 1935.

48. UASC box 17, file 15, telegram, Kuiper to Struve, October 28, 1935.

49. UCSC, Kuiper to Wright, October 30, 1935.

50. UCSC, Kuiper to Wright, October 30, 1935; Wright to Kuiper, November 4, 1935.

51. UASC box 17, file 15, telegram, Voûte to Kuiper, November 1, 1935.

52. UASC box 17, file 15, telegram, Kuiper to Struve, November 1, 1935.

53. UASC box 17, file 15, telegram, Kuiper to Struve, November 7, 1935.

54. W. W. Morgan, "Frank Elmore Ross 1874–1960," *BMNAS* 39 (1967): 389–402; S. B. Nicholson, "Frank Elmore Ross 1874–1960," *PASP* 73 (1961): 182–84; W. Osborn, "Frank Elmore Ross and His Variable Star Discoveries," *Journal of the American Association of Variable Star Observers* 40 (2012): 133–40.

55. Barnard was looking for variable stars in globular clusters. In the process he discovered the star of high proper motion now known as Barnard's Star.

56. UASC box 17, file 15, telegram, Ross to Kuiper, January 25, 1936. See also UASC box 17, file 15, Kuiper to Ross, January 29, 1936.

57. UCSC: Kuiper to Wright, November 14, 1935.

58. UASC box 17, file 15, Kuiper to Struve, December 23, 1935, UASC, box 17, file 15.

59. UASC, box 17, file 15, Struve to Kuiper, January 6, 1936.

60. UASC box 17, file 15, Kuiper to Struve, January 1, 1936.

61. Dean Gale was a deep-dyed racist (not unusual in those days), regarded Indians like Chandra as "colored people," and would not allow them on campus. William Sheehan, personal communication to author, July 9, 2017.

62. UASC box 17, file 15, Kuiper to Struve, February 6, 1936.

63. UChSC box 19, folder 20, Kuiper to Chandrasekhar, March 15, 1936.

64. UASC box 17, file 15, Struve to Kuiper, March 9, 1936.

CHAPTER 4

1. D. E. Osterbrock, *Yerkes Observatory, 1892–1950: The Birth, Near Death, and Resurrection of a Scientific Institution* (Chicago: University of Chicago Press, 1997).

2. Hale was the only son of Chicago elevator magnate William Hale, a precocious amateur scientist, and became the greatest astronomical entrepreneur of his time. The first telescope used at Yerkes was a 12-inch Clark that his father had acquired for him and set up in young Hale's own private observatory at Kenwood on Chicago's south side. (The Hale mansion still exists, though the observatory—and Hale's library—are long gone. They have been turned into high-end condos). William Sheehan, personal communication to author, July 9, 2017.

3. Lick and Yerkes Observatories were inspired by the first astrophysical observatory at Potsdam, founded in 1874. William Sheehan, personal communication to author, July 9, 2017.

4. "Henry Ives Cobb," Wikipedia.org, last updated August 10, 2018, https://en.wikipedia.org/wiki/Henry_Ives_Cobb.

5. William Sheehan, personal communication to author, July 9, 2017.

6. E. E. Barnard, "The Bruce Photographic Telescope of the Yerkes Observatory," *ApJ* 21 (1905): 35–48.

7. William Sheehan, personal communication to author, July 9, 2017.

8. Dale Cruikshank provided me with some reminiscence of his time at Yerkes (summers of 1958, 1959, and 1960):

> Roy Wickham was the modest and kindly custodian, and the way he tipped his head was almost a signature characteristic of his. Marguerite Van Biesbroeck lived with her brother Georges and sister-in-law Julia at 60 Constance, where they ran the boarding house for itinerate astronomers and students. I took French language lessons from Marguerite ("Miss Van B") one summer. She had been a music conservatory student in Belgium, and had a large library of bound volumes of piano and vocal music. She was disposing of these, and offered some or all of them to me. At the end of that first summer, I drove home from Wisconsin with a car trunk full of those music volumes, some written as copy books in her own hand. They have been a treasure to me ever since. Of course, I still have them. Georges ("Dr. Van B") taught me how to measure double stars with the filar micrometer on the 40-inch refractor, and when I continued similar measurements back home with the Drake University 8-inch refractor, he kindly commented on my results in a hand-written letter that I still have. Regarding Charles Ridell, he was no longer

at YO when I was there, but Kuiper often referred to him as a master instrument maker. (Dale Cruickshank, personal communication to author, August 22, 1917)

9. T. Hockney and T. R. Williams, "Van Biesbroeck, Georges-Achille," *BEA* 2 (2007): 1168–69.

10. "Yerkes Observatory Staff," The University of Chicago Photographic Archive, accessed November 11, 22016, http://photoarchive.lib.uchicago.edu/db.xqy?one=apf6-04226.xml.

11. UChSC, box 19, folder 20, Kuiper to Chandrasekhar, October 12, 1937.

12. UASC box 17, file 16, Struve to Kuiper, February 20, 1940.

13. Yerkes Study Group, "Final Report of the Yerkes Study Group: 30 November 2007," http://astro.uchicago.edu/yerkes/ysg/YSG_Final_Report.pdf.

14. D. S. Evans and J. D. Mulholland, *Big and Beautiful: A History of the McDonald Observatory* (Austin: University of Texas Press, 1986).

15. Frost to Benedict, April 1, 1926, cited by Evans and Mulholland, *Big and Beautiful*, 13.

16. G. A. Wegner, "Elvey, Cristian Thomas," *BEA* 2 (2007): 337.

17. Evans and Mulholland, *Big and Beautiful*, 38.

18. O. Struve, "Research Program of the Yerkes and the McDonald Observatories," *PA* 41 (1933): 543–48.

19. The 12-inch refractor was the maiden instrument that Hale brought from his Kenwood Observatory to Yerkes Observatory in the early days, before the 40-inch refractor became available. William Sheehan, personal communication to author, July 7, 2017.

20. D. Osterbrock, "Obituary: Franklin Evans Roach, 1905–1993," *Bulletin of the American Astronomical Society* 26 (1994): 1608–10. On a personal level, Dale Cruikshank has written, "I became acquainted with Franklin Roach in Hawaii, where he lived during the years I was there. He had written a book on the light of the night sky (airglow, aurora, etc.) and had taken a keen interest in the history of astronomy. We discussed that subject many times, and he told me of his first encounter (as a student at Yerkes Observatory) with Otto Struve—the story revealed Struve's commanding presence from day one as Director. Roach was a kindly and lively scientist, who, at least in his later years, wore his hair in such a way, which together with his general appearance, made him look just like Benjamin Franklin." He also adds, "At dinner with him and his wife one evening in his Waikiki condominium (which had a huge pair of ship's binoculars mounted on a pillar on his lanai overlooking the ocean), a phone call came in saying that a naked-eye nova had been spotted in Cygnus. We all trooped outside to get a clear view of the sky, and sure enough, there was a new star. This was V1500 Cygni, which was discovered in late August, 1975, and was visible to the naked-eye for about a week. The dinner party was therefore at the very end of August." Dale Cruickshank, personal communication to author, August 22, 2017.

21. Evans and Mulholland, *Big and Beautiful*, 90.

22. Evans and Mulholland, *Big and Beautiful*, 80.

23. UASC box 17, file 16, Kuiper to Struve, May 30, 1941.

24. UASC box 17, file 16, Kuiper to Struve, July 8, 1941.

25. UChSC box 19, folder 20, Sarah Kuiper to Chandrasekhar, June 10, 1940.

26. UChSC box 19, folder 20, Kuiper to Chandrasekhar, November 9, 1940.

27. UChSC box 19, folder 20, Sarah Kuiper to Chandrasekhar, January 29, 1941.

28. UASC box 17, file 16, Struve to Kuiper, December 27, 1940.

29. UChSC box 19, folder 20, Kuiper to Chandrasekhar, February 3, 1941.

30. UChSC box 19, folder 20, Kuiper to Chandrasekhar, January 16, 1941.

31. G. P. Kuiper, "List of Known White Dwarfs," PASP 53 (1941): 248–52.

32. H. Ludendorff, "Uber die Radialbewengung von ε Aurigae," *AN* 171 (1912): 50–56.

33. G. P. Kuiper, O. Struve, and B. Strömgren, "The Interpretation of ε Aurigae," *ApJ* (1937): 86, 570–612.

34. Su-Shu Huang, "An Interpretation of ε Aurigae," *ApJ* 141 (1965): 976–84.

35. G. P. Kuiper, "On the Interpretation of β Lyrae and Other Close Binaries," *ApJ* 93 (1941): 133–77.

36. "Disability History Month: John Goodricke the Deaf Astronomer," *BBC News Magazine*, December 18, 2012, http://www.bbc.com/news/magazine-20725639. See also C. Gilman, "John Goodricke and His Variable Stars," *Sky and Telescope* 56 (1978): 400–403. William Sheehan adds,

> He became deaf in childhood after an illness, but he was fortunate to have come from a good family (the Goodricke baronets of Ribston Hall). He was actually born in Groningen. Because of the family means, he received a good education (Thomas Braidwood's Academy for deaf pupils in Edinburgh and Warrington Academy), and when he returned to live with his parents in York, he happened to become acquainted with a worthy mentor, Edward Pigott, whose father had built a private observatory and who was already interested in variable stars. He gave Goodricke a list, and Goodricke—perhaps because of his deafness even more sensitive to nuances in stellar brightness—made the most of the opportunity given to him. The location of Goodricke's observatory, in the Treasurer's House right next to York Minster, has been established, and one can even determine by recreating Goodricke's observations that he observed from the eastern-most window of the second floor, looking south towards (over) the Minster. No other astronomer, not even Horrocks, accomplished so much at such an early age. (William Sheehan, personal communication to author, July 9, 2017)

37. F. Carrier, G. Burki, and M. Burnet, "Search for Duplicity in Periodic Variable Be Stars," *Astronomy and Astrophysics* 385 (2002): 488–502.

38. G. P. Kuiper, "The Empirical Mass-Luminosity Relation," *ApJ* 88 (1938): 472–507.

39. J. Halm, "Stars, Motion in Space, etc. Further Considerations Relating to the Systematic Motions of the Stars," *MNRAS* 71 (1911): 610–39.

40. E. Hertzsprung, "Über die Verwendung photographischer effektiver Wellenlaengen zur Bestimmung von Farbenäquivalenten," *Publikationen des Astrophysikalischen Observatoriums zu Potsdam* 22 (1911): 1–40.

41. A. Van Maanen, "The Masses and Absolute Magnitudes of Visual Binaries," *PASP* 31 (1919): 231–33.

42. K. Lundmark, "Luminosities, Colours, Diameters, Densities, Masses of the Stars" in *Handbuch der Astrophysik* vol. 5, *Das Sternsystem*, 210–573 (Berlin/Heidelberg: Springer, 1932); A. S. Eddington, "The Connection of Mass with Luminosity for Stars," *The Observatory* 53 (1930): 342–44.

43. G. P. Kuiper, "The Empirical Mass-Luminosity Relation," *ApJ* 88 (1938): 472–507.

44. P. P. Parenago, "Issledovaniia izmenenii bleska 208 peremennykh zvezd (1920–1937)," *Trudy Gosudarstvennogo astronomicheskogo instituta im. P.K. Sternberga* 12 (1938): 130–32.

45. UChSC box 19, folder 20, Sarah Kuiper to Chandrasekhar, June 6, 1941.

46. Catherine and Daniel Popper married in 1940 and had one son. Daniel was the residential astronomer at McDonald Observatory from 1939 to 1942, when he joined the war effort at UC Berkeley, his alma mater. After the war, he returned to Yerkes Observatory until he moved to become a founding member in of the astronomy department at UCLA 1947. He became a leading observer of binary stars. See C. Lacy, "Popper, Daniel Magnes," *BEA* 2 (2007): 924–25.

47. Dale Cruikshank, personal communication to author, September 9, 2017.

48. UChSC box 19, folder 20, Kuiper to Chandrasekhar, December 10, 1941.

49. UChSC box 19, folder 20, Sarah Kuiper to Chandrasekhar, June 6, 1941.

50. UARSC, S. des Tombe, "Oral History Recorded at Half Moon Bay."

51. Evans and Mulholland, *Big and Beautiful*, 96.

52. Evans and Mulholland, *Big and Beautiful*, 97.

53. Evans and Mulholland, *Big and Beautiful*, 97.

54. Kuiper to Chandrasekhar, December 10, 1941, UChSC, box 19, folder 20.

55. UASC box 17, file 16, Kuiper to Struve, January 28, 1942.

56. UASC box 17, file 16, Kuiper to Struve, January 28, 1942.

57. UASC box 17, file 16, Kuiper to Struve, January 15, 1942.

58. UASC box 17, file 16, Struve, Mary to Kuiper (postcard), January 20, 1942.

59. UASC box 17, file 16, Struve to Kuiper, January 22, 1942.

60. UASC box 17, file 16, Kuiper to Struve, January 26, 1942.

61. UASC box 17, file 16, Struve to Kuiper, March 20, 1942.

62. UASC box 17, file 16, Kuiper to Struve, April 4, 1942, and several others at this time.

63. UASC box 17, file 16, Struve to Kuiper, June 8, 1942.

64. UChSC box 19, folder 20, Kuiper, Sarah to Chandrasekhar, June 20, 1943.

65. UChSC box 19, folder 20, Kuiper to Chandrasekhar, March 30, 1943.

66. Morgan's early life had been traumatically erratic: he moved roughly thirteen times between his birth and the end of high school. His father was unstable. In fact, he had a great fear of travel—and probably did not want to be uprooted. He lived in the same house at Yerkes for approximately sixty-eight years, and though he became somewhat more willing to travel in later years, we know from the diary books he kept for therapy after his breakdown in 1951 (shortly after he announced the discovery of the spiral arm structure of the Milky Way) that traveling was always a challenge for him. He was a very dependent personality who struggled mightily to avoid these tendencies—and in particular struggled against the strong male types whom he regarded as dominant, like Struve, Hiltner, and Kuiper. William Sheehan, personal communication to author, July 9, 2017.

67. UASC box 17, file 16, Struve to Kuiper, May 14, 1942.

68. UASC box 17, file 16, Struve to Kuiper, May 21, 1943.

69. UASC box 17, file 16, Struve to Kuiper, June 4, 1944.

70. E. B. Frost, "Edward Emerson Barnard 1857–1923," *BMNAS* 21 (1927): 1–23.

71. UChSC box 19, folder 20, Kuiper to Chandrasekhar, October 17, 1943. The comment about the "Hun" and library at Naples refers to events during four days in September 1943 as the allies advanced on Naples. In attempting to suppress a violent and ultimately successful insurrection by the citizens of Naples, the Germans set fire to the National Library and opened fire on the crowd. See "Four Days of Naples," Wikipedia.org, last updated June 28, 2018, https://en.wikipedia.org/wiki/Four_days_of_Naples.

72. UChSC box 19, folder 20, Sarah Kuiper to Chandrasekhar, June 20, 1943.

CHAPTER 5

1. "Oral History: Clarence Nichols Hickman," Engineering and Technology History Wiki, accessed November 26, 2016, http://www.ieeeghn.org/wiki/index.php/Oral-History: Clarence_Nichols_Hickman. See also the Mechanical Music Digest Archives, accessed November 26, 2016, http://www.mmdigest.com/Archives/Authors/Aut1171.html.

2. UChSC, box 19, folder 20, Kuiper to Chandrasekhar, March 30, 1943.

3. UCHSC, box 19, folder 20, Kuiper to Chandrasekhar, March 30, 1943.

4. L. W. Alvarez, "Alfred Lee Loomis 1887–1975," *BMNAS* 51 (1980): 307–41. William Sheehan comments, "Alfred Loomis. A lot of interesting material on Wikipedia about him. He was first cousin to Henry Stimson, the Secretary of State, whose biography Mac Bundy later wrote, and after graduating from Harvard Law School began practicing corporate law in the firm of Winthrop and Stinson. He was obviously very successful, and could afford to buy a 17,000 acre spread on Hilton Head for his private enjoyment— riding, boating, fishing and hunting. His hobbies included automobiles and yachting,

including racing America Cup yachts against the Vanderbilts and Astors." William Sheehan, personal communication to author, July 9, 2017.

5.   F. E. Terman, *Electronic and Radio Engineering*, 4th ed. (New York: McGraw-Hill, 1955).

6.   UCHSC, box 19, folder 20, Kuiper to Chandrasekhar, August 5, 1943.

7.   UCHSC, box 19, folder 20, Kuiper to Chandrasekhar, September 19, 1943.

8.   UCSC, Kuiper to Moore, February 29, 1944. UASC, box 15, file 24, Kuiper to Russell, January 18, 1944. The ten brightest satellites were Io, Europa, Ganymede, Callisto (which orbit Jupiter), Titan (Saturn), Ariel, Umbriel, Titania, Oberon (Uranus), and Triton (Neptune).

9.   UCHSC, box 19, folder 20, Kuiper to Chandrasekhar, January 28, 1944.

10. UASC, box 15, folder 24, Kuiper to Russell, January 17, 1944; Russell to Kuiper, January 17, 1944.

11. UCSC, Kuiper to Moore, February 29, 1944.

12. H. N. Russell, "The Atmospheres of the Planets," *Nature* 135 (1935): 219–26.

13. J. H. Jeans, *The Dynamical Theory of Gases* (London: Cambridge University Press, 1916).

14. R. Wildt, "The Geochemistry of the Atmosphere and the Constitution of the Terrestrial Planets," *RMP* 14 (1942): 151–59.

15. G. P. Kuiper, "Planetary Atmospheres and Their Origin," in *The Atmospheres of the Earth and Planets* (Chicago: University of Chicago Press, 1949).

16. UCHSC, box 19, folder 20, Kuiper to Chandrasekhar, January 28, 1944.

17. UCHSC, box 19, folder 20, Kuiper to Chandrasekhar, January 28, 1944.

18. G. P. Kuiper, "Titan: A Satellite with an Atmosphere," *ApJ* 100 (1944): 378–83.

19. A. E. Milne, *Sir James Jeans: A Biography* (1952; repr., Cambridge: Cambridge University Press, 2013).

20. Examples are *Stars in their Courses* (1931; repr., Cambridge: Cambridge University Press, 2009) and *Through Space and Time* (1934; repr., Cambridge: Cambridge University Press, 2009).

21. Lomonosov's observations of an atmosphere on Venus, based on limb darkening, have been disputed by several authors, and some of their reservations apply to the Titan case: See, for example, W. P. Sheehan and J. M. Pasachoff, "Atmosphere of Venus: Problems in Perception," *Physics Today* 66 (2013): 9–10; J. M. Pasachoff and W. Sheehan, "Lomonosov, the Discovery of Venus's Atmosphere, and the Eighteenth-Century Transits of Venus," *Journal Astronomical History and Heritage* 15 (2012): 3–14; J. Westfall and W. Sheehan, *Celestial Shadows: Eclipses, Transits, and Occultations* (New York: Springer, 2015), 312–21.

22. UASC, box 15, file 24, Russell to Kuiper, January 25, 1944.

23. J. Comas Solá, "Observationes des Satellites Principaux de Jupiter et de Titan," *AN* 179 (1908): 289–90.

24. A. Coustenis, "Titan," in *Encyclopedia of the Solar System*, ed. T. Spohn, D. Breuer, and T. Johnson, 831–49 (Amsterdam: Elsevier, 2014).

25. G. P. Kuiper, "Titan: A Satellite with an Atmosphere," *ApJ* 100 (1944): 378–83.

26. R. V. Jones, *Most Secret War: British Scientific Intelligence 1939–1945* (London: Penguin, 1978).

27. "Diary—Gerard P. Kuiper," UASC, box 57, folder 16.

28. Jones, *Most Secret War*. See also "Telecommunications Research Establishment," Wikipedia .org, last updated August 17, 2018, https://en.wikipedia.org/wiki/Telecommunications _Research_Establishment.

29. Kuiper makes repeated reference to Pinetree and Widewing in the diary. "Diary— Gerard P. Kuiper," UASC, box 57, folder 17, OSRD Liaison Office reference number WA- 3457 1C. See also "Bushy Park," Wikipedia.org, last updated August 16, 2018, https:// en.wikipedia.org/wiki/Bushy_Park; and "Eight Air Force," Wikipedia.org, last updated September 1, 2018, https://en.wikipedia.org/wiki/Eighth_Air_Force.

30 UASC, box 57, folder 15, Kuiper to Terman, March 13, 1943.

31. UASC, box 57, folder 15, Kuiper to Terman, March 13, 1943.

32. UChSC, box 19, folder 20, Kuiper to Chandrasekhar, December 10, 1944. The U.S. military used a system, called V-Mail (victory mail), of batch microfilming letters and printing them once the film reached the United States.

33. B. T. Pash, *The Alsos Mission* (New York: Charter, 1970).

34. A. Jacobsen, *Operation Paperclip: The Secret Intelligence Program that Brought Nazi Scientists to America* (New York: Little, Brown, 2014).

35. Pash, *The Alsos Mission*.

36. B. Benderson, "Samuel Abraham Goudsmit 1902–1978," *BMNAS* (2008): 1–28, http://www.nasonline.org/publications/biographical-memoirs/memoir-pdfs/goudsmit -samuel.pdf.

37. UASC, box 57, folder 15, Goudsmit to Kuiper, April 1, 1945.

38. UASC, box 57, folder 16, thirteen untitled sheets stapled together and beginning "It is not intended . . ."

39. UASC, box 57, folder 15, Kuiper to Goudsmit, May 12, 1945.

40. UASC, box 57, folder 21, two sheets torn from a small notebook.

41. UASC, box 57, folder 15, Kemble to Alsos personnel, April 13, 1945.

42. These techniques included adding iron filings in the paint used on the exterior of the submarines, which failed because they are too conductive, and covering the submarines with magnetized nets, which failed because water swept the nets away.

43. K. Landrock, "Friedrich Georg Houtermans (1903–1966)—Ein bedeutender Physiker des 20. Jahrhunderts," *Naturwissenschaftliche Rundschau* 56 (2003): 187–99.

44. UASC, box 17, folder 17, Struve to Kuiper, February 7, 1946.

45. UASC box 17, folder 16, Kuiper to Struve, February 19, 1946.

46. UASC, box 17, folder 17, Struve to Kuiper, February 7, 1946.

47. UASC, box 57, folder 15, Kuiper to Fisher, May 9, 1945.

48. UASC, box 57, folder 15, Kuiper to Fisher, June 30, 1945.

49. UASC, box 57, folder 15, Kuiper to Goudsmit, July 2, 1945. This is an eight-page report describing visits to Liège (to see Paugren about a stereoplanigraph for the US Navy), Eindhoven (to see Casimir, from whom he learned about various anti-Nazi German scientists, scientists who wanted to go to the United States, a temporary university in Eindhoven, locals executed by Germany, the concentration camp at Vugt), Amsterdam (visited Gorter, Clay, Pannekoek, and Sizoo and learned about their health and looting), Leiden (equipment taken by Germans, but no deaths, university closed after protests, Hertzsprung criticized for not being sufficiently anti-German, Oort fled), Delft (some looting, two professors executed, many students fled after refusing to pledge allegiance to Germany), Groningen (nothing of consequence), Utrecht (Minnaert imprisoned and in poor health, witnessed many shootings). He then described the poor conditions in Holland such that 85 percent of babies born in March and April 1945 died.

50. G. P. Kuiper, "Reports from French astronomers," *ApJ* 101 (1945): 254.

51. G. P. Kuiper, "German Astronomy During the War," *PA* 6 (1946): 263–86.

52. UASC, box 57, folder 16, Goudsmit to Kuiper, April 10, 1945.

53. UASC, box 57, folder 19, Kuiper to Clark, November 19, 1945. UASC, box 57, folder 19, Clark to Kuiper, November 19, 1945.

54. UASC, box 17, folder 17, Struve to Kuiper, August 2, 1944.

55. B. H. Bodenheimer, "Louis George Henyey 1910–1970," *BMNAS* 66 (1995): 167–89.

56. UASC, box 17, folder 6, Sarah Kuiper to Struve, June 4, 1945.

57. UASC, box 57, folder 24, Kuiper to Terman, December 17, 1945.

58. D. Bruckner, "30 Former Top-Secret U.S. Spies Hold a Reunion Here," *Chicago Sun Times*, September 18, 1960, p. 32.

59. It is not clear to me whether the "Major Fisher" (with no initials) to which Kuiper sent his reports on German scientists (UASC, box 57, folder 15, Kuiper to Fisher, May 9, 1945, and UASC, box 57, folder 15, Kuiper to Fisher, June 30, 1945) is the same person as referred to here.

60. UASC, box 57, folder 18, Kuiper to Goudsmit, September 24, 1960.

61. UCSC, box 94, folder 1, Kuiper to Moore, April 12, 1945.

CHAPTER 6

1. M. Walker, *The Cold War: A History* (New York: Holt Paperbacks, 1995).

2. Kuiper had interviewed Eric Regener, who (with Karl-Otto Kiepenhaur) was the first to do upper-atmosphere cosmic ray research with V2 rockets. Regener was active throughout the 1930s measuring the properties of cosmic rays at high altitude, using balloons, and

measuring their attenuation in lakes. Examples of his work are E. Regener, "Spectrum of Cosmic Rays," *Nature* 127 (1931): 233–34; E. Regener, "Intensity of Cosmic Radiation in the High Atmosphere," *Nature* 130 (1932): 364; E. Regener, "Energy of Cosmic Rays," *Nature* 131 (1933): 130; and E. Regener and G. Pfotzer, "Vertical Intensity of Cosmic Rays by Threefold Coincidences in the Stratosphere," *Nature* 136 (1935): 718–19.

3. E. W. Kutzscher, "Review of Detectors of Infrared Radiation," *Electro-optical Systems Design* 5 (1973): 30.

4. Much of my information on the early history of infrared detectors is taken from A. Rogalski, "History of Infrared Detectors," *Opto-electronics Review* 20 (2012): 279–308.

5. D. J. Lovell, "The Development of Lead Salt Detectors," *American Journal of Physics* 37 (1969): 467–78.

6. UASC, box 57, file 16, fourteen pages that could be tri-folded to fit in a breast pocket that systematically listed the technologies of interest to the allies. I refer to them here as the "briefing sheets."

7. E. W. Kutzscher and A. R. Vogeler, *The Odyssey of a Scientist: The Career of Dr. Edgar Kutzscher in Germany and America* (oral history program, California State University, Fullerton, 1986).

8. L. M. Brown and J. B. Ketterson, "Robert J. Cashman," *PT* 43 (1990): 148.

9. UASC, Greenstein to Dale Cruikshank, August 30, 1982.

10. UASC, box 17, folder 16, Kuiper to Struve, November 23, 1945.

11. UASC, box 17, folder 15, Kuiper to Struve, September 28, 1945.

12. UASC, box 17, folder 15, Kuiper to Struve, November 13, 1945.

13. UASC, box 17, folder 15, Kuiper to Struve, November 23, 1945.

14. G. P. Kuiper, W. Wilson, and R. J. Cashman, "An Infrared Stellar Spectrometer," *ApJ* 106 (1947): 243–51.

15. I believe he published alone because he worked alone. For most of the projects he tackled, he did not need help, and since he put research so much ahead of, for example, training others, his focus ensured a solo approach. The spectrometer is a very good example of Kuiper arranging for coauthors when the work demanded it. There are many later more examples of necessary coauthors, such as the lunar atlases. On this topic William Sheehan commented to me, "He was very different from Morgan, for example, who gave credit even to grad students who helped. Don Osterbrock often told me that the most cited paper of his career was the one on the spiral arms with Morgan and Stewart Sharpless, even though their role was actually quite minor—getting wide-angle images with the Greenstein-Henyey camera of the HII regions of the MW." W. Sheehan, personal communication to author, September 3, 2017.

16. R. J. Cashman, "New Photo-Conductive Cells," *Journal of the Optical Society of America* 36 (1946): 356. See also R. J. Cashman, "Photodetectors for Ultraviolet, Visible and Infrared Radiation," *Proceedings of the National Electronics Conference* (1946): 171–80.

17. Kuiper, Wilson, and Cashman, "An Infrared Stellar Spectrometer."

18. UChSC, box 19, folder 20, Kuiper to Bartky, January 1, 1946.

19. UChSC, box 19, folder 20, Bartky to Kuiper, January 6, 1946.

20. UChSC, box 19, folder 20, proposal to the U.S. Navy titled "Infra Red Project Proposed as a Navy Task Order."

21. R. L. Walker, "Arthur Adel (1908–1994)," accessed December 11, 2016, https://aas.org /obituaries/arthur-adel-1908-1994.

22. A. Adel and V. M. Slipher, "The Constitution of the Atmospheres of the Giant Planets," *Physical Review* 46 (1934): 902–6.

23. Interview of Arthur Adel by Robert Smith, August 12, 1987, AIP, https://www.aip.org /history-programs/niels-bohr-library/oral-histories/5000.

William Sheehan has added some comments:

The Lowell Observatory was definitely in the doldrums during this period, and Slipher—perhaps because he had had to scrimp by during the Depression, after Percival's widow Constance drained down the endowment in lawsuits and the Stock Market crashed—notoriously spent much of his time in his office at Lowell doing real estate and other business, not astronomy. (Something that drove C. O. Lampland nuts!). The anti-Semitism involved in Adel's being passed over not only for the directorship but also dismissed is one of the darker chapters of the observatory's history; though it was not unusual by the standards of the time. Percival Lowell's brother Abbott Lawrence Lowell as president of Harvard tried to limit Jewish enrollment to 15% of the student body, and to ban African-American students from living in the Freshman Halls. (In both cases the Harvard Board of Overseers overruled him.) He also led the Brahmin opposition to the nomination of Louis Brandeis to the Supreme Court, claiming Brandeis—who was Jewish— lacked the requisite "legal temperament and capacity."

The Lowell trustee who dismissed Adel, William Lowell Putnam II, was a typical WASP of his time—though I suspect V.M. Slipher may have been more important to the decision to part with Adel. Slipher by now was scientifically stagnant, jealous of his prerogatives and wary of anyone scientifically more up-to-date. He was a more or less self-taught spectroscopist/observer and knew nothing of atomic and nuclear physics. (Lampland, on the other hand, was quite sophisticated in those areas). So Slipher had become very comfortable in his directorship and allowed to do as he pleased; it had become a sinecure; he spent much of his time as I said doing his real estate, and became very wealthy; in the event, he remained in the position until 1954, when he was 75. Brad Smith, who knew the Sliphers from when he worked under Clyde Tombaugh for the natural satellites of the Moon project in the 1950s, said that V.M. was not a very nice man; he was a businessman, neither more nor less.

Brandeis said of A.L. Lowell and his ilk in 1916 that they were men "who have been blinded by privilege, have no evil purpose, and many of whom have a distinct public spirit, but whose environment—or innate narrowness—have obscured all vision and sympathy with the masses." (William Sheehan, personal communication to author, July 9, 2017)

24. Interview of Arthur Adel by Robert Smith.

25. UChSC, box 19, folder 20, undated notes titled "Telephone Conversation with Dr. Arthur Adel, June 26, 1946, 2 p.m." See also UChSC, box 19, folder 20, Kuiper to Adel, June 26, 1946.

26. UChSC, box 19, folder 20, Adel to Kuiper, June 29, 1946.

27. UASC, box 17, folder 16, Struve to Kuiper, February 4, 1941.

28. UASC, box 17, folder 16, Struve to Kuiper, July 28, 1941.

29. UASC, box 17, folder 16, Struve to Kuiper, May 16, 1943.

30. UASC, box 17, folder 16, Struve to Kuiper, May 20, 1944.

31. UASC, box 17, folder 16, Struve to Kuiper, July 12, 1944.

32. UASC, box 17, folder 16, Struve to Kuiper, March 13, 1945.

33. UASC, box 17, folder 16, Struve to Kuiper, February 11, 1946.

34. G. P. Kuiper, "Infrared Spectra of Planets," *ApJ* 106 (1947): 251–54.

35. R. Wildt, "On the Chemistry of the Atmosphere of Venus," *ApJ* 96 (1942): 312–14.

36. P. Cattermole, "Chapter 4: The Venusian Atmosphere" In *Venus: The Geological Story* (Baltimore: Johns Hopkins University Press, 1994).

37. Lomonosov's observations have been reappraised by Jay Pasachoff and William Sheehan, who do not think the blurring at the limb had anything to do with the atmosphere of Venus. See Sheehan and Pasachoff, "Atmosphere of Venus"; Pasachoff and Sheehan, "Lomonosov"; Westfall and Sheehan, *Celestial Shadows*, 312–21.

38. J. J. Schroeter, "Observations on the Atmospheres of Venus and the Moon, Their Respective Densities, Perpendicular Heights, and the Twi-Light Occasioned by Them. By John Jerome Schroeter, Esq. of Lilienthal, in the Dutchy of Bremen. Translated from the German," *Philosophical Transactions of the Royal Society of London* 82 (1792): 309–61.

39. F. E. Ross, "Photographs of Venus," *ApJ* 68 (1928): 57–92.

40. Kuiper, "Infrared Spectra of Planets."

41. G. P. Kuiper, "Determination of the Rotation Rate of Venus," *ApJ* 120 (1953): 603–5; R. S. Richardson, "Observations of Venus Made at Mount Wilson in the Winter of 1954–55," *PASP* 67 (1955): 304.

42. C. Boyer and P. Guérin, "Etude de la rotation rétrograde, en 4 jours, de la couche extérieure nuageuse de Vénus," *Icarus* 11 (1969): 338–55; C. Boyer, "The 4-Day Rotation of the Upper Atmosphere of Venus," *Planetary and Space Science* 21, no. 9 (1973): 1559–61. William Sheehan adds, "Charles Boyer was Magistrate of Brazzaville, The Congo, and an amateur astronomer who used a 10-inch reflector. The idea of photographing Venus in

the ultraviolet was suggested to him by Henri Camichel, a ham radio aficionado at Pic du Midi. It was his small-scale images of Venus in the ultraviolet that led to the identification of a four-day rotation of the upper atmosphere—though few professional astronomers credited the result, which was only confirmed by Mariner 10 in 1974." For details, see W. Sheehan and T. Dobbins, "Charles Boyer and the Clouds of Venus," *Sky and Telescope* (June 1999): 56–60.

43. R. Wildt and E. J. Meyer, "Das Spektrum des Planeten Jupiter," *ZfA* 3 (1931): 354–68.

44. T. Dunham, "Note on the Spectra of Jupiter and Saturn," *PASP* 45 (1933): 42–44.

45. Kuiper, "Infrared Spectra of Planets."

46. UASC, box 17, folder 16, Colwell to astronomy department, April 28, 1947.

47. UASC, box 17, folder 16, minutes of faculty meeting, March 26, 1947.

48. D. E. Osterbrock, *Yerkes Observatory, 1892–1950* (Chicago: University of Chicago Press, 1997).

49. O. Struve, "The Story of an Observatory (The Fiftieth Anniversary of the Yerkes Observatory)," *PA* 55 (1947): 227–44, 283–94. The same article appeared as O. Struve, "The Yerkes Observatory: Past, Present, and Future Science," *Science* 106 (1947): 217–20.

50. At Struve's stellar symposium, papers were presented by Strömgren, Greenstein, Lawrence Aller (Indiana University), Alfred Joy (Mount Wilson), and Struve. At Kuiper's planetary symposium, talks were given by Greenstein, Swings, Herzberg, van de Hulst, Kuiper, Carl-Gustav Rossby (meteorologist), and James Franck and Harrison Brown (chemists), Rollin Chamberlin (geologist), Spitzer (Princeton), and Whipple (Harvard). Kuiper invited Harry Wexler (Weather Bureau), Raymond Seeger (ONR), and James Van Allen (V2 program) at Whipple's suggestion.

51. G. P. Kuiper, *The Atmospheres of the Earth and Planets.*

52. J. A. Hynek, ed., *Astrophysics: A Topical Symposium* (Chicago: University of Chicago Press, 1951).

53. G. P. Kuiper, *The Atmospheres of the Earth and Planets*, rev. ed. (Chicago: University of Chicago Press, 1952). The increase in length by sixty-eight pages compared to the first edition was almost entirely because Kuiper doubled the length of his chapter. A two-page chapter by James Frank on photosynthesis on Mars was dropped.

54. G. P. Kuiper, "Survey of Planetary Atmospheres," in *The Atmospheres of the Earth and Planets*, 304–45.

55. G. P. Kuiper, "Planetary and Satellite Atmospheres," *RPP* 13 (1950): 247–75.

56. Evans and Mulholland, *Big and Beautiful*, 118.

57. E. C. Slipher, "Observations of Mars in 1924 Made at the Lowell Observatory: I. Visual and Photographic Observations of the Surface," *PASP* 36 (1924): 255–60.

58. G. P. Kuiper, "Planetary and Satellite Atmospheres," *RPP* 13 (1950): 266.

59. A reviewer has pointed out that there are no "green areas" on Mars and what Kuiper observed were simply darker areas that only seemed green because of their contrast with

the brighter areas. He (or she) also pointed out that Sagan and Pollack much later showed that the dark areas are underlying rock and the light areas are deposits of windblown red dust.

60. R. E. Doel, *Solar System Astronomy in America: Communities, Patronage, and Interdisciplinary Research, 1920–1960* (Cambridge: Cambridge University Press, 1996).

61. G. P. Kuiper, "New Absorptions in the Uranian Atmosphere," *ApJ* 109 (1949): 540–41.

62. G. P. Kuiper, "The Diameter of Neptune," *ApJ* 110 (1949): 93–94.

63. G. P. Kuiper, "The Diameter of Pluto," *PASP* 62 (1950), 133–37.

64. G. P. Kuiper, "Determination of the Pole of Rotation of Venus," *ApJ* 120 (1954): 603–5.

65. G. P. Kuiper, "The Fifth Satellite of Uranus," *PASP* 61 (1949): 129; G. P. Kuiper, "The Second Satellite of Neptune," *PASP* 61 (1949): 175–76.

66. UChSC, box 19, folder 20, Burhoe to Chandrasekhar and Struve, November 12, 1948.

67. G. P. Kuiper, "Note on the Origin of the Asteroids," *PNAS* 39 (1953): 1159–61; G. P. Kuiper, "On the Origin of the Lunar Surface Features," *PNAS* 40 (1954): 1096–12; G. P. Kuiper, "On the Origin of the Solar System," *Proceedings of a Topical Symposium, Commemorating the 50th Anniversary of the Yerkes Observatory and Half a Century of Progress in Astrophysics*, ed. J. A. Hynek, 357–424 (New York: McGraw-Hill, 1951).

CHAPTER 7

1. UARSC, Derek Sears papers, Page to Cruickshank, August 16, 1982.

2. UChSC, box 19, file 20, Sarah Kuiper to Chandrasekhar, June 13, 1941.

3. UASC, box 19, file 20, Kuiper to Chandrasekhar, April 21, 1948; March 17, 1948; March 17, 1948.

4. UARSC, S. des Tombe, "Oral History Recorded at Half Moon Bay."

5. E. A. Whitaker, *The University of Arizona's Lunar and Planetary Laboratory: Its Founding and Early Years* (Tucson: University of Arizona Lunar and Planetary Laboratory, 1985).

6. UARSC, S. des Tombe, "Oral History Recorded at Half Moon Bay."

7. Evans and Mulholland, *Big and Beautiful*.

8. Carolyn M. Bertins, "Livestock Crossing," in *Killer Buyer: Adventures of a New Mexico Horse Dealer*, accessed December 1, 2016, http://horsestories.com/kb.shtml. I am grateful to Bertins for permission to include this passage.

9. UASC, box 17, folder 17, Struve to Kuiper, February 6, 1950.

10. UASC, box 17, folder 17, Struve to Kuiper, February 6, 1950.

11. UChSC, box 19, folder 20, Kuiper to Struve, February 7, 1950.

12. UASC, box 17, folder 17, Struve to Kuiper, February 11, 1950.

13. UASC, box 17, folder 17, Struve to Kuiper, February 11, 1950.

14. Bertins, "Livestock Crossing."

15. UChSC, box 19, folder 20, Keys to Kuiper, January 15, 1948.

16. UChSC, box 19, folder 20, program.

17. UChSC, box 19, folder 20, Kuiper to Keys/Morgenstern, undated, but a copy is attached to Kuiper to Chandrasekhar, February 20, 1948.

18. UChSC, box 19, folder 20, Kuiper to Chandrasekhar, February 20, 1948.

19. UASC, box 17, folder 17, Kuiper to Struve, February 2, 1948.

20. UASC, box 17, folder 17, Struve to Kuiper, February 10, 1948.

21. UChSC, box 19, folder 20, Kuiper to Chandrasekhar, February 20, 1948.

22. UChSC, box 19, folder 20, Kuiper to Chandrasekhar, February 20, 1948.

23. UChSC, box 19, folder 20, Kuiper to Chandrasekhar, February 20, 1948.

24. UChSC, box 19, folder 20, Chandrasekhar to Kuiper, March 8, 1948.

25. UChSC, box 19, folder 20, program of Tuesday, February 17, 1948.

26. The *Life* magazine article appeared in the June 28, 1948, issue under the title "Mars in Color," the article being mostly a full-page color image of Mars.

27. Evans and Mulholland, *Big and Beautiful*, 119.

28. UASC, box 17, folder 17, Kuiper to Struve, February 28, 1950.

29. UASC, box 17, folder 17, Page to Struve, September 20, 1949.

30. UASC, box 17, folder 17, Chandrasekhar to Struve, September 29, 1949.

31. Krisciunas, "Otto Struve 1897–1963."

32. B. Gustafsson, "Bengt Strömgren's Approach to the Galaxy," in *The Galaxy Disk in Cosmological Context*, ed. J. Andersen, J. Bland-Hawthorn, and B. Nordström, 3–16, *Proceedings IAU Symposium* 254 (Cambridge: Cambridge University Press, 2009).

33. W. Herschel, "Account of a Comet," *Philosophical Transactions of the Royal Society of London* 71 (1781): 492–501.

34. A. I. Lexell, "Recherches sur la nourvelle planete, decouverte par M. Herschel et nominee Georgium Sidus," *Acta Academiae scientiarum Imperialis Petropolitanae* 1 (1783): 303–29.

35. "Uranus," Wikipedia.org, last updated August 20, 2018, https://en.wikipedia.org/wiki/Uranus.

36. W. Herschel, "An Account of the Discovery of Two Satellites Revolving Round the Georgian Planet," *Philosophical Transactions of the Royal Society of London* 77 (1787): 125–29.

37. William Sheehan, personal communication to author, July 9, 2017.

38. J. McFarland, "Lassell, William," *BEA* 1 (2007): 681–82.

39. The equatorial mount was invented by Joseph Ritter von Fraunhofer for refracting telescopes. William Sheehan, personal communication to author, July 9, 2017.

40. W. Lassell, "On the Interior Satellites of Uranus," *MNRAS* 12 (1851): 15–18.

41. W. Lassell, "On the Interior Satellites of Uranus," *MNRAS* 12 (1851): 15–18.

42. J. Herschel, *Outlines of Astronomy* (1852; repr. Charleston, S.C: BiblioBazaar, 2009).

43. U. J. Le Verrier, "Recherches sur les mouvements d'Uranus," *AN* 25 (1846): 85–91.

44. He had apparently been indisposed, perhaps of a strained ankle, in the lead-up to the discovery of Neptune; apparently he received some communication about the subject from his friend Dawes. He possessed a large equatorially mounted reflector and was very keen to show its mettle, so he was on the lookout for satellites as soon as the planet turned up at Berlin. Interestingly, he also thought he saw a ring, which later has been shown to have been an illusion, apparently due to distortions in the optics. William Sheehan, personal communication to author, July 9, 2017.

45. W. Lassell, "Discovery of Supposed Ring and Satellite of Neptune," *MNRAS* 7 (1846): 157.

46. C. Flammarion, *Astronomie Populaire* (1880; repr. Cambridge: Cambridge University Press, 1880).

47. This is not to say that astronomers were not busy discovering the moons of other planets. Phobos and Deimos were visually discovered in 1877 by Asaph Hall, E. E. Barnard discovered Amalthea, the fifth satellite of Jupiter, visually in 1892; and William H. Pickering discovered Phoebe, the ninth satellite of Saturn, photographically in 1898, after which all additional satellites had been discovered photographically. Also, Perrine, Melotte, and Nicholson added satellites of Jupiter photographically before Kuiper discovered Miranda and Nereid.

48. G. P. Kuiper, "The Fifth Satellite of Uranus," *PASP* 61 (1949): 129.

49. D. L. Harris, "The Satellite System of Uranus" (PhD diss., University of Chicago, 1949).

50. UChSC, box 19, folder 20, Chandrasekhar to Kuiper, March 8, 1948.

51. UChSC, box 19, folder 20, Kuiper to Chandrasekhar, March 9, 1948.

52. G. P. Kuiper, "The Second Satellite of Neptune," *PASP* 1 (1949): 175–76.

53. Evans and Mulholland, *Big and Beautiful*.

54. Evans and Mulholland, *Big and Beautiful*.

55. This poem and source is quoted in Evans and Mulholland, *Big and Beautiful*. We have tried without success to trace the original and are uncertain of its source. I assume the poet is the Carl S. Junge, who lived in Chicago and made beautiful art deco book illustrations, posters, and designs in the 1930s. Luc Devroye, "Carl S. Junge," December 14, 2002, http://luc.devroye.org/junge/.

56. S. S. Sheppard, D. Jewitt, and J. Kleyna, "An Ultradeep Survey for Irregular Satellites of Uranus: Limits to Completeness," *AJ* 129 (2005): 518–25.

57. M. R. Showalter et al., "New Satellite of Neptune: S/2004 N 1," Central Bureau Electronic Telegrams no. 3586, #1 (2013).

58. G. P. Kuiper, "The Diameter of Neptune," *ApJ* 110 (1949): 93–94.

59. G. P. Kuiper, "The Diameter of Pluto," *PASP* 62 (1950): 133–37.

60. C. X. Huang and G. Á. Bakos, "Testing the Titius-Bode Law Predictions for Kepler Multi-planet Systems," *MNRAS* 442 (2014): 674–81.

61. F. Nölke, *Der Entwicklungsgang unseres Planetensystems* (Berlin and Bonn: Dummers verlag, 1930); G. P. Kuiper, "Boekbespreking: Prof. Dr. friedrich Nölke, Der Entwicklungsgang unseres Planetsystems," *Hemel en Dampkring* 28 (1930): 310–15.

62. G. P. Kuiper, "Is er een wet van Titus-Bode?," *Hemel en Dampkring* 28 (1930): 306–15.

63. G. P. Kuiper, "The Law of Planetary and Satellite Distances," *ApJ* 109 (1949): 308–13.

64. G. P. Kuiper, "No. 1. Organization and Programs of the Laboratory," *COMM* 1 (1962): 1–20.

65. B. Suirles-Jeffreys, "Obituary—Barbara Middlehurst," *QJRAS* 36 (1995): 461–62.

66. G. P. Kuiper, "Reports from French Astronomers," *ApJ* 101 (1945), 254 with two plates.

67. UASC, box 17, file 16, Kuiper to Struve, September 28, 1946. A double-edged micrometer is a mechanical device for precisely measuring small distances. Polarization is the action of restricting the vibrations of a transverse wave, especially light, wholly or partially to one direction.

68. UChSC, box 19, folder 21, Chandrasekhar to Kuiper, September 29, 1946. When asked to comment on a rough draft of Cruikshank's biography of Kuiper for the NAS, Chandrasekhar made the same comment. See UARSC, Chandrasekhar to Cruikshank, August 16, 1982.

69. UCSD, Kuiper to Griggs, March 31, 1955.

70. F. J. Vine and D. H. Matthews, "Magnetic Anomalies over Oceanic Ridges," *Nature* 199 (1963): 947–49; U. Marvin, *Continental Drift—The Evolution of a Concept* (Washington, D.C.: Smithsonian Institution Press, 1973).

71. This description may be a little hard on the astronomers of the day. William Sheehan (personal communication to author, July 7, 2017) suggests that the era was typified by the planetary work of amateurs such as W. F. Denning, E. M. Antoniadi, T. E. R. Phillips, and B. M. Peek, who tended to publish in journals such as the British Astronomical Association journal. Professionals focused on extrasolar system astronomy. Tim Swindle (personal communication to author, July 26, 2017) has suggested to me that the moribund state of planetary astronomy prior to Kuiper was more a result of the state of instrument development. The nineteenth century's thrust into astrophysics enabled major leaps in extrasolar system astronomy but did little for planetary work. However, rocketry's appearance led to major leaps in planetary science. Many factors dictated the nature of the second era.

72. B. J. Bok, "Book Reviews: The Earth as a Planet," *Scientific Monthly* 81 (1955), 42; J. B. Irwin, "Book Reviews: Planets and Satellites," *Science* 134 (1961): 1356–57. J. F. Heard, "Review of Publications—Planets and Satellites," *Journal of the Royal Astronomical Society of Canada* 56 (1962): 86; J. B. Irwin, "Book Reviews: The Moon, Meteorites, and Comets," *Science* 143 (1964): 946–47.

73. See, for example, D. W. Sears, K. Bryson, and D. Ostrowski, "The Internal Structure of Earth-Impacting Meteoroids: The View from the Microscope, the Laboratory Bench,

and the Telescope," *54th AIAA Aerospace Sciences Meeting, AIAA SciTech, (AIAA 2016–0997)* (2016), https://arc.aiaa.org/doi/10.2514/6.2016-0997.

74. W. W. Morgan, "Gerard Kuiper, 1905–1973," *The University of Chicago Record*, April 14, 1974.

75. UChSC, box 19, folder 20, Kuiper to Chandrasekhar, November 10, 1941.

76. We have no indication of what was "different" about Shatzel's background, but we are told that he was opinionated, and this will not have appealed to the autocratic Struve. Perhaps a clue lies in Shatzel's German name, which may have been a problem for those familiar with the German wartime atrocities.

77. UASC, box 17, folder 17, Struve to Kuiper, November 14, 1949.

78. UASC, box 17, folder 17, Shatzel to Struve, December 6, 1949.

79. A. V. Shatzel, "Photometric Studies of Asteroids III. The Light-Curve of 44 Nysa," *ApJ* 120 (1954): 547–50.

80. D. L. Harris III, "Comet Notes: Comet 1929 II (Forbes); Comet 1948 d (Pajdusakova-Mrkos); Comet Schwassmann-Wachmann 1; Comet 1946 k (Bester)," *PA* 56 (1948): 277–78.

81. D. L. Harris and S. Chandrasekhar, "On the Radiative Equilibrium of a Stellar Atmosphere XXIV," *ApJ* 108 (1948): 92–111.

82. D. Harris and G. Van Biesbroeck, "Positions and Preliminary Orbit of Nereid, Neptune's second satellite," *Astronomical Journal* 54 (1949): 197–98.

83. D. L. Harris, "The Satellite System of Uranus" (PhD diss., University of Chicago, 1949).

84. T. Gehrels, *On the Glassy Sea: In Search of a Worldview* (Charleston, S.C.: BookSurge, 2007).

85. Strictly speaking, albedo—the amount of light the asteroid reflects—also needs to be known, but this can reasonably be determined from the spectrum and, in any event, is a secondary effect.

86. W. Poundstone, *Carl Sagan: A Life in the Cosmos* (New York: Henry Holt, 1999).

87. L. Reiffel, "A Study of Lunar Research Flights, Vol. I," TR 59–39, Research Directorate, Air Force Special Weapons Center, Kirkland Air Force Base, New Mexico.

88. UARSC, Sagan to Cruikshank, September 29, 1980.

CHAPTER 8

1. S. Seager, *Exoplanets* (Tucson: University of Arizona Press, 2010).

2. B. C. Sproul, *Primal Myths: Creation Myths Around the World* (New York: Harper One [Harper Collins], 1979).

3. UASC, box 17, file 17, Kuiper to Struve, October 5, 1949.

4. R. Descartes, ed. and trans. S. Gaukroger, *The World and Other Writings* (Cambridge: Cambridge University Press, 1998).

5. D. Wootten, *Galileo: Watcher of the Skies* (New Haven, Conn.: Yale University Press, 2013).

6. I. Kant, trans. I. Johnston, *Universal Natural History and Theory of Heavens* (Arlington, Va.: Richer Resources, 2009).

7. M. M. Woolfson, "The Solar System—Its Origin and Evolution" *QJRAS* 34 (1993): 1–20.

8. J. Jeans, "The Part Played by Rotation in Cosmic Evolution," *Memoirs of the Royal Astronomical Society* 77 (1917): 186–99.

9. T. C. Chamberlin and R. D. Salisbury, *Geology* (New York: Holt, 1906); F. R. Moulton, "On the Evolution of the Solar System," *ApJ* 22 (1905): 165–81.

10. M. M. Woolfson, "A Capture Theory of the Origin of the Solar System," *Proceedings of the Royal Society* A282 (1964): 485–504.

11. K. E. Edgeworth, "The Evolution of Our Planetary System," *Journal of the British Astronomical Association* 53 (1943): 181–88.

12. J. McFarland, "Kenneth Essex Edgeworth—Victorian Polymath and Founder of the Kuiper Belt?," *Vistas in Astronomy* 40 (1996): 343–54.

13. C. F. von Weizsäcker, "Über die Entstehung des Planetensystems," *ZfA* 22 (1943): 319–55.

14. UChSC, box 19, folder 20, Kuiper to Chandrasekhar, undated (probably January/February 1945).

15. Pash, *The Alsos Mission*.

16. "Ernst von Weizsäcker," Wikipedia.org, last updated August 3, 2018, https://en.wikipedia.org/wiki/Ernst_von_Weizsäcker.

17. A hotel is currently located at this address, and we may assume that it did in 1949.

18. A copy of this letter is in UChSC, box 19, folder 20.

19. UChSC, box 19, folder 20, Sarah Kuiper to Chandrasekhar, February 2, 1950. Gerard published a letter in the *London Times* from his temporary home at 14 Henrietta Street, Covent Garden, London, that appeared on December 8 or 9, 1949.

20. "Ernst von Weizsäcker."

21. G. P. Kuiper, "Discourse Following Award of the Kepler Gold Medal at the AAAS Meeting, Franklin Institute, Philadelphia," *COMM* 9, no. 183 (1971): 404.

22. Kuiper, "On the Origin of the Solar System," *Proceedings*.

23. UASC, box 17, file 17, Kuiper to Struve, October 5, 1949. The *Astrophysical Journal* paper Kuiper mentions is Kuiper, "The Law of Planetary and Satellite Distances" (see *ApJ* 109, p. 555 for a correction).

24. UASC, box 17, file 17, Struve to Kuiper, February 6, 1950.

25. G. P. Kuiper, "On the Origin of the Solar System," *PNAS* 37 (1951): 1–14.

26. D. Ter Haar, "Further Studies on the Origin of the Solar System," *ApJ* 111 (1950): 179–90, with an appendix by S. Chandrasekhar and D. Ter Haar.

27. A planet is a celestial body that is in orbit around the Sun, has sufficient mass for its self-gravity to overcome rigid body forces so that it assumes a hydrostatic equilibrium (nearly round) shape, and has cleared the neighborhood around its orbit.

28. Seager, *Exoplanets*.

29. Kuiper, "On the Origin of the Solar System," *PNAS*.

30. Kuiper, "On the Origin of the Solar System," *PNAS*.

31. These papers were H. P. Berlage, "On the Electrostatic Field of the Sun due to Its Corpuscular Rays," *Proceedings of the Royal Academy Amsterdam* 33 (1930): 614–18; "On the Electrostatic Field of the Sun as a Factor in the Evolution of the Planets," *Proceedings of the Royal Academy Amsterdam* 33 (1930): 719–22; "On the Structure and Internal Motion of the Gaseous Disc Constituting the Original State of the Planetary System," *Proceedings of the Royal Academy Amsterdam* 35 (1932): 553–62; "Viscosity and Steady States of the Disc Constituting the Embryo of the Planetary System," *Proceedings of the Royal Academy Amsterdam* 37 (1934): 221–32; "A Study of the Systems of Satellites from the Standpoint of the Disc—Theory of the Origin of the Planetary System," *Annalen v.d. Bosscha-Sterrenwacht* 4 (1934): 79–94; "The Theorem of Minimal Loss of Energy due to Viscosity in Steady Motion and the Origin of Planetary System from a Rotating Gaseous Disc," *Proceedings of the Royal Academy Amsterdam* 38 (1935): 857–63; "Spontaneous Development of a Gaseous Disc Revolving Round the Sun into Rings and Planets," pts. 1 and 2, *Proceedings of the Royal Academy Amsterdam* 43 (1940): 534–40; 558–65.

32. UASC, Gehrels to Cruikshank, May 11, 1979.

33. H. P. Berlage, *The Origin of the Solar System* (Oxford and New York: Pergamon, 1968).

34. UARSC, Gehrels to Cruikshank to, May 11, 1979.

35. UARSC, Cruikshank to Gehrels, May 15, 1979.

36. Kuiper, "On the Origin of the Solar System," *PNAS*.

37. M. A. Barucci et al., *The Solar System Beyond Neptune* (Tucson: University of Arizona Press, 2008).

38. D. H. Levy, *Clyde Tombaugh: Discoverer of Planet Pluto* (Cambridge, Mass.: Sky, 1991); D. P. Cruikshank and W. Sheehan, *Discovering Pluto: Exploration at the Edge of the Solar System* (Tucson: University of Arizona Press, 2018). Tim Swindle has suggested that Tombaugh found Pluto near *a* place that Lowell had predicted—he had predicted lots of places. Lowell's main contribution was inspiring the search.

39. A. O. Leuschner, "The Astronomical Romance of Pluto," *PASP* 44 (1932): 197–214.

40. I. Yamamoto, "Prof. Yamamoto's Suggestion on the Origin of Pluto," *Bulletin of the Kwagan Observatory* 3, no. 288 (1934): 288–89.

41. R. A. Lyttleton, "On the Possible Results of an Encounter of Pluto with the Neptunian System," *MNRAS* 97 (1936): 108–15.

42. G. P. Kuiper, "The Formation of the Planets," pts. 1, 2, and 3, *Royal Astronomical Society of Canada* 50 (1956): 57–68; 105–21; 158–76.

43. W. J. Luyten, "Pluto Not a Planet?," *Science* 123 (1956): 896. He was referring to Lyttle-ton, "On the Possible Results."

44. G. P. Kuiper, "The Planet Pluto," *Science* 124 (1956): 322.

45. "Pluto and the Developing Landscape of Our Solar System," International Astronomical Union, accessed November 11, 2016, http://www.iau.org/public/themes/pluto/. The IAU resolution is "Resolution B5," International Astronomical Union, accessed November 11, 2016, http://www.iau.org/static/resolutions/Resolution_GA26-5-6.pdf.

46. Edgeworth, "The Evolution of our Planetary System," 186.

47. Kuiper, "On the Origin of the Solar System," *PNAS*, 13.

48. J. H. Oort, "The Structure of the Cloud of Comets Surrounding the Solar System, and a Hypothesis Concerning Its Origin," *BAN* 11 (1950): 91–408; F. L. Whipple, "A Comet Model I. The Acceleration of Comet Encke," *ApJ* 111 (1950): 375–94.

49. "Resolution B5."

CHAPTER 9

1. E. A. Whitaker, "Clash of the Titans (or Why and How the LPL was Born)" (Power-Point presentation, Symposium to Celebrate the 50th Anniversary of the University of Arizona's Lunar and Planetary Laboratory, Tucson, Arizona, October 2, 2010), https://www.lpl.arizona.edu/sites/default/files/history/lpl50/lectures.pdf.

2. R. E. Doel, *Solar System Astronomy in America: Communities, Patronage, and Interdisci-plinary Research, 1920–1960* (Cambridge: Cambridge University Press, 1996).

3. UARSC, Sagan to Cruikshank, December 1, 1982.

4. W. P. Sheehan and T. A. Dobbins, *Epic Moon: A History of Lunar Exploration in the Age of the Telescope* (Richmond, Va.: Willmann-Bell, 2001). See also G. Racki et al., "Ernst Julius Öpik's (1916) Note on the Theory of Explosion Cratering on the Moon's Surface—The Complex Case of a Long-Overlooked Benchmark Paper," *MAPS* 49 (2017): 1861–74.

5. U. B. Marvin, "Oral Histories in Meteoritics and Planetary Science: X. Ralph B. Bald-win," *Meteoritics and Planetary Science* 38 (2003): A163–A175.

6. R. B. Baldwin, "The Meteoritic Origin of Lunar Craters," *PA* 50 (1942): 356–69; R. B. Baldwin, "The Meteoritic Origin of Lunar Structures," *PA* 51 (1943): 117–27.

7. R. B. Baldwin, *The Face of the Moon* (Chicago: University of Chicago Press, 1949).

8. R. B. Baldwin, *The Measure of the Moon* (Chicago: University of Chicago Press, 1963).

9. G. P. Kuiper, "On the Origin of the Lunar Surface Features," *PNAS* 40 (1954): 1096–1112.

10. J. R. Arnold, J. Bigeleisen, and C. A. Hutchison, "Harold Clayton Urey 1893–1981," *BMNAS* 68 (1995): 361–411.

11. R. Rhodes, *The Making of the Atomic Bomb* (New York: Simon and Schuster, 1986).

12. S. L. Miller and H. C. Urey, "Organic Compound Synthesis on the Primitive Earth," *Science* 130 (1959): 245–21.

13. H. C. Urey, *The Planets: Their Origin and Development* (New Haven, Conn.: Yale University Press, 1952).

14. H. N. Russell, "The Solar Spectrum and the Earth's Crust," *Science* 39 (1914): 791–94. See also A. E. Nordenskiöld, "Nordenskiöld's Arctic Investigations," *Science* 5 (1885): 430–32.

15. H. C. Urey, "The Origin and Development of the Earth and Other Terrestrial Planets," *Geochimica et Cosmochimica Acta* 1 (1951): 209–77; H. C. Urey, "Meteorites and the Moon," *Science* 147 (1965): 1262–65.

16. H. Jeffreys, *The Earth*, 4th ed. (Cambridge: Cambridge University Press, 1959).

17. Doel, *Solar System Astronomy in America*.

18. G. P. Kuiper, "On the Origin of the Solar System," *Proceedings of the New York Academy of Sciences* 37 (1951): 1–233.

19. S. J. Peale, "Generalized Cassini's Laws," *Astronomical Journal* 74 (1969): 483–87.

20. Kuiper, "On the Origin of Lunar Surface Features," 1102.

21. H. C. Urey, "Some Criticisms of 'On the Origin of the Lunar Surface Features,' by G. P. Kuiper," *Proceedings of the National Academy of Sciences of the United States of America* 41 (1955): 423–28.

22. G. P. Kuiper, "The Lunar Surface-Further Comments," *Proceedings of the National Academy of Sciences of the United States of America* 41 (1955): 820–23.

23. "Elger (1985)" is T. G. Elger, *The Moon: A Full Description and Map of Its Principal Physical Features* (1895; repr. HardPress, 2012).

24. D. J. Stevenson, "Origin of the Moon—The Collision Hypothesis," *Annual Review of Earth and Planetary Sciences* 15 (1987): 271–315.

25. Baldwin, *The Face of the Moon*.

26. UCSD, Kuiper to Urey, December 1, 1949.

27. UCSD, Urey to Kuiper, June 13, 1951.

28. UCSD, Urey to Kuiper, November 13, 1951.

29. UCSD, Kuiper to Urey, November 14, 1951.

30. UCSD, Kuiper to Urey, January 23, 1952.

31. Francis William Aston (1877–1945), known for discovering, by means of his mass spectrograph, isotopes in many radioactive elements.

32. Hans Eduard Suess (1909–93) did research in the field of cosmochemistry investigating the abundances of elements in meteorites with Urey at the University of Chicago and UC San Diego.

33 UCSD, Bullard to Urey, August 8, 1955.

34. UCSD, Urey to Page, September 27, 1955.

35. UCSD, Kuiper to IAU, June 3, 1954.

36. UCSD, Urey to Kuiper, June 23, 1954.

37. UCSD, Kuiper to Oosterhoff, August 3, 1955.

38. UCSD, Oosterhoff to Urey, September 2, 1955.

39. UCSD, Urey to Waterman, February 2, 1955.

40. G. P. Kuiper, ed., *The Sun* (Chicago: University of Chicago Press, 1953); G. P. Kuiper, ed., *The Earth as a Planet* (Chicago: University of Chicago Press, 1954); G. P. Kuiper and B. M. Middlehurst, eds., *Planets and Satellites* (Chicago: University of Chicago Press, 1961); G. P. Kuiper and B. M. Middlehurst, eds., *The Moon, Meteorites and Comets* (Chicago: University of Chicago Press, 1963).

41. Kuiper, *The Atmospheres of the Earth and Planets*; Kuiper, *The Atmospheres of the Earth and Planets*, rev. ed.

42. UCSD, box 51, folder 6–7, Urey to Waterman, February 9, 1955.

43. J. T. Wilson, "The Development and Structure of the Crust," in *The Earth as a Planet*, ed. G. P. Kuiper, 138–214 (Chicago: University of Chicago Press, 1954).

44. UCSD, box 51, folder 6–7, Urey to Griggs, March 31, 1955.

45. UCSD, box 51, folder 6–7, Urey to Kimpton, August 2, 1955.

46. UCSD, box 51, folder 6–7, Kimpton to Urey, September 29, 1955.

47. G. P. Kuiper, "On the Origin of the Satellites and Trojans," *Vistas in Astronomy* 2 (1956): 1631–66.

48. UCSD, Urey to the editors of *The Observatory*, April 17, 1957.

49. UCSD, Porter to Urey, June 4, 1957.

50. UCSD, Carpenter to Urey, January 8, 1960.

51. UCSD, Urey to Carpenter, January 19, 1960.

52. UCSD, Carpenter to Urey, January 25, 1960.

53. UCSD, Urey to Meinel, February 1, 1960.

54. UCSD, Meinel to Urey, February 3, 1960.

55. UCSD, Urey to Hartmann, November 2, 1962.

56. H. C. Urey, "The Origin of the Moon's Surface Features," pts. 1 and 2, *Sky and Telescope* 15 (1956): 108–11; 161–63.

57. UCSD, Hartmann to Urey, November 6, 1962.

58. UCSD, Urey to Hartmann, November 13, 1962.

59. UCSD, Urey to Spradley, November 1, 1962.

60. L. H. Spradley, "Lunar Globe Photography," *Communications of the Lunar and Planetary Laboratory* 1 (1962): 31–34.

61. UCSD, Urey to Sarah Kuiper, December 28, 1973.

CHAPTER 10

1. A. Von Humboldt, *Cosmos: A Sketch of a Physical Description of the Universe* (repr., Seattle, Wash., and Charleston, S.C.: CreateSpace, 2014), 1:111.

2. "Cataline Sky Survey," Wikipedia.org, last updated September 2, 2018, https://en .wikipedia.org/wiki/Catalina_Sky_Survey.

3. O. P. Popova et al., "Chelyabinsk Airburst, Damage Assessment, Meteorite Recovery and Characterization," *Science* 342 (2013): 1069–73.

4. W. Alvarez et al., "Iridium Anomaly Approximately Synchronous with Terminal Eocene Extinctions," *Science* 216 (1982): 886–88.

5. G. P. Kuiper, "The Moon," *Journal of Geophysical Research* 64 (1959): 1714.

6. W. K. Hartmann and D. R. Davis, "Satellite-Sized Planetesimals and Lunar Origin," *Icarus* 24 (1975): 504–15.

7. The flying dinosaurs escaped this fate, and their descendants are the familiar birds that exist today.

8. L. W. Alvarez et al., "Extraterrestrial Cause for the Cretaceous-Tertiary Extinction," *Science* 208 (1980): 1095–1108.

9. "Extinction Even," Wikipedia.org, "List of Extinction Events" section, last updated August 22, 2018, https://en.wikipedia.org/wiki/Extinction_event#List_of_extinction _events.

10. E. L. Krinov, trans. J. S. Romankiewicz, *Giant Meteorites* (Oxford and New York: Pergamon, 1966).

11. C. F. Chyba, P. J. Thomas, and K. J. Zahnle, "The 1908 Tunguska Explosion: Atmospheric Disruption of a Stony Asteroid," *Nature* 361 (1993): 40–44.

12. Popova et al., "Chelyabinsk Airburst."

13. H. Povenmire, "The Sylacauga, Alabama Meteorite: The Impact Locations, Atmosphere Trajectory, Strewn Field and Radiant," *Abstracts of the Lunar and Planetary Science Conference* 26 (1995): 1133–34.

14. P. Jenniskens et al., "The Mbale Meteorite Shower," *Meteoritics and Planetary Science* 29 (1994): 246–54.

15. K. Yau, P. Weissman, and D. Yeomans, "Meteorite Falls in China and Some Related Human Casualty Events," *Meteoritics* 29 (1994): 864–71.

16. D. P. Cruikshank and E. A. Whitaker, "The Yerkes Observatory Meteorite Hoax of July 1960," *Meteorite Magazine* 16 (2010): 5–6.

17. G. Foderà Serio, A. Manara, and P. Sicoli, "Giuseppe Piazzi and the Discovery of Ceres," in *Asteroids III*, ed. W. F. Bottke Jr. et al., 17–24 (Tucson: University of Arizona Press, 2002). When asteroids are first detected they are given a temporary label reflecting year and order of discovery, hence 1 Ceres, 2 Pallas, 3 Juno, and 4 Vesta. Once their orbit has been determined well-enough to enable future recovery, they are given a permanent number and name.

18. For sources for this paragraph see D. W. G. Sears, "The Explored Asteroids: Science and Exploration in the Space Age," *Space Science Reviews* 194 (2015): 139–235.

19. C. T. Kowal, "A Solar System Survey," *Icarus* 77 (1989): 118–23, https://airandspace.si
.edu/stories/editorial/finding-pluto-blink-comparator.

20. The photoelectric effect discovered by Einstein, for which he received the Nobel Prize,
led to the development of a large range of devices to measure the intensity of light culminating in the photomultiplier tube that would eventually have the ability to amplify
the signal by billions. It was only a year after the invention of the photomultiplier tube
by RCA in 1934 that William Calder of Harvard College Observatory attached to a
telescope and applied the technique to the study of asteroids. See "The Nobel Prize in
Physics 1921," Nobel Media AB, 2014, http://www.nobelprize.org/nobel_prizes/physics
/laureates/1921/; and W. A. Calder, "Photoelectric Photometry of Asteroids," *Harvard
College Observatory Bulletin* 904 (1936): 11–18. At Yerkes Observatory in the 1950s, the
local expert in the astronomical photometry was William Albert Hiltner, who liked to
be called Al and who worked closely with Kuiper and his colleagues for many years.

21. G. P. Kuiper, "Satellites, Comets, and Interplanetary Material," *Proceedings of the New
York Academy of Science* 39 (1953): 1153–58.

22. G. P. Kuiper, "Note on the Origin of the Asteroids," *Proceedings of the New York Academy
of Science* 39 (1953): 1159–61.

23. "Ingrid van Houten-Groeneveld," Wikipedia.org, last updated February 24, 2018,
https://en.wikipedia.org/wiki/Ingrid_van_Houten-Groeneveld.

24. I. Groeneveld and G. P. Kuiper, "Photometric Studies of Asteroids. I," *ApJ* 120 (1954):
200–220.

25. 4 Vesta, 7 Iris, 9 Metis, 15 Eunomia, 17 Thetis, 25 Phocaea, 39 Latitia, 97 Klotho, and 511
Davida.

26. The target asteroids were 1 Ceres, 40 Harmonia, 2 Pallas, 3 Juno, 9 Metis, 10 Hygeia, 14
Irene, 39 Laetitia, 44 Nysa, 354 Elenora, 433 Eros, and 532 Herculina. See I. Groeneveld
and G. P. Kuiper, "Photometric Studies of Asteroids. II," *ApJ* 120 (1954): 529–46.

27. A. V. Shatzel, "Photometric Studies of Asteroids. III. The light curve of 44 Nysa," *ApJ*
120 (1954): 547–50.

28. *Chicago Tribune*, "Planetarium Defends its View on Research," June 25, 1959, p. 75 (pt. 5,
p. 10).

29. I. I. Ahmad, "Photometric Studies of Asteroids IV. The Light-Curves of Ceres, Hebe,
Flora, and Kalliope," *ApJ* 120 (1954): 551–59.

30. T. Gehrels, "Photometric Studies of Asteroids. V. The Light-Curve and Phase Function
of 20 Massalia," *ApJ* 123 (1956): 331–38.

31. T. Gehrels, "Photometric Studies of Asteroids VI. Photographic Magnitudes," *ApJ* 125
(1957): 550–70.

32. UASC, box and file unknown, Oort to Kuiper, July 30, 1953.

33. "Cornelis Johannes van Houten," Wikipedia.org, last updated May 5, 2018, https://en
.wikipedia.org/wiki/Cornelis_Johannes_van_Houten.

34. I. Van Houten-Groeneveld and C. J. Van Houten, "Photometric Studies of Asteroids VII," *ApJ* 127 (1958): 253–73.

35. 7 Iris, 8 Flora, 11 Parthenope, 15 Eunamia, 16 Psyche, 17 Thetis, 25 Phocaea, and 39 Laetitia.

36. T. Gehrels and D. Owings, "Photometric Studies of Asteroids IX. Additional Light-Curves," *ApJ* 135 (1962): 906–24; H. J. Wood and G. P. Kuiper, "Photometric studies of Asteroids X," *ApJ* 137 (1962): 1279–85.

37. G. P. Kuiper et al., "Survey of Asteroids," *ApJ*, sup. 3 (1958): 289–427.

38. Strictly speaking, it is necessary to know the albedo of an asteroid—the amount of incident light that it reflects—but since albedos vary over less than a factor of 10 while asteroids vary over many orders of magnitude, it is not a big effect.

39. Kuiper et al., "Survey of Asteroids."

40. Evans and Mulholland, *Big and Beautiful.*

41. In the late nineteenth century the blink comparator became obsolete because computers can basically do the same thing digitally—it's the same concept that modern search programs use.

42. C. J. Van Houten et al., "The Palomar-Leiden Survey of Faint Minor Planets," *AA Supplement Series* 2 (1970): 339–448.

43. Research Institute Leiden Observatory, "Annual Report 2005," 6.

44. UASC, box and file unknown, Van Houten to Kuiper, November 3, 1970; Kuiper to Van Houten, November 9.

45. Sears, "The Explored Asteroids."

46. W. Alvarez et al., "Iridium Anomaly Approximately Synchronous with Terminal Eocene Extinctions," *Science* 216 (1982): 886–88.

47. H. B. Hammel et al., "HST Imaging of Atmospheric Phenomena Created by the Impact of Comet Shoemaker-Levy 9," *Science* 267 (1995): 1288–96.

48. S. Larson, "The Catalina Sky Survey for NEOs," *Bulletin of the American Astronomical Society* 30 (1998), 1037; A. J. Drake, "First Results from the Catalina Real-Time Transient Survey," *ApJ* 696 (2009): 870–84.

CHAPTER 11

1. UChSC, box 19, folder 21. The graph shows data for twelve major telescopes, from the 9.5-inch Dorpat telescope, which became operative in 1820, to the 200-inch Palomar, which was constructed in 1948. The correlation between log size and date is very strong. The graph was probably originally attached to the memorandum UChSC, box 19, folder 21, Kuiper to Chandrasekhar, March 26, 1953.

2. UChSC, box 19, folder 21, Kuiper to Chandrasekhar, March 26, 1953. The supplement is dated March 31, 1953.

3. UChSC, box 19, folder 21, Chandrasekhar to Kuiper to, April 1, 1953.

4. This is an exaggeration of Kuiper's. Without modern adaptive optics, it would not be possible to see stellar discs or asteroid shapes with terrestrial telescopes.

5. W. L. Putnam, "A Yankee Image: The Life and Times of Roger Lowell Putnam," (Flagstaff, Ariz.: Lowell Observatory, 1991).

6. UASC, box 25, file 8, Kuiper to Brunk, October 30, 1968. By "the hump" Kuiper meant that layer of the atmosphere above which there is very little water. It also means being above the clouds.

7. Johnson-type telescopes were low-cost, medium-size reflectors specially designed for infrared astronomy. The relatively poor optical quality was good enough to feed most of incoming infrared radiation into the large entrance aperture of the photometer or spectrometer. See G. De Vacouleurs, "Harold Lester Johnson 1921–1980," *BMNAS* 67 (1991): 241–61.

8. UASC, box 25, file 8, Kuiper to Brunk, June 24, 1967.

9. Victor Blanco, "Brief History of the Cerro Tololo Inter-American Observatory," Cerro Tololo Inter-American Observatory, February 1993, http://www.ctio.noao.edu/noao /content/ctio-history.

10. Blanco, "Brief History of the Cerro Tololo Inter-American Observatory."

11. D. Lorenzen, "Jürgen Stock and His Impact on Modern Astronomy in South America," *Revista Mexicana de Astronomía y Astrofísica (Serie de Conferencias)* 25 (2006): 71–72.

12. UASC, box 50, file 9, Stock to Kuiper, March 9, 1960; Kuiper to Brunk, June 24, 1967. Also in this file are thirteen multipage reports during 1960 and 1961 labelled "Stock Report" and later "Chile Site Survey Report." See also J. Stock, "Procedures for Location of Astronomical Observatory Sites," *Chile Site Survey Technical Report* 1 (1962).

13. UASC, Kuiper to Zachariason, August 19, 1960, box and file unknown.

14. S. Vasilevskis and D. E. Osterbrock, "Charles Donald Shane 1895–1983," *BMNAS* 62 (1989): 487–511.

15. "Charles Donald Shane: The Lick Observator," Regional History Project, UCSC LibraryShane, 1969, http://escholarship.org/uc/item/4sb4j79p.

16. Interview of Frank Edmondson by David DeVorkin, April 21, 1977, and February 2, 1978, AIP, https://www.aip.org/history-programs/niels-bohr-library/oral-histories/4588-1 and https://www.aip.org/history-programs/niels-bohr-library/oral-histories/4588-2.

17. W. H. Haas, "In Memoriam: Alika K. Herring," *Journal of the Association of Lunar and Planetary Observers* 40 (1996): 30–31. His middle name was Kamalaniokeaukaha, although his connection with Hawaii is unclear. William Sheehan, personal communication to author, July 7, 2017.

18. UASC, box and file unknown, Rutlland to Kuiper, July 17, 1960.

19. UASC, box and file unknown, Kuiper to Rutlland, July 23, 1960.

20. UASC, box and file unknown, Kuiper to Stock, March 16, 1961.

21. UASC, box and file unknown, W. and S. Mathias to Shane, October 9, 1961.

22. UASC, box and file unknown, W. and S. Mathias, Paul Kuiper, "General Report and Evaluation of the Chile Project," undated, but probably October 9, 1961.

23. UASC, box and file unknown, Kuiper to Keller, October 30, 1961.

24. UASC, box and file unknown, Kuiper to Whaley, October 10, 1961.

25. UASC, box and file unknown, Kuiper to Harrell, January 11, 1962.

26. UASC, box and file unknown, Shane to Whitaker, February 1, 1974.

27. UASC, box and file unknown, Kuiper to Newell, March 9, 1963.

28. M. M. Waldrop, "Mauna Kea (I): Halfway to Space," *Science* 214 (1981): 1010–13; M. M. Waldrop, "Mauna Kea (II): Coming of Age," *Science* 214 (1981): 1110–14.

29. Waldrop, "Mauna Kea (I)"; Waldrop, "Mauna Kea (II)"; John T. Jeffries, "Astronomy in Hawai`i 1964–1970," University of Hawaii Institute for Astronomy, accessed November 19, 2016, http://www.ifa.hawaii.edu/users/jefferies/Preface.htm.

30. Waldrop, "Mauna Kea (I)."

31. Whitaker, *The University of Arizona's Lunar and Planetary Laboratory*.

32. Whitaker, *The University of Arizona's Lunar and Planetary Laboratory*.

33. The information in this paragraph is from Whitaker, *The University of Arizona's Lunar and Planetary Laboratory*; and Waldrop, "Mauna Kea (I)."

34. W. H. Lambright, *Powering Apollo: James E. Webb of NASA* (Baltimore: Johns Hopkins University Press, 1995).

35. Evans and Mulholland, *Big and Beautiful*.

36. Whitaker, *The University of Arizona's Lunar and Planetary Laboratory*.

37. Interview of William Brunk by Joseph N. Tatarewicz, July 21, 1983, and August 9, 1983, AIP, https://www.aip.org/history-programs/niels-bohr-library/oral-histories/28198-1 and https://www.aip.org/history-programs/niels-bohr-library/oral-histories/281981-2.

38. Interview of John Jefferies by Spencer Weart, July 29, 1977, AIP, https://www.aip.org/history-programs/niels-bohr-library/oral-histories/4693.

39. Interview of William Brunk by Joseph N. Tatarewicz, July 21, 1983, and August 9, 1983, AIP, https://www.aip.org/history-programs/niels-bohr-library/oral-histories/28198-1 and https://www.aip.org/history-programs/niels-bohr-library/oral-histories/28198-2.

40. Waldrop, "Mauna Kea (I)."

41. Whitaker, *The University of Arizona's Lunar and Planetary Laboratory*.

42. UASC, box 50, file 11, small grey notebook labeled "Chilean Site Survey." Pages 19 and 45–67 concern his flight to Mount Tequila, December 13 to December 18, 1973.

43. Between the time it occupied the balcony of the castle (i.e., Chapultepec Castle) and moved to the western outskirts of the city, it was based at the Palacio del ex Arzobispado, in Tacubaya. It was to a site near the observatory that Percival Lowell moved the newly unveiled Clark refractor from Flagstaff at the opposition of Mars in December 1896. William Sheehan, personal communication to author, July 7, 2017.

44. https://en.wikipedia.org/wiki/Category:Astronomical_observatories_in_Mexico (retrieved December 24, 2016).

45. UARSC, S. des Tombe, "Oral History Recorded at Half Moon Bay."

46. Whitaker, *The University of Arizona's Lunar and Planetary Laboratory*.

47. G. P. Kuiper, 1972. No. 172. The Lunar and Planetary Laboratory and its Telescopes. COMM 6, 155–170.

48. Details of these efforts can be found in UASC, box 50 folders 1–27 and box 51, folders 1–5.

49. UASC, box 25, folder 8, Kuiper to Brunk, January 31, 1969; box 25, folder 8, Kuiper to Brunk, February 14, 1969; box 25, folder 8, Kuiper to Brunk, March 21, 1969; box 25, folder 8, Kuiper to Brunk, October 2, 1969; box 25, folder 8, Kuiper to Brunk, October 13, 1969; box 25, folder 8, Kuiper to Brunk, November 10, 1969.

50. William Sheehan, personal communication to author, July 7, 2017.

51. T. Gehrels, "Ultraviolet Photometry Using High Altitude Balloons," *Applied Optics* 6 (1967): 231–33.

52. G. P. Kuiper, F. F. Forbes, and H. L. Johnson, "No. 93. A Program of Astronomical Infrared Spectroscopy from Aircraft," *COMM* 6 (1967): 155–70.

53. Kuiper, Forbes, and Johnson, "No. 93. A Program of Astronomical Infrared Spectroscopy."

54. H. H. Aumann, C. M. Gillespie, and F. J. Low, "The Internal Powers and Effective Temperatures of Jupiter and Saturn," *ApJ* 157 (1969): L69–L72.

55. J. P. Gardner et al., "The James Webb Space Telescope," *Space Science Reviews* 123 (2006): 485–606.

56. Kuiper, Forbes, and Johnson, "No. 93. A Program of Astronomical Infrared Spectroscopy."

57. UASC, box 29 folder 5 contains considerable documentation dealing with Kuiper's airborne astronomy program and how it was funded. Kuiper wrote proposals to NASA for time on the aircraft and obtained funds for the work from the NASA institutional award made to the University of Arizona. Kuiper was required to submit regular proposals and progress reports to the Space Science Committee, which managed those funds. He was a member of the committee, but the chair was Al Weaver, the university administrator responsible for research.

58. Kuiper, Forbes, and Johnson, "No. 93. A Program of Astronomical Infrared Spectroscopy."

59. Kuiper, Forbes, and Johnson, "No. 93. A Program of Astronomical Infrared Spectroscopy."

60. D. P. Cruikshank, F. A. de Wiess, and G. P. Kuiper, "No. 126. A High-Resolution Solar Spectrometer for Air-Borne Infrared Observations," *COMM* 7 (1968): 233–38.

61. G. P. Kuiper and F. F. Forbes, "No. 95. High Altitude Spectra from NASA CV 990 Jet. I: Venus, 1–2.5 Microns, Resolution 20 cm-1," *COMM* 6 (1967): 177–89; G. P. Kuiper et al., "High Altitude Spectra from NASA CV 990 Jet. II. Water Vapor on Venus," *COMM* 6 (1969): 209–28; G. P. Kuiper, "No. 173. Interpretation of the Jupiter Red Spot, I," *COMM* 9 (1972): 247–313; F. J. Low and G. H. Rieke, "The Instrumentation and

Techniques of Infrared Photometry," in *Methods of Experimental Physics—Astrophysics. Part A, Optical*, ed. N. Carleton, 415–62 (Cambridge: Academic Press, 1974).

62. The solar spectrum series of articles is G. P. Kuiper and D. L. Steinmetz, "No. 94. Solar Comparison Spectra, 1.0–2.5 μm," *COMM* 6 (1967): 171–76; G. P. Kuiper et al., "No. 124. Arizona-NASA Atlas of Infrared Solar Spectrum—Report II," *COMM* 7 (1968): 197–220; G. P. Kuiper and L. A. Bijl, "No. 125. Arizona-NASA Atlas of Infrared Solar Spectrum—Report III," *COMM* 7 (1968): 221–32; L. A. Bijl, G. P. Kuiper, and D. P. Cruikshank, "No. 160. Arizona-NASA Atlas of Infrared Solar Spectrum, Report IV," *COMM* 9 (1969): 1–27; L. A. Bijl, G. P. Kuiper, and D. P. Cruikshank, "No. 161. Arizona-NASA Atlas of Infrared Solar Spectrum, Report V," *COMM* 9 (1969): 29–51; G. P. Kuiper et al., "No 162. Arizona-NASA Atlas of Infrared Solar Spectrum, Report VI," *COMM* 9 (1969): 53–63; L. A. Bijl, G. P. Kuiper, and D. P. Cruikshank, "No. 163. Arizona-NASA Atlas of Infrared Solar Spectrum, Report VII," *COMM* 9 (1969): 65–92; L. A. Bijl, G. P. Kuiper, and D. P. Cruikshank, "No. 164. Arizona-NASA Atlas of Infrared Solar Spectrum, Report VIII," *COMM* 9 (1969): 93–120; L. A. Bijl, G. P. Kuiper, and D. P. Cruikshank, "No. 165. Arizona-NASA Atlas of Infrared Solar Spectrum, Report IX," *COMM* 9 (1969): 121–53; D. C. Benner, L. Randic, and A. B. Thomson, "No. 166. Arizona-NASA Atlas of Infrared Solar Spectrum, Report X," *COMM* 9 (1970): 155–69.

Dale Cruikshank writes (personal communication, July 8, 2017), "I was on all of the solar observing flights, several of which were flown out of Hawaii. During a one-day break from the flying, Kuiper and I went to the Big Island to visit Kilauea Volcano and the USGS Hawaiian Volcano Observatory. The volcano was quiet that day, but there was an on-going series of episodic eruptions in Halemaumau that was quite spectacular. After the solar flights were concluded and we returned to LPL in Tucson, the Kilauea eruptions resumed, and Kuiper sent me back to Hawaii to study them. He was thrilled with the photos and description I brought back to LPL."

63. "List of NASA Aircraft," Wikipedia.org, last updated August 19, 2018, https://en.wikipedia.org/wiki/List_of_NASA_aircraft.

64. Official program, "Dedication of the Gerard P. Kuiper Airborne Observatory, May 21, 1975. In memory of Gerard P. Kuiper, former director of the Lunar and Planetary Laboratory of The University of Arizona."

CHAPTER 12

1. Doel, *Solar System Astronomy in America*.

2. E. A. Whitaker, *Mapping and Naming the Moon: A History of Lunar Cartography and Nomenclature* (Cambridge: Cambridge University Press, 1999).

3. W. P. Sheehan and T. A. Dobbins, *Epic Moon: A History of Lunar Exploration in the Age of the Telescope* (Richmond, Va.: Willmann-Bell, 2001), 59–72.

4. The IAU is composed of thirty-three commissions, of which Commission 16 was tasked with arbitrating lunar nomenclature. There were thirty-four members with Kuiper as president. Kuiper included Urey in the list, but when Urey received Commission 16 mail he wrote to Kuiper, "What is Commission 16 of the International Astronomical Union? I received your letter from June 3 addressed to Commission members, but as far as I know I am not a member." UCSD, Urey to Kuiper, June 23, 1954.

5. The letter is reproduced in Whitaker, "Clash of the Titans."

6. Whitaker, *The University of Arizona's Lunar and Planetary Laboratory*, 8–11.

7. R. J. McKim, "Obituary of Alan Pennell Lenham 1930–1996," *Journal of the British Astronomical Association* 107 (1997): 50.

8. "Dai" is a familiar form of "David" used in Wales.

9. "David 'Dai' Arthur (Lunar Scientist)," accessed December 8, 2016, https://the-moon .wikispaces.com/D.W.G.+Arthur#David%20%22Dai%22%20Arthur.

10. J. Ashbrook, "Surveying the Moon," *Sky and Telescope* 15 (1956): 452–53.

11. UASC, box and file unknown, "Application for Extension of No. 1 of Contract 19(604)-3873," January 1, 1958. Four research associates were to be supported: Joseph W. Tapscott ($8,000/year), J. Harlen Bretz ($700/7 weeks), D. W. G. Arthur ($4,200/7 months), and Ewen Whitaker ($4,200/7 months). Also supported were a photographic specialist, Frank E. Manning ($2,000, part time), and secretary, Lucille Schott ($1650/6 months), and three students: Dale Cruikshank (undergraduate, $350/1.5 months), Carl Huzzen (graduate student, $1,450/9.66 months), Clayton Smith (graduate student, $1,350/9.5 months). These figures can be multiplied by about ten to bring them up to 2018 dollars. There were three photographic assistants, Carter, Williams, and Newman (requiring $1,000 between them) and then $8,750 for Kuiper, amounting to five months of support irregularly spread throughout the year. With a reserve of $5,300, this came to a total budget of $39,000, or $390,000 in 2018 dollars. This can be contrasted with the previous NSF budget for two years, July 1, 1956, to June 30, 1958, with no salary for Kuiper, $2,000 for A. P. Lenham ($2,000/6 months), $100 for equipment, $250 for supplies, and $150 for travel. With 15 percent overhead, this came to a total of $5,750 ($58,000 in 2018 dollars).

12. G. P. Kuiper, *Photographic Lunar Atlas based on photographs taken at the Mount Wilson, Lick, Pic du Midi, McDonald and Yerkes Observatories, Edition A* (Chicago: University of Chicago Press, 1960).

13. G. P. Kuiper, *Photographic Lunar Atlas based on photographs taken at the Mount Wilson, Lick, Pic du Midi, McDonald and Yerkes Observatories, Edition B* (Arlington, Va.: U.S. Air Force Cambridge Research Laboratories, 1961).

14. J. Brouet, "Photographic Lunar Atlas," *Ciel et Terre* 77 (1961): 350.

15. "Frank K. Edmondson," Wikipedia.org, last updated January 1, 2018, http://en.wikipedia .org/wiki/Frank_K._Edmondson.

16. F. K. Edmondson, "Book Review: Photographic Lunar Atlas," *Science* 132 (1960): 290–91.

17. G. P. Kuiper, ed., D. W. G. Arthur and E. A. Whitaker, comps., *Orthographic Atlas of the Moon, Edition A, Supplement 1 to the Photographic Lunar Atlas (Contributions of the Lunar and Planetary Laboratory, No. 1)* (Tucson: University of Arizona Press, 1960). G. P. Kuiper, ed., D. W. G. Arthur and E. A. Whitaker, comps., *Orthographic Atlas of the Moon, Edition B, Supplement 1 to the Photographic Lunar Atlas (Contributions of the Lunar and Planetary Laboratory, No. 1)* (Arlington, Va.: U.S. Air Force Cambridge Research Laboratories, 1960).

18. F. K. Edmondson, "Book Review: Moon Maps," *Science* 134 (1961): 323.

19. E. A. Whitaker et al., *Rectified Lunar Atlas. Supplement Number Two to the Photographic Lunar Atlas* (Tucson: University of Arizona Press, 1964). E. A. Whitaker et al., *Rectified Lunar Atlas. Supplement Number Two to the USAF Lunar Atlas* (Arlington, Va.: U.S. Air Force Cambridge Research Laboratories, 1960).

20. F. K. Edmondson, "Book Review: Rectified Lunar Atlas. Supplement 2 to the Photographic Lunar Atlas," *Science* 145 (1964): 1289–90.

21. B. K. Byers, "Destination Moon: A History of the Lunar Orbiter Program," NASA Technical Memorandum NASA TM X-3847, 1977.

22. "Digital Lunar Orbiter Photographic Atlas of the Moon," Lunar and Planetary Institute, accessed December 12, 2016, http://www.lpi.usra.edu/resources/lunar_orbiter/.

23. T. P. Hansen, "Guide to Lunar Orbiter Photographs," NASA SP-242, Scientific and Technical Information Office, NASA, Washington, D.C., 1970.

24. G. P. Kuiper et al., eds., *Consolidated Lunar Atlas. Supplement Numbers 3 and 4 to the USAF Photographic Lunar Atlas* (Tucson: University of Arizona Lunar and Planetary Laboratory, 1967).

25. The four "System of Lunar Craters" papers are D. W. G. Arthur et al., "No. 30 The system of Lunar Craters, Quadrant I," *COMM* 2 (1963): 71–78 (plus 73-page catalog); D. W. G. Arthur et al., "No. 40 The System of Lunar Craters, Quadrant II," *COMM* 3 (1963): 1–2 (plus 72-page catalog); D. W. G. Arthur, A. P. Agnieray et al., "No. 50 The System of Lunar Craters, Quadrant III," *COMM* 3 (1965): 1–2 (plus 159-page catalog); D. W. G. Arthur, R. H. Pellicori, and C. A. Wood, "No. 70 The System of Lunar Craters, Quadrant IV," *COMM* 5 (1966): 1–2 (plus 220-page catalog).

26. UASC, box 25, folder 8, Kuiper to Brunk, April 1, 1968.

27. UASC, box 25, folder 8, Lunar Orbiter Project Office to Brunk, March 6, 1968.

28. UASC, box 25, folder 8, Kuiper to Brunk, April 2, 1968.

29. UASC, box 25, folder 8, Kuiper to Brunk, March 7, 1973.

30. UASC, box 25, folder 8, Kuiper to Brunk, May 14, 1973.

31. R. E. Doel, "Evaluating Soviet Lunar Science in Cold War America," *Osiris* 7 (1992): 238–64.

32. UARSC, S. des Tombe, "Oral History Recorded at Half Moon Bay."

33. There are a very large number of letters between Kuiper and his CIA contacts in UASC, box 42, folders 26 and 27.

34. Dale Cruikshank (personal communication to author, August 11, 2017) has noted, "Regarding Mr. Hunsacker, in an interview I had with him upon my return from the USSR in the summer of 1969, he told me that Q was not an abbreviation, but actually a name. I don't know if he was serious, but Mr. Hunsacker was quite bald, and he told me his parents named him Q because they knew he would eventually have no hair and would resemble a cue ball in billiards."

35. UARSC, S. des Tombe, "Oral History Recorded at Half Moon Bay."

36. UASC, box and file unknown, Fagerstrom to Kuiper, August 20, 1954. See also Sessions to Kuiper, January 19, 1962, in same folder.

37. UASC, box and file unknown, Kuiper to Sessions, January 24, 1962.

38. "Leonard Reiffel," Wikipedia.org, last updated April 25, 2018, https://en.wikipedia.org/wiki/Leonard_Reiffel.

39. L. Reiffel, "A Study of Lunar Research Flights Vol I," Air Force Special Weapons Center, AFSWC-TR-59–39, 1959.

CHAPTER 13

1. Interview of Frank K. Edmondson by David DeVorkin on April 21, 1997, AIP, https://www.aip.org/history-programs/niels-bohr-library/oral-histories/4588-1 (retrieved November 21, 2016).

2. Doel, *Solar System Astronomy in America*.

3. UASC, UASC, box and file unknown, Kuiper to Oort, July 1, 1958.

4. Evans and Mulholland, *Big and Beautiful*.

5. UASC, UASC, box and file unknown, Kuiper to Oort, July 1, 1958.

6. UASC, UASC, box and file unknown, "Yerkes, the end" folder, "Proposal for a 'Center' or 'Institute' of Planetary and Lunar Studies, Gerard P. Kuiper, Universities of Chicago and Texas."

7. G. Burbidge, "An Accidental Career," *Annual Review of Astronomy and Astrophysics* 45 (2007): 1–41.

8. E. M. Burbidge, "Synthesis of the Elements in Stars," *RMP* 29 (1957): 547–650.

9. S. Chandrasekhar and K. H. Prendergast, "The Equilibrium of Magnetic Stars," *PNAS* 42 (1956): 5–9.

10. "D. Nelson Limber," *PT* 31 (1978): 93.

11. UASC, UASC, box and file unknown, Morgan to Kuiper, November 10, 1958.

12. UASC, UASC, box and file unknown, Kuiper to Morgan, November 21, 1958.

13. UASC, UASC, box and file unknown, "Yerkes the End" file, Burbidge and Chamberlin to Kuiper, May 11, 1959.

14. UASC, UASC, box and file unknown, "Yerkes the End" file, Kuiper to Burbidge and Chamberlin, May 12, 1959.

15. UASC, box and file unknown, "Notes in Preparation for a Senior Staff Meeting," June 8, 1959.

16. John Spencer, "William Merz Sinton (1925–2004)," American Astronomical Society, accessed December 12, 2016, https://aas.org/obituaries/william-merz-sinton-1925-2004.

17. UASC, box 17, folder 14, Burbidge, Burbidge and Prendergast to Kuiper, August 14, 1959.

18. UASC, box 13, folder 1, Kuiper to Johnson, August 18, 1959.

19. UASC, box 17, folder 15, staff meeting minutes, August 17, 1959.

20. UASC, Kuiper, "Summary of Statements at Departmental Meeting of August 17, 1959."

21. UASC, statements with respect to the Burbidge, Burbidge, and Prendergast complaint by Chandrasekhar, Morgan, Limber, Hiltner, Kraft, and Chamberlain.

22. UASC, Kuiper to Burbidge, and Burbidge, Prendergast, August 18, 1959.

23. UASC, Burbidge, Burbidge, and Prendergast to Kuiper, August 21, 1959.

24. UASC, Kuiper's handwritten notes dated October 30, 1959, "Department of Astronomy, Meeting on Friday, October 30, 2 pm."

25. UASC, Kuiper to Zachariasen, November 12, 1959.

26. For example, Kuiper wrote a seven-page confidential memorandum to Chancellor Kimpton on November 1, 1959, with a follow-up memo the next day asking that it not become part of the university's official record. See also UASC, Kuiper to Zachariasen, December 6, 1959, and Kuiper to Harrison, April 17, 1960.

27. UASC, Kuiper, agenda and draft for departmental policy committee, November 11, 1959; UASC, Harrison to Kuiper, April 26, 1960.

28. UASC, Kuiper to Kimpton, November 1, 1959.

29. UASC, Kuiper, "Report on Meeting, Nov. 9th, 1959."

30. UASC, November 11, 1959, Kuiper to Zachariasen.

31. UASC, box and file unknown, Zachariasen to astronomy faculty, October 7, 1959.

32. UASC, Morgan to astronomy faculty, December 14, 1959.

33. A. Roland, *Model Research*, vols. 1–2 (Washington, D.C.: NASA SP-4103, 1985), http://history.nasa.gov/SP-4103/sp4103.htm.

34. I. Stewart, *Organizing Scientific Research for War: The Administrative History of the Office of Scientific Research and Developments* (Boston: Little, Brown, 1948), https://ia800300.us.archive.org/33/items/organizingscient00stew/organizingscient00stew.pdf.

35. Doel, *Solar System Astronomy in America*.

36. "Sul Ross State University," Wikipedia.org, last updated June 5, 2018, https://en.wikipedia.org/wiki/Sul_Ross_State_University.

37. "In Memoriam: Aden B. Meinel, 1922–2011," The Optical Society, October 2, 2011, http://www.osa.org/en-us/about_osa/newsroom/obituaries/adenmeinel/.

38. Whitaker, *The University of Arizona's Lunar and Planetary Laboratory*.

39. UASC, Harrison to Kuiper, February 15, 1960.

40. UChSC, box 19, folder 20, newspaper clipping, "Kuiper, Astronomy Head at University of Chicago, to Leave Post," *Chicago Daily Tribune*, February 13, 1960, p. D10.

41. "U.A. Hires Famous Astronomer," *Tucson Daily Citizen*, February 13, 1960, p. 1. See also "UofA to have Country's only Lunar and Planetary Lab," *Arizona Daily Citizen*, May 12, 1960, p. 46.

42. UASC, Kuiper to Oort, February 19, 1960.

43. Interview of Subrahmanyan Chandrasekhar by Spencer Weart, Monday, October 31, 1977, AIP, https://www.aip.org/history-programs/niels-bohr-library/oral-histories/4551 -1, https://www.aip.org/history-programs/niels-bohr-library/oral-histories/4551-2, and https://www.aip.org/history-programs/niels-bohr-library/oral-histories/4551-3.

44. Harvill's career is briefly described in the program for the recognition dinner for his retirement, which is in UASC, box 8, folder 34. The amazing growth of the university during his tenure as president is also documented there.

45. Dale Cruikshank remarks (personal communication to author, August 11, 2017), "Harvill was indeed a fan of Kuiper. When I gave a speech at the 50-year anniversary event at LPL, Harvill attended and we chatted afterward. He was retired, of course, but was still proud of the vision of Kuiper that he had enabled at University of Arizona, and the heights to which it had risen."

46. Edwin Francis Carpenter, Wikipedia.org, last updated November 3, 2017, https://en .wikipedia.org/wiki/Edwin_Francis_Carpenter.

47. "A. Richard 'Dick' Kassander: September 10, 1920–July 27, 2017," Tempe Mortuary, accessed September 26, 2018, https://www.meaningfulfunerals.net/obituary/4319608 ?fh_id=14347.

48. Whitaker, *The University of Arizona's Lunar and Planetary Laboratory*.

49. UASC, box and file unknown.

50. Whitaker, *The University of Arizona's Lunar and Planetary Laboratory*.

51. UASC, box 17, folder 17, Struve to Kuiper, January 28, 1960.

52. UASC, box 17, folder 17, Kuiper to Struve, February 2, 1960.

53. UASC, box 17, folder 17, Struve to Kuiper, February 5, 1960.

54. UASC, box 17, folder 17, Struve to Kuiper, June 30, 1960.

55. UASC, box 17, folder 17, Kuiper to Cowling, August 7, 1963.

56. Whitaker, *The University of Arizona's Lunar and Planetary Laboratory*.

57. Whitaker, *The University of Arizona's Lunar and Planetary Laboratory*.

CHAPTER 14

1. M. Sevigny, *Under Desert Skies: How Tucson Mapped the Way to the Moon and Planets* (Tucson: University of Arizona Press, 2016).

2. UASC, biofile, "University of Arizona Staff Re Commendation."

3. Whitaker, *The University of Arizona's Lunar and Planetary Laboratory*.

4. UARSC, S. Larson, Oral History recorded at Tucson, Arizona, November 9, 2014.

5. The UBVRI nomenclature indicates that the star was viewed through five filters (ultraviolet, blue, violet, red, and infrared), and the relative brightness of the star through these was used to characterize the star.

6. G. De Vacouleurs, "Harold Lester Johnson 1921–1980," *BMNAS* 67 (1995), 241–61.

7. D. W. G. Sears, "Oral Histories in Meteoritics and Planetary Science—XX: Dale Cruikshank," *Meteoritics and Planetary Science* 48 (2013): 700–711. Cruikshank has explained to me that the mentor who wrote a letter of recommendation for him to be summer assistant at Yerkes Observatory in 1958 was Philip S. Riggs, an astronomer who taught astronomy and mathematics at Drake University in Des Moines, Iowa, where he grew up. Riggs was a strong influence on Cruikshank as a schoolboy and to some degree was a role model. When Cruikshank returned home from his first summer at Yerkes, Riggs (normally quite formal) asked him, "Has old Gerry learned to get along with people yet?"

8. D. W. G. Sears, "Oral Histories in Meteoritics and Planetary Science—XXIV: William K. Hartmann," *Meteoritics and Planetary Science* 49 (2014): 1119–38.

9. "Dr. Alan Binder," LinkedIn profile, accessed August 25, 2017, https://www.linkedin.com/in/dr-alan-binder-65278074.

10. Sears, "Oral Histories in Meteoritics and Planetary Science—XXIV: William K. Hartmann."

11. O. Richard Norton, "Master Optician, Master Observer," Cave-Astrola.com, accessed December 15, 2016, http://www.cave-astrola.com/history/thomascave/optician/index.html.

12. UARSC, interview of Spencer Titley by Melissa Sevigny, October 9, 2007.

13. Interview of Eugene Shoemaker by Ron Doel, January 30, 1986, June 16, 1987, June 17m 1987, and September 8, 1988, Niels Bohr Library and Archives, AIP, https://www.aip.org/history-programs/niels-bohr-library/oral-histories/5082-1, https://www.aip.org/history-programs/niels-bohr-library/oral-histories/5082-2, https://www.aip.org/history-programs/niels-bohr-library/oral-histories/5082-3, https://www.aip.org/history-programs/niels-bohr-library/oral-histories/5082-4.

14. UARSC, interviews by Melissa Sevigny of Spencer Titley, October 9, 2007; William Hartmann, December 1, 2006; Alan Binder, May 18, 2007; and Dale Cruikshank, October 22, 2007.

15. See the collection of short memories of Kuiper by his students at LPL's website, https://www.lpl.arizona.edu/history/points-of-light/founding/early-grads.

16. UARSC, Gehrels to Cruikshank, May 11, 1979.

17. Dale Cruikshank, personal communication to author, July 8, 2017.

18. UASC, box 8, folder 33, Harvill to Kassander, July 25, 1961.

19. Kuiper, "No. 1. Organization and Programs of the Laboratory."

20. G. P. Kuiper, "The Lunar and Planetary Laboratory," pts. 1 and 2, *Sky and Telescope* 27 (1964): 4–7; 88–92. The text of the *Sky and Telescope* articles was similar to the earlier

article in the *Communications*, See Kuiper, "No. 1. Organization and Programs of the Laboratory."

21. G. P. Kuiper and D. P. Cruikshank, "No. 95 High Altitude Spectra, I: Venus, 1–2.5 Microns," *COMM* 6 (1967): 177–90.

22. T. C. Owen and G. P. Kuiper, "No. 32 A Determination of the Composition and Surface Pressure of the Martian Atmosphere," *COMM* 2 (1964): 113–32.

23. G. P. Kuiper, "No. 23. Infrared Spectra of Stars and Planets II. Water Vapor in Omicron Ceti," *COMM* 1 (1963): 179–88.

24. G. P. Kuiper, "No. 25. Infrared Spectra of Stars and Planets III. Reconnaissance of A0-B8 Stars, 1–2.5 Microns," *COMM* 1 (1963): 179–88.

25. G. P. Kuiper, "No. 26. Spectrometric Records from 0.3–0.55 m for Some A and B Stars with Special References to the Balmer Series," *COMM* 2 (1963): 33–47.

26. G. P. Kuiper et al., "No. 16. An Infrared Stellar Spectrometer," *COMM* 1 (1962): 119–27.

27. G. P. Kuiper, "No. 15. Infrared Spectra of Stars and Planets. I. Photometry of the Infrared Spectrum of Venus, 1–2.5 Microns," *COMM* 1 (1962): 83–117.

28. Kuiper and Cruikshank, "No. 95 High Altitude Spectra, I."

29. UASC, box 25, folder 8, Kuiper to Brunch, December 10, 1969; G. P. Kuiper, "No. 101. Identification of the Venus Cloud Layers," *COMM* 6 (1969): 229–50.

30. G. P. Kuiper, "No. 31. Infrared Spectra of Stars and Planets. IV: The Spectrum of Mars, 1–2.5 Microns, and the Structure of Its Atmosphere," *COMM* 2 (1964): 79–112; T. C. Owen and G. P. Kuiper, "A Determination of the Composition and Surface Pressure of the Martian Atmosphere," *COMM* 2 (1964): 113–32.

31. L. Mertz and I. Coleman, "Infrared Spectrum of Saturn's Ring," *AJ* 71 (1966): 748–49.

32. G. P. Kuiper, D. P. Cruikshank, and U. Fink, "The Composition of Saturn's Rings," *Sky and Telescope* (January 14, 1970; February 80, 1970); C. B. Pilcher et al., "Saturn's Rings: Identification of Water Frost," *Science* 167, no. 3923 (1970): 1372–73.

33. C. B. Pilcher, "Saturn's Rings: Identification of Water Frost," *Science* 167 (1970): 1372–73.

34. G. P. Kuiper, "Infrared Observations of Planets and Satellites," *AJ* 62 (1957): 245.

35. G. P. Kuiper, "No. 108. Comments on the Galilean Satellites," *COMM* 10 (1973): 28–34.

36. Interview of Homer Newell by Richard F. Hirsh, July 17, 1980, and October 20, 1980, https://www.aip.org/history-programs/niels-bohr-library/oral-histories/4795-1 and https://www.aip.org/history-programs/niels-bohr-library/oral-histories/4795-2.

37. W. H. Lambright, *Powering Apollo: James E. Webb of NASA* (Baltimore: Johns Hopkins Press, 1995); P. Bizony, *The Man Who Ran the Moon: James E. Webb, NASA, and the Secret History of Project Apollo* (New York: Thunder's Mouth, 2006).

38. R. S. Kraemer, *Beyond the Moon: The Golden Age of Planetary Exploration 1971–1978* (Washington, D.C.: Smithsonian Institution Press, 2000).

39. Interview of William Brunk by Joseph N. Tatarewicz, July 21, 1983, and August 9, 1983, https://www.aip.org/history-programs/niels-bohr-library/oral-histories/28198-1 and https://www.aip.org/history-programs/niels-bohr-library/oral-histories/28198-2.

40. UASC, box 29, folder 9, "A Proposal to the National Aeronautics and Space Administration for the Construction of an Interdisciplinary Space Sciences Center at the University of Arizona, Tucson, Arizona, May 15, 1964."

41. Whitaker, *The University of Arizona's Lunar and Planetary Laboratory.*

42. Official program, "Dedication of the Gerard P. Kuiper Airborne Observatory."

43. UASC, Box 29, folder 9, "Symposium: The Lunar Surface, Fri, Sat, Jan 27–28, 1967, 9–12 AM, 2–5 PM, Steward Observatory, Room 102."

44. UASC, box 29, folder 3, Harvill to Hall, Kuiper, Meinel, and Weaver, December 17, 1965.

45. UASC, box 29, folder 3, "Space Sciences Sub Committee Meeting," October 5, 1967.

46. UASC, box 29, folder 3, Kuiper, "Report to Space Sciences Committee on Utilization of Space in Space Studies Building," September 19, 1967.

47. Whitaker, *The University of Arizona's Lunar and Planetary Laboratory.*

48. UARSC, interview of Robert Strom by Melissa Sevigny, October 26, 2006.

49. P. Moore, *Guide to the Moon* (New York: Norton, 1953).

50. T. Hockey and T. R. Williams, "Van Biesbroeck, Georges-Acille," *BEA* 2 (2007): 1168–69.

51. J. V. Scotti, "Wieslaw Z. Wisniewski (1931–1994)," *Icarus* 112 (1994): 300–301.

52. F. J. Low, G. H. Rieke, and R. D. Gehrz, "History of Modern Infrared Astronomy—1960 to 1983," University of Arizona Department of Astronomy/Steward Observatory, accessed December 15, 2016, http://ircamera.as.arizona.edu/Astr_518/flowhist.pdf.

53. H. H. Aumann, C. M. Gillespie, and F. J. Low, "The Internal Powers and Effective Temperatures of Jupiter and Saturn," *ApJ* 157 (1969): L69–L72.

54. G. H. Rieke et al., "The Multiband Imaging Photometer for Spitzer (MIPS)," *ApJ Supplement Series* 154 (2004): 25–28 (with 1 plate).

55. The UBVRIJKLMN nomenclature adds five filters in the infrared (labelled alphabetically J to N) to the UBVRI system described earlier. The relative brightness of the star through these ten filters is used to characterize the star.

56. G. P. Kuiper, "The Lunar and Planetary Laboratory II," *Sky and Telescope* 27 (1964): 88–92.

57. "Elizabeth Roemer," Wikipedia.org, last updated July 22, 2018, https://en.wikipedia.org/wiki/Elizabeth_Roemer.

58. UASC, box 25, folder 8, Kuiper to Brunk, July 14, 1966.

59. UARSC, interview of George Rieke by Melissa Sevigny, October 23, 2009.

60. Low and Rieke, "The Instrumentation and Techniques of Infrared Photometry."

61. UARSC, interview of William Hubbard by Melissa Sevigny, September 19, 2007.

CHAPTER 15

1. G. P. Kuiper, "No. 172. The Lunar and Planetary Laboratory and its Telescopes," *COMM* 9 (1972): 199–247; Whitaker, *The University of Arizona's Lunar and Planetary Laboratory.*

2. Whitaker, *The University of Arizona's Lunar and Planetary Laboratory.*

3. De Vacouleurs, "Harold Lester Johnson 1921–1980."

4. "UBV" is a system for classifying stars based on their intensity as seen through ultraviolet, blue and violent filters.

5. UASC, box and file unknown, Kuiper to Olson, February 12, 1973.

6. These are now available online as pdfs at https://www.lpl.arizona.edu/sic/collection /journal.

7. UChSC, box 19, folder 21, Chandrasekhar to Kuiper, October 1, 1963. See also Chandrasekhar to Kuiper, October 28, 1963.

8. UChSC, box 19, folder 21, Kuiper to Chandrasekhar, October 9, 1963.

9. UASC, box 25, folder 8, Kuiper to Brunk, June 30, 1966.

10. UASC, box 25, folders 7, 8, and 9, Kuiper-Brunk correspondence 1965–73.

11. G. P. Kuiper, "No. 93. A Program of Astronomical Infrared Spectroscopy from Aircraft," *COMM* 6 (1962): 155–70; G. P. Kuiper, "The Surface Structure of the Moon," in *The Nature of the Lunar Surface: Proceedings of the 1965 IAU-NASA Symposium, April 15–16, 1965*, ed. W. N. Hess, J. A. Menzel, and J. A. O'Keefe, 99–105 (Baltimore: John Hopkins University Press, 1965); G. P. Kuiper, "No. 58. Interpretation of Ranger VII records," *COMM* 4 (1965): 1–70.

12. UASC, box 25, folder 8, Kuiper to Brunk, June 28, 1969. Halides are compounds of chlorine, fluorine, bromine, iodine, and astatine.

13. G. P. Kuiper, "No. 1 Organization and Program of the Laboratory," *COMM* 1 (1962): 1–20; G. P. Kuiper, "No. 142 High Altitudes Sites and IR Astronomy," *COMM* 8 (1962): 121–64.

14. B. M. Middlehurst, "Transient Lunar Events: Possible Causes," *Nature* 209 (1966): 602; B. M. Middlehurst and P. Moore, "Lunar Transient Phenomena: Topographical Distribution," *Science* 155 (1967): 449–51.

15. UASC, box 25, file 8, Kuiper to Middlehurst, March 27, 1967.

16. B. M. Middlehurst et al., "Chronological Catalog of Reported Lunar Events," NASA Technical Report NASA TR R-277, 1968.

17. B. S. Jeffreys, "Obituary. Barbara Middlehurst," *QJRAS* 35 (1995): 461–62.

18. B. M. Middlehurst, "No. 21 Surface Photometry of Extended Images," *COMM* 2 (1964): 161–66.

19. B. M. Middlehurst, "No. 28 Helium Lines in Early Type Stars: Systematic Spectra and Possible Causes of the Trumpler Shift," *COMM* 2 (1964): 59–64.

20. Nicolai Alexandrovich Kozyrev's spectroscopic observations of Alphonsus were inspired by photographs by Dinsmore Alter at the Griffith Observatory of apparent haze over the floor in violet images compared to red, which clearly had to do with the light-scattering properties of the filter used in the different images rather than, as he believed, a cloud. His supposed spectrographic image showing an emission of gas from Alphonsus was very convincing to the transient lunar phenomena (TLP) contingent, led by Patrick Moore, the popular British TV astronomer and writer. The whole story of Kozyrev—including

his rather tragic experiences in one of Stalin's gulags (he is mentioned in Solzhenitzyn's Gulag Archipelago)—is described in Sheehan and Dobbins, *Epic Moon*, 312–16. Moore continued to hold forth about the TLP until 1999, with even lunar specialists tending to tiptoe around the subject (especially in Britain; such was Moore's influence and standing). Then Dobbins and Sheehan published "The TLP Myth: A Case for the Prosecution," *Sky and Telescope* (September 1999): 118–23. As we now know from Martin Mobberley's biography of Moore, that was the straw that broke the camel's back. He gave up and soon started acknowledging even the meteorite impact theory of the lunar craters' formation. (Before that he had been an increasingly lonely vulcanist.) William Sheehan, personal communication to author, July 7, 2017.

21. Doel, "Evaluating Soviet Lunar Science in Cold War America."

22. R. J. Turner, "No. 195 A Model of the Eastern Portion of Schroter's Valley," *COMM* 10 (1973): 81–93; R. J. Turner et al., "No. 149 The Northeast Rim of Tycho," *COMM* 8 (1970): 203–34.

23. Kuiper, "The Lunar and Planetary Laboratory II."

24. For example, G. Eberhard, *Grundlagen der Astrophysik: Zweiter Teil II (Handbuch der Astrophysik)* (Berlin: Springer, 1931).

25. UASC, box 17, folder 17, Kuiper to Struve, November 29, 1954.

26. UASC, box 17, folder 17, Kuiper to Struve, January 5, 1956.

27. UASC, box and file unknown, Clemence to NSF, January 7, 1955.

28. Nine volumes were planned, but volume 4, *Clusters and Binaries*, appears never to have been published. Kuiper and Middlehurst acted as general editors, while each volume had its own editor or editors. All were published by the University of Chicago Press. In the order they appeared, the volumes were G. P. Kuiper and B. M. Middlehurst, eds., *Volume 1: Telescopes* (1960); J. L. Greenstein, ed., *Volume 6: Stellar Atmospheres* (1960); W. A. Hiltner, ed., *Volume 2: Astronomical Techniques* (1962); K. A. Strand, ed., *Volume 3: Basic Astronomical Data* (1963); A. Blaauw and M. Schmidt, eds., *Volume 5: Galactic Structure* (1965); L. H. Aller and D. B. McLaughlin, eds., *Volume 8: Stellar Structure* (1965); B. M. Middlehurst and L. H. Aller, *Volume 7: Nebulae and Interstellar Matter* (1968); A. Sandage, M. Sandage, and J. Kristian, eds., *Volume 9: Galaxies and the Universe* (1975).

29. Kuiper and Middlehurst stuck to their original numbering system for the volumes, so the volumes did not come out in order. Also, being published over such an extended time meant inconsistencies in information because some were more recent than others.

30. J. B. Irwin, "Review: Instruments and Techniques," *Science* 138 (1962): 127–28; J. B. Irwin, "Review: Astronomical Data," *Science* 142 (1963): 217–18. See also A. J. Deutsch, "Review: Tools of the Astronomer," Science 133 (1961): 1416.

31. G. P. Kuiper, "German Astronomy During the War," *PA* 54 (1946): 263–86; G. P. Kuiper, "Reports from French Astronomers," *ApJ* 101 (1945): 254 (with 2 plates); G. P. Kuiper, "Meeting of 1945 July 13," *MNRAS* 105 (1945): 192.

32. UASC, box 57, folder 18, Kuiper to Goudsmit, September 24, 1960.

33. UASC, biofile, Kuiper to editor of the *Arizona Daily Star*, July 23, 1969.

34. UASC, box and file unknown, Bithos to Kuiper, September 15, 1971.

35. UASC, box and file unknown, Kuiper to Weaver, October 28, 1971.

36. UASC, box and file unknown, Kuiper to Udall, November 3, 1971.

37. UASC, box 8. folder 34, Harvill to Kuiper, November 10, 1971.

38. UASC, box 8, folder 34, Kuiper to Harvill, February 1, 1971. The building is now Aloft Tucson University. See http://www.alofttucsonuniversity.com/.

39. UASC, box 8, folder 34, Kuiper to Harvill, August 16, 1972.

40. UASC, box 8, folder 34, Kuiper to Harvill, August 18, 1972.

41. "Journalists Criticized by Steiger," *Arizona Daily Star*, Tucson, March 14, 1971.

42. UASC, box 8, folder 34, Kuiper to Harvill, March 15, 1971.

43. UASC, box 44, folder 8, to box, 46, folder 3, inclusive.

44. "About Us," University of Arizona Department of Hydrology and Atmospheric Sciences, accessed September 28, 2018, https://has.arizona.edu/about-us.

45. W. K. Hartmann, "Pollution: Patterns of Visibility Reduction in Tucson," *Journal of the Arizona Academy of Science* 7 (1972): 101–8.

46. Hartmann, "Pollution."

47. UASC, biofile, Kuiper to the Editor of the Tucson Daily Star, June 5, 1968.

48. According to LPL folklore, Kuiper's desk was on a raised platform, with a flag next to the desk. Visitors would be seated below so that they would be looking up at him.

49. Whitaker, *The University of Arizona's Lunar and Planetary Laboratory*.

50. UASC, Kuiper to staff, January 12, 1973; Kuiper to staff, January 25, 1973; Kuiper to Planetary Photography, May 16, 1973; Kuiper to faculty and staff, October 1, 1973; Kuiper to 61-inch observers, October 31, 1973; Kuiper to staff, June 15, 1972; Kuiper to staff, December 13, 1971; Kuiper to staff, May 11, 1971.

51. UASC, Kuiper to staff, December 26, 1968.

52. UASC, Kuiper to LPL personnel, undated.

53. S. A. Stern, "Forging a New Solar System—GPK," *Astronomy* (March 1999): 40–45.

54. UASC, box 25, folder 8, Kuiper to Brunk, May 7, 1968.

55. UASC, box 8, folder 34, Kuiper to Harvill, March 24, 1972.

56. UASC, box 8, folder 34, Harvill to Kuiper, April 11, 1972.

57. UASC, box 8, folder 34, Harvill to Kuiper, April 14, 1972.

58. UASC, box 8, folder 33, Harvill to Kuiper, September 10, 1964.

59. UASC, box 25, folder 8, Brunk to Kuiper, February 22, 1973.

60. John M. Logsdon, ed., "Biographical Appendix," in *Exploring the Unknown* (Washington, D.C.: National Aeronautics and Space Administration, 1995), 802, http://history.nasa.gov/SP-4407vol7App.pdf.

61. For this narrative, I have relied on the oral histories of LPL members collected by Melissa Sevigny: UARSC, interview of Alan Binder by Melissa Lamberton, May 18, 2007; interview of George V. Coyne by Melissa Lamberton, August 31, 2007; interview of Dale P. Cruikshank by Melissa Lamberton, October 22, 2007; interview of Charles Sonett by Melissa Lamberton, December 14, 2006; interview of Ewen A. Whitaker by Melissa Lamberton, December 14, 2006; interview of Laurel L. Wilkening by Melissa Lamberton, February 20, 2007; interview of Charles A. Wood by Melissa Lamberton, May 22, 2007.

62. UASC, box 16, file 8, Kuiper to Schaefer, December 11, 1973.

63. UASC, box 16, file 8, Schaefer to Kuiper, June 2, 1973.

64. UASC, box 16, file 8, Kuiper to Schaefer, December 11, 1973.

65. Kuiper's apparent reluctance to make stationary freely available might have been caused by an incident involving the well-known popularizer of astronomy, Patrick Moore. Ewen Whitaker and Chuck Wood sent a letter to the British Astronomical Association making serious criticisms of Patrick Moore. They wanted the letter published in the Association's journal, but the journal refused and forwarded the letter to Patrick Moore. This resulted in a very calm and patient letter from Moore to Kuiper. Kuiper apologized to Moore and reprimanded Whitaker and Wood with a complaint about their rudeness and the use of LPL letterhead. UASC box and file unknown, Moore to Kuiper, September 7, 1966; Kuiper to Moore, September 14, 1966; Kuiper to Moore, October 18, 1966.

66. UARSC, interview of Charles Sonett by Melissa Sevigny, December 14, 2006.

67. UARSC, interview of Alan Binder by Melissa Lamberton, May 18, 2007.

68. C. P. Sonett, "Electromagnetic Induction in the Moon," *Reviews of Geophysics* 20 (1982): 411–55.

69. Daniel Stolte, "Charles P. Sonett: The Legacy of a Pioneering Space Scientist," UA News, University of Arizona, October 5, 2011, https://uanews.arizona.edu/story/charles-p-sonett-the-legacy-of-a-pioneering-space-scientist.

CHAPTER 16

1. At the U.S. Space and Rocket Center in Huntsville, Alabama, are a V2 and a Redstone rocket lying alongside each other. Stripped of their skins they are seen to be identical. Also underscoring their similarity is the distinctive circular stand that was used for launch. This conspicuous ring can also be seen on any photograph of the V2 launches by the Germans during World War II, and the structure is easily visible in photographs of the launch of astronauts Alan Shepard and Gus Grissom on their Redstone-Mercury launch assemblies.

2. Whitaker, *The University of Arizona's Lunar and Planetary Laboratory*. Tim Swindle has noted that, since the launch occurred midday in Chicago, it is surprising Kuiper would not have heard about it before meeting Whitaker at the airport seven or eight hours later.

3. Whitaker, *The University of Arizona's Lunar and Planetary Laboratory*.

4. Doel, *Solar System Astronomy in America*.

5. R. C. May, "Lunar Impact: A History of Project Ranger," NASA SP-4210, 1977.

6. Byers, "Destination Moon."

7. "Surveyor Program 1969, Surveyor Results," NASA SP-184, 1969.

8. UASC, box 29, folder 14, Silverstein to Kuiper, May 9, 1961.

9. Whitaker, *The University of Arizona's Lunar and Planetary Laboratory*.

10. This is an often-quoted anecdote, but I cannot find it documented anywhere.

11. G. P. Kuiper, "No. 58 Interpretation of Ranger VII Records," *COMM* 4 (1966): 1–70.

12. Whitaker, *The University of Arizona's Lunar and Planetary Laboratory*.

13. Whitaker, *The University of Arizona's Lunar and Planetary Laboratory*.

14. Hess, Menzel, and O'Keefe, *The Nature of the Lunar Surface*.

15. UASC, box 8, folder 33, Kuiper to Harvill, March 27, 1965.

16. UASC, box 8, folder 33, Kuiper to Harvill, March 18, 1963.

17. UASC, box 8, folder 33, Kuiper to Harvill, March 27, 1965; box 8, folder 34, Kuiper to Harvill, October 11, 1969.

18. T. Gold, "The Lunar Surface," *MNRAS* 115 (1955): 585–604.

19. H. Bondi, "Thomas Gold. 22 May 1920—22 June 2004: Elected FRS 1964," *Biographical Memoirs of Fellows of the Royal Society* 52 (2006): 117–10; G. Burbidge and M. Burbidge, "Thomas Gold 1920–2004," *BMNAS* 88 (2006): 1–14.

20. UASC, box 25, folder 8, telegram, Kuiper to Newell, April 4, 1965.

21. UASC, box 8, folder 33, Smith to Harvill, April 24, 1962.

22. UASC, box 8, folder 34, Harvill to Kuiper, March 30, 1968.

23. J. Lear, "What the Moon Ranger Couldn't See," *Saturday Review*, September 9, 1964, pp. 35–40.

24. UASC, biofile, Kuiper to Monroe, September 3, 1964.

25. UASC, box 8, folder 33, Harvill to Kuiper, September 10, 1964.

26. UASC, box 29, folder 14, Silverman to Kuiper, May 9, 1961; box 29, folder 14, Kuiper to Silverman, May 18, 1961. The entire proposal is in the same folder.

27. UASC, box 8, folder 33, Kuiper to Harvill, May 18, 1961.

28. UASC, box 25, folder 8, Kuiper to Brunk, unfinished and never sent, July 22, 1969.

29. D. W. G. Sears, "Oral Histories in Meteoritics and Planetary Science—XVI: Donald D. Bogard," *Meteoritics and Planetary Science* 47 (2012): 416–33.

30. "The age of the samples would be critical": A Urey-type chondritic Moon would be as old as the solar system, which was then known to be around 4.6 billion years, whereas a volcanic Moon could be younger, and the rocks could have a variety of ages.

31. UCSD, Urey to Sarah Kuiper, December 28, 1973.

32. UASC, box 25, folder 8, Kuiper to Newell, April 12, 1965, with addendum.

33. Presumably referring to the 1964 General Assembly of the IAU.

34. UASC, box 25, folder 8, Kuiper to Brunk, March 7, 1973.

35. UASC, box 25, folder 8, Kuiper to Brunk, May 14, 1973.

36. UASC, box 8, folder 34, Harvill to Kuiper, July 11, 1973.

37. UASC, box 25, folder 8, Brunk to Kuiper, November 29, 1973.

CHAPTER 17

1. Godfrey Sill was a chemistry teacher at the high school level who obtained a research position at LPL. He worked with Cruikshank on a series of laboratory studies of spectra of materials that drew on his experience in synthesis and handling of hazardous chemicals. Godfrey was a Catholic priest in the Carmelite order, and he signed his name on scientific papers with G. T. Sill and O. Carm. Dale Cruikshank, personal communication with author, July 8, 2017.

2. UASC, box 50, folder 11, small grey notebook labeled "Chilean Site Survey." Pages 19 and 45–67 concern his flight to Mount Tequila, December 13 to December 18, 1973.

3. UARSC, interview of Laurel Wilkening by Melissa Sevigny, February 20, 2007. Wilkening mentions her marriage to Godfrey Sill and discusses their flight to Mexico with Kuiper. Another oral history by Wilkening recounting many of these events can be found at https://www.jsc.nasa.gov/history/oral_histories/NASA_HQ/Herstory/WilkeningLL/Wilkening_11-15-01.htm.

4. D. Jesús de Alba Martinez, "Harro Barraza, Guillermo," *BEA* (2007): 1471–72.

5. "El Observatorio Astrofísico Guillermo Haro en Cananea, Sonora," Instituto Nacional de Astrofísica, Optica y Electrónica, December 8, 2009, http://www.inaoep.mx/~astrofi/cananea/. See also "Guillermo Haro Observatory," Wikipedia.org, last updated November 8, 2017, https://en.wikipedia.org/wiki/Guillermo_Haro_Observatory.

6. J. Wilcock, *Mexico on Five Dollars a Day* (New York: Arthur Frommer, 1964).

7. "Xochicalco," Wikipedia.org, last updated January 12, 2018, https://en.wikipedia.org/wiki/Xochicalco.

8. UARSC, Sarah Kuiper-Roth to Cruikshank, September 16, 1982.

9. UARSC, S. des Tombe, "Oral History Recorded at Half Moon Bay."

10. "Jakarta Intercultural School," Wikipedia.org, last updated September 10, 2018, https://en.wikipedia.org/wiki/Jakarta_Intercultural_School.

11. Arizona Daily Star for December 25, 1973.

12. UASC, biofile, "Memorial Service, December 30, 1973, for Gerard P. Kuiper, 1905–1973."

13. UASC, biofile, "Text of Remarks by B. J. Bok."

14. Whitaker, *The University of Arizona's Lunar and Planetary Laboratory*.

15. E. A. Whitaker, "Kuiper Archives," *Science* 184 (1974): 1231.

16. D. L. Shirley, "The Mariner 10 Mission to Venus and Mercury," *Acta Astronautica* 53 (2003): 375–85.

17. P. D'Incecco, "Kuiper Crater on Mercury—An Opportunity to Study Recent Surface Weathering Trends with Messenger," *Abstracts of the Lunar and Planetary Science Conference* 1815 (2012).

18. "Planetary Name: Welcome," Gazetteer of Planetary Nomenclature, IAU Working Group for Planetary System Nomenclature, accessed April 25, 2017, https://planetarynames.wr .usgs.gov/.

19. To my certain knowledge, Nicaragua and Monaco have issued postage stamps with Kuiper's image and Guinea has twice issued such stamps. There are probably others.

20. UARSC, Perton and Huisman to Cruikshank, June 11, 2001; S. des Tombe, "Oral History Recorded at Half Moon Bay."

21. D. P. Cruikshank, "Gerard Peter Kuiper," *BMNAS* 62 (1993): 258–95.

22. "So it was really amazing to have him need an audience, and I was the audience." UARSC, interview of Charles Wood by Melissa Sevigny, May 22, 2007.

23. UASC, Greenstein to Dale Cruikshank, August 30, 1982.

INDEX

ABOUT THE AUTHOR

Derek W. G. Sears was a professor at the University of Arkansas for thirty years and is now a senior research scientist at NASA. He has published widely on meteorites, lunar samples, asteroids, and the history of planetary science. *Gerard P. Kuiper and the Rise of Modern Planetary Science* is his sixth book.